Theoretical Surface Science

Advanced Texts in Physics

This program of advanced texts covers a broad spectrum of topics which are of current and emerging interest in physics. Each book provides a comprehensive and yet accessible introduction to a field at the forefront of modern research. As such, these texts are intended for senior undergraduate and graduate students at the MS and PhD level; however, research scientists seeking an introduction to particular areas of physics will also benefit from the titles in this collection.

Springer
*Berlin
Heidelberg
New York
Hong Kong
London
Milan
Paris
Tokyo*

Physics and Astronomy   ONLINE LIBRARY

http://www.springer.de/phys/

Axel Groß

Theoretical Surface Science

A Microscopic Perspective

With 105 Figures,
11 Tables and 32 Exercises

 Springer

Professor Dr. Axel Groß
Technische Universität München
Physik-Department T30
85747 Garching
Germany
e-mail: agross@ph.tum.de

Library of Congress Cataloging-in-Publication Data: Gross, Axel, 1961–. Theoretical surface science: a microscopic perspective/ Axel Gross. p. cm. – (Advanced texts in physics, ISSN 1439-2674) Includes bibliographical references and index. ISBN 354043903X (alk. paper) 1. Surfaces (Physics) I. Title. II. Series. QC173.4.S94 G74 2002 530.4'27–dc21 2002026851

ISSN 1439-2674
ISBN 3-540-43903-X Springer-Verlag Berlin Heidelberg New York

This work is subject to copyright. All rights are reserved, whether the whole or part of the material is concerned, specifically the rights of translation, reprinting, reuse of illustrations, recitation, broadcasting, reproduction on microfilm or in any other way, and storage in data banks. Duplication of this publication or parts thereof is permitted only under the provisions of the German Copyright Law of September 9, 1965, in its current version, and permission for use must always be obtained from Springer-Verlag. Violations are liable for prosecution under the German Copyright Law.

Springer-Verlag Berlin Heidelberg New York
a member of BertelsmannSpringer Science+Business Media GmbH

http://www.springer.de

© Springer-Verlag Berlin Heidelberg 2003
Printed in Germany

The use of general descriptive names, registered names, trademarks, etc. in this publication does not imply, even in the absence of a specific statement, that such names are exempt from the relevant protective laws and regulations and therefore free for general use.

Typesetting: Data prepared by the author using a Springer TeX macro package
Data conversion: Frank Herweg, Leutershausen
Cover design: *design & production* GmbH, Heidelberg

Printed on acid-free paper   SPIN 10872108   57/3141/ba   5 4 3 2 1 0

In memoriam
David and Sabeth

Preface

Recent years have seen tremendous progress in the theoretical treatment of surface structures and processes. While a decade ago most theoretical studies tried to describe surfaces either on a qualitative level using empirical parameters or invoked rather severe approximative models, there is now a large class of surface system that can be addressed quantitatively based on first-principles electronic structure methods. This progress is mainly due to advances in the computer power as well as to the development of efficient electronic structure algorithms. However, ab initio studies have not only been devoted to microscopic aspects. Instead, starting from a description of the electronic structure and total energies of surfaces, a hierarchy of methods is employed that allows the theoretical treatment of surfaces from the microscopic length and time scales up to the macroscopic regime. This development has led to a very fruitful cooperation between theory and experiment which is reflected in the large number of research papers that result from a close collaboration between experimental and theoretical groups.

Still, in my opinion, this progress had not been reflected in the available surface science textbooks. I felt that there was a need for an advanced textbook on theoretical surface science. Rather than following a macroscopic thermodynamic approach, the textbook should be based on a microscopic point of view, so-to-say in a bottom-up approach. This provided the motivation to start the project this book resulted from. The text is based on a class on theoretical surface science held at the Technical University in Munich. The class and the manuscript evolved simultaneously, taking into account the feedback from the students attending the class.

I have tried to give a comprehensive overview of most fields of modern surface science. However, instead of listing many different data I have rather picked up some benchmark systems whose description allows the presentation and illustration of fundamental concepts and techniques in theoretical surface science. The theoretical results are compared to experiments where possible, but experimental techniques are not introduced. Still this book is not only meant for students and researchers in theoretical surface science, but also for experimentalists who either are interested in the basic concepts underlying the phenomena at surfaces or want to get an introduction into the methods their theoretical colleagues are using. Of course not every aspect of surface science could have been covered, for example surface magnetism is hardly touched upon.

I have tried to present derivations for most of the theoretical methods presented in this book. However, I did not intend to overburden this book with lengthy calculations. A detailed list of references is provided for the reader who wants to get more detailed information on specific methods or systems. In particular, I have tried to select excellent comprehensive review papers that can serve as a basis for further reading. For that reason, the reference list does not necessarily reflect the scientific priority, but rather the usefulness for the reader. In this context I would also like to apologize to all colleagues who feel that their own work is not properly presented in this book. It is important to note that in fact I am not the first to use a bottom-up approach in the presentation of first-principles calculations. This concept was developed by Matthias Scheffler for his class on theoretical solid-state physics at the Technical University in Berlin, and in using this concept for the present textbook I am deeply indebted to him.

Such a book would indeed not be possible without interaction with colleagues. I am grateful to my students Arezoo Dianat, Christian Bach, Markus Lischka, Thomas Markert, Christian Mosch, Ataollah Roudgar and Sung Sakong for stimulating discussions in the course of the preparation of this book and for their careful proofreading of the manuscript. Special thanks go to my colleagues and friends Wilhelm Brenig, Peter Kratzer, Eckhard Pehlke and again Matthias Scheffler for their careful and competent reading of the manuscript and their helpful suggestions in order to further improve the book.

This book is also a product of the insight gained in the discussions and collaborations with numerous colleagues, in particular Steve Erwin, Bjørk Hammer, Ulrich Höfer, Dimitrios Papaconstantopoulos, Helmar Teichler, Steffen Wilke, Martin Wolf, and Helmut Zacharias. In addition, I am indebted to Claus Ascheron from Springer-Verlag who supported this book project from the early stages on.

Finally I would like to thank my wife Daniella Koopmann, my son Noah and my yet unborn daughter for their patience and understanding for all the time that I devoted to writing this book instead of taking care of them.

München, April 2002 *Axel Groß*

Contents

1. Introduction 1
2. **The Hamiltonian** 5
   2.1 The Schrödinger Equation 5
   2.2 Born–Oppenheimer Approximation 7
   2.3 Structure of the Hamiltonian 8
   Exercises 18
3. **Electronic Structure Methods and Total Energies** 21
   3.1 Hartree–Fock Theory 21
   3.2 Quantum Chemistry Methods 31
   3.3 Density Functional Theory 36
   3.4 Pseudopotentials 41
   3.5 Implementations of Density Functional Theory 44
   3.6 Further Many-Electron Methods 49
   3.7 Tight-Binding Method 50
   Exercises 52
4. **Structure and Energetics of Clean Surfaces** 57
   4.1 Electronic Structure of Surfaces 57
   4.2 Metal Surfaces 63
   4.3 Semiconductor Surfaces 72
   4.4 Ionic Surfaces 80
   4.5 Interpretation of STM Images 85
   4.6 Surface Phonons 87
   Exercises 92
5. **Adsorption on Surfaces** 95
   5.1 Potential Energy Surfaces 95
   5.2 Physisorption 97
   5.3 Newns–Anderson Model 104
   5.4 Atomic Chemisorption 110
   5.5 Effective Medium Theory and Embedded Atom Method 116
   5.6 Reactivity Concepts 123

	5.7	Adsorption on Low-Index Surfaces	128
	5.8	Adsorption on Precovered Surfaces	136
	5.9	Adsorption on Structured Surfaces	139
	5.10	Reactions on Surfaces	142
		Exercises	144
6.	**Gas-Surface Dynamics**	147
	6.1	Classical Dynamics	147
	6.2	Quantum Dynamics	149
	6.3	Parametrization of ab initio Potentials	154
	6.4	Scattering at Surfaces	157
	6.5	Atomic and Molecular Adsorption on Surfaces	162
	6.6	Dissociative Adsorption and Associative Desorption	171
		Exercises	189
7.	**Kinetic Modelling of Processes on Surfaces**	191
	7.1	Determination of Rates	191
	7.2	Diffusion	195
	7.3	Kinetic Lattice Gas Model	198
	7.4	Kinetic Modelling of Adsorption and Desorption	199
	7.5	Growth	205
	7.6	Reaction Kinetics on Surfaces	217
		Exercises	221
8.	**Electronically Non-adiabatic Processes**	225
	8.1	Determination of Electronically Excited States	225
	8.2	Electronic Excitation Mechanisms at Surfaces	228
	8.3	Electronic Friction Effects	229
	8.4	Reaction Dynamics with Electronic Transitions	232
	8.5	Electronic Transitions	234
		Exercises	239
9.	**Perspectives**	241
	9.1	Solid-liquid Interface	241
	9.2	Nanostructured Surfaces	243
	9.3	Biologically Relevant Systems	244
	9.4	Industrial Applications	246

References 249

Index 271

1. Introduction

It has always been the goal of theoretical surface science to understand the fundamental principles that govern the geometric and electronic structure of surfaces and the processes occuring on these surfaces like gas-surface scattering, reactions at surfaces or growth of surface layers. Processes on surfaces play an enormously important technological role. We are all surrounded by the effects of these processes in our daily life. Some are rather obvious to us like rust and corrosion. These are reactions that we would rather like to avoid. Less obvious are surface reactions that are indeed very advantageous. Many chemical reactions are promoted tremendously if they take place on a surface that acts as a catalyst. Actually most reactions employed in the chemical industry are performed in the presence of a catalyst. Catalysts are not only used to increase the output of a chemical reaction but also to convert hazardous waste into less harmful products. The most prominent example is the car exhaust catalyst.

The field of modern surface science is characterized by a wealth of microscopic experimental information. The positions of both substrate and adsorbate atoms on surfaces can be determined by scanning microscopes, the initial quantum states of molecular beams hitting a surface can be well-controlled, and desorbing reaction products can be analysed state-specifically. This provides an ideal field to establish a microscopic theoretical description that can either explain experimental findings or in the case of theoretical predictions can be verified by experiment.

And indeed, recent years have seen a tremendous progress in the microscopic theoretical treatment of surfaces and processes on surfaces. While some decades ago a phenomenological thermodynamic approach was prevalent, now microscopic concepts are dominant in the analysis of surface processes. A variety of surface properties can now be described from first principles, i.e. without invoking any empirical parameters.

In fact, the field of theoretical surface science is no longer limited to explanatory purposes only. It has reached such a level of sophistication and accuracy that reliable predictions for certain surface science problems have become possible. Hence both experiment and theory can contribute on an equal footing to the scientific progress. In particular, computational surface science may act as a virtual chemistry and physics lab at surfaces. Computer

Fig. 1.1. The virtual chemistry and physics lab at surfaces: simulation of surface structures and processes at surfaces on the computer

experiments can thus add relevant information to the research process. Such a computer experiment in the virtual lab is illustrated in Fig. 1.1.

In this book the theoretical concepts and computational tools necessary and relevant for theoretical surface science will be introduced. I will present a microscopic approach towards the theoretical description of surface science. Based on the fundamental theoretical entity, the Hamiltonian, a hierarchy of theoretical methods will be introduced in order to describe surface processes. But even for the largest time and length scales I will develop a statistical rather than a thermodynamic approach, i.e., all necessary parameters will be derived from microscopic properties.

Following this approach, theoretical methods used to describe static properties such as surface structures and dynamical processes such as reactions on surfaces will be presented. An equally important aspect of the theoretical treatment, however, is the proper analysis of the results that leads to an understanding of the underlying microscopic mechanisms. A large portion of the book will be devoted to the establishment of theoretical concepts that can be used to catagorize the seemingly immense variety of structures and processes at surfaces. The discussion will be rounded up by the presentation of case studies that are exemplary for a certain class of properties. Thus I will address subjects like surface and adsorbate structures, reactivity concepts, dynamics and kinetics of processes on surfaces and electronically nonadiabatic effects. All chapters are supplemented by exercises in which the reader is invited either to reproduce some important derivations or to determine typical properties of surfaces.

The theoretical tools employed for surface science problems are not exclusively used for these particular problems. In fact, surface science is a research field at the border between chemistry and solid state physics. Consequently, most of the theoretical methods have been derived either from quantum chemistry or from condensed matter physics. It is outside of the scope of this book to derive all these methods in every detail. However, most of the methods used commonly in theoretical surface science will be addressed and discussed. Hence this book can be used as a reference source for theoretical methods. It is not only meant for graduate students doing research in theoretical surface science but also for experimentalists who want to get an idea about the methods their theoretical colleagues are using.

However, it is fair to say that for a certain class of systems theoretical surface science is still not accurate enough for a reliable description. These problems will be mentioned throughout the book. The open problems and challenges will be presented, but also the opportunities will be illustrated that open up once the problems are solved.

In detail, this book is structured as follows. In the next chapter we first introduce the basic Hamiltonian appropriate for surface science problems. We will consider general properties of this Hamiltonian that are important for solving the Schrödinger equation. At the same time the terminology necessary to describe surface structures will be introduced.

A large part will be devoted to the introduction of electronic structure methods because they are the foundation of any ab initio treatment of surface science problems. Both wave-function and electron-density based methods will be discussed. In addition, the most important techniques used in implementations of electronic structure methods will be addressed.

The structure and energetics of metal, semiconductor and insulators surface is the subject of the following chapter. Using some specific substrates as examples, the different mechanisms determining the structure of these surfaces will be introduced. In this context, the theoretical treatment of surface phonons is also covered. In the succeeding chapter, the theoretical description of surfaces will be extended to atomic and molecular adsorption systems. Reactivity concepts will be discussed which provide insight into the chemical trends observed in adsorption on clean, precovered and structured surfaces. These concepts are also applied to simple reactions on surfaces.

Dynamics of scattering, adsorption and desorption at surfaces is the subject of the next chapter. Classical and quantum methods to determine the time evolution of processes on surfaces will be introduced. The determination of reaction probabilities and distributions in gas-surface dynamics will be illustrated. For processes such as diffusion, growth or complex reactions at surfaces, a microscopic dynamical simulation is no longer possible. A theoretical treatment of such processes based on ab initio calculations is still possible using a kinetic approach, as will be shown in the next chapter.

While most of the processes presented in this book are assumed to occur in the electronic ground state, there is an important class of electronically non-adiabatic processes at surfaces. The theoretical description of nonadiabatic phenomena has not reached the same level of maturity as the treatment of electronic ground-state properties, as the following chapter illustrates, but there are promising approaches. Finally, I will sketch future research directions in surface science where theory can still contribute significantly to enhance the understanding, and I will give examples of first successful applications.

2. The Hamiltonian

Any theoretical description has to start with the definition of the system under consideration and a determination of the fundamental interactions present in the system. This information is all contained in the Hamiltonian which is the central quantity for any theoretical treatment. All physical and chemical properties of any system can be derived from its Hamiltonian. Since we are concerned with microscopic particles like electrons and atoms in surface science, the proper description is given by the laws of quantum mechanics. This requires the solution of the Schrödinger equation.

In this chapter we will first describe the Hamiltonian entering the Schrödinger equation appropriate for surface science problems. One general approximation that makes the solution of the full Schrödinger equation tractable is the decoupling of the electronic and nuclear motion which is called the Born–Oppenheimer or adiabatic approximation. We will then have a closer look at the specific form of the Hamiltonian describing surfaces. We will discuss the symmetries present at surfaces. Taking advantage of symmetries can greatly reduce the computational cost in theoretical treatments. Finally, we will introduce and illustrate the nomenclature to describe the structure of surfaces.

2.1 The Schrödinger Equation

In solid state physics as well as in chemistry, the only fundamental interaction we are concerned with is the electrostatic interaction. Furthermore, relativistic effects are usually negligible if only the valence electrons are considered. To start with, we treat core and valence electrons on the same footing and neglect any magnetic effects. Then a system of nuclei and electrons is described by the nonrelativistic Schrödinger equation with a Hamiltonian of a well-defined form,

$$H = T_{\text{nucl}} + T_{\text{el}} + V_{\text{nucl}-\text{nucl}} + V_{\text{nucl}-\text{el}} + V_{\text{el}-\text{el}} \,. \tag{2.1}$$

T_{nucl} and T_{el} are the kinetic energy of the nuclei and the electrons, respectively. The other terms describe the electrostatic interaction between the positively charged nuclei and the electrons. As long as it is not necessary, we will not take the spin into account for the sake of clarity of the equations.

Consequently, neglecting spin the single terms entering the Hamiltion are explicitly given by

$$T_{\text{nucl}} = \sum_{I=1}^{L} \frac{\boldsymbol{P}_I^2}{2M_I}, \tag{2.2}$$

$$T_{\text{el}} = \sum_{i=1}^{N} \frac{\boldsymbol{p}_i^2}{2m}, \tag{2.3}$$

$$V_{\text{nucl}-\text{nucl}} = \frac{1}{2} \sum_{I \neq J} \frac{Z_I Z_J e^2}{|\boldsymbol{R}_I - \boldsymbol{R}_J|}, \tag{2.4}$$

$$V_{\text{nucl}-\text{el}} = -\sum_{i,I} \frac{Z_I e^2}{|\boldsymbol{r}_i - \boldsymbol{R}_I|}, \tag{2.5}$$

and

$$V_{\text{el}-\text{el}} = \frac{1}{2} \sum_{i \neq j} \frac{e^2}{|\boldsymbol{r}_i - \boldsymbol{r}_j|}. \tag{2.6}$$

Throughout this book we will use CGS-Gaussian units as it is common practice in theoretical physics textbooks. Atoms will usually be numbered by capital letter indices. Thus, Z_I stands for the charge of the I-th nuclei. The factor $\frac{1}{2}$ in the expressions for $V_{\text{nucl}-\text{nucl}}$ and $V_{\text{el}-\text{el}}$ ensures that the interaction between the same pair of particles is not counted twice.

In principle we could stop here because all what is left to do is to solve the many-body Schrödinger equation using the Hamiltonian (2.1)

$$H\Phi(\boldsymbol{R}, \boldsymbol{r}) = E\Phi(\boldsymbol{R}, \boldsymbol{r}). \tag{2.7}$$

The whole physical information except for the symmetry of the wave functions is contained in the Hamiltonian. In solving the Schrödinger equation (2.7), we just have to take into account the appropriate quantum statistics such as the Pauli principle for the electrons which are fermions. The nuclei are either bosons or fermions, but usually their symmetry does not play an important role in surface science. Often relativistic effects can also be neglected. Only if heavier elements with very localized wave functions for the core electron are considered, relativistic effects might be important since the localization leads to high kinetic energies of these electrons.

Note that only the kinetic and electrostatic energies are directly present in the Hamiltonian. We will later see that the proper consideration of the quantum statistics leads to contributions of the so-called *exchange-correlation energy* in the effective Hamiltonians. However, it is important to realize that the energy gain or cost according to additional effective terms has to be derived from the energy gain or cost in kinetic and electrostatic energy that is caused by the quantum statistics.

Unfortunately, the solution of the Schrödinger equation in closed form is not possible. Even approximative solutions are far from being trivial. In the rest of the book we will therefore be concerned with a hierarchy of approximations that will make possible the solution of (2.7) at least within reasonable accuracy. The first step in this hierarchy will be the so-called Born–Oppenheimer approximation.

2.2 Born–Oppenheimer Approximation

The central idea underlying the Born–Oppenheimer [1] or adiabatic approximation is the separation in the time scale of processes involving electrons and atoms. Except for hydrogen and helium, atoms have a mass that is 10^4 to 10^5 times larger than the mass of an electron. Consequently, at the same kinetic energy electrons are 10^2 to 10^3 times faster than the nuclei. Hence one supposes that the electrons follow the motion of the nuclei almost instantaneously. Most often one simply assumes that the electrons stay in their ground state for any configuration of the nuclei. The electron distribution then determines the potential in which the nuclei moves.

In practice, one splits up the full Hamiltonian and defines the electronic Hamiltonian H_{el} for fixed nuclear coordinates $\{\boldsymbol{R}\}$ as follows

$$H_{el}(\{\boldsymbol{R}\}) = T_{el} + V_{nucl-nucl} + V_{nucl-el} + V_{el-el} \,. \tag{2.8}$$

In (2.8) the nuclear coordinates $\{\boldsymbol{R}\}$ do not act as variables but as parameters defining the electronic Hamiltonian. The Schrödinger equation for the electrons for a given fixed configuration of the nuclei is then

$$H_{el}(\{\boldsymbol{R}\})\Psi(\boldsymbol{r},\{\boldsymbol{R}\}) = E_{el}(\{\boldsymbol{R}\})\Psi(\boldsymbol{r},\{\boldsymbol{R}\}) \,. \tag{2.9}$$

Again, in (2.9) the nuclear coordinates $\{\boldsymbol{R}\}$ are not meant to be variables but parameters. In the Born–Oppenheimer or adiabatic approximation the eigenenergy $E_{el}(\{\boldsymbol{R}\})$ of the electronic Schrödinger equation is taken to be the potential for the nuclear motion. $E_{el}(\{\boldsymbol{R}\})$ is therefore called the Born–Oppenheimer energy surface. The nuclei are assumed to move according to the atomic Schrödinger equation

$$\{T_{nucl} + E_{el}(\boldsymbol{R})\}\Lambda(\boldsymbol{R}) = E_{nucl}\Lambda(\boldsymbol{R}) \,. \tag{2.10}$$

Often the quantum effects in the atomic motion are neglected and the classical equation of motion are solved for the atomic motion:

$$M_I \frac{\partial^2}{\partial t^2}\boldsymbol{R}_I = -\frac{\partial}{\partial \boldsymbol{R}_I}E_{el}(\{\boldsymbol{R}\}) \,. \tag{2.11}$$

The force acting on the atoms can be conveniently evaluated using the Hellmann–Feynman theorem [2,3]

$$\boldsymbol{F}_I = -\frac{\partial}{\partial \boldsymbol{R}_I}E_{el}(\{\boldsymbol{R}\}) = \langle \Psi(\boldsymbol{r},\{\boldsymbol{R}\})|\frac{\partial}{\partial \boldsymbol{R}_I}H_{el}(\{\boldsymbol{R}\})|\Psi(\boldsymbol{r},\{\boldsymbol{R}\})\rangle \,. \tag{2.12}$$

In principle, in the Born–Oppenheimer approximation electronic transition due to the motion of the nuclei are neglected. One can work out the Born–Oppenheimer approximation in much more detail (see, e.g., [4]), however, what it comes down to is that the small parameter m/M is central for the validity of the adiabatic approximation (see Exercise 2.1). In fact, the Born–Oppenheimer approximation is very successful in the theoretical description of processes at surfaces. Still its true validity is hard to prove because it is very difficult to correctly describe processes that involve electronic transition (see Chap. 8).

If it takes a finite amount of energy to excite electronic states, i.e., if the adiabatic electronic states are well-separated, then it can be shown that electronically nonadiabatic transitions are rather improbable (see, e.g., [5]). In surface science this applies to insulator and semiconductor surfaces with a large band gap. At metal surfaces no fundamental band gap exists so that electronic transitions with arbitrarily small excitations energies can occur. Still, the strong coupling of the electronic states in the broad conduction band leads to short lifetimes of excited states and thus to a fast quenching of these states.

On the other hand, there are very interesting processes in which electronic nonadiabatic processes are induced, as we will see in Chap. 8. The theoretical treatment of these systems requires special techniques that will also be discussed later in this book.

2.3 Structure of the Hamiltonian

Employing the Born–Oppenheimer approximation means first to solve the electronic structure problem for fixed atomic coordinates. The atomic positions determine the external electrostatic potential in which the electrons move. Furthermore, they determine the symmetry properties of the Hamiltonian.

Surface science studies are concerned with the structure and dynamics of surfaces and the interaction of atoms and molecules with surfaces. If not just ordered surface structures are considered, then the theoretical surface scientists has to deal with the interaction of a system with only few degrees of freedom, the atom or molecule, with a system, the surface or substrate, that has in principle an infinite number of degrees of freedom. Thus the substrate exhibits a quasi-continuum of states. Usually different methods are used to deal with the single subsystems: molecules are treated by quantum chemistry methods while surfaces are handled by solid-state methods.

To deal with both subsystems on an equal footing represents a real challenge for any theoretical treatment, but it also makes up the special attraction of theoretical surface science. We will focus on this issue in more detail in the next chapter. But before considering a strategy to solve the Schrödinger equation it is always important to investigate the symmetries of the Hamiltonian. Not only rigorous results can be derived from symmetry considerations, but these considerations can also reduce the computational effort dramatically. This can be demonstrated very easily [6]. Let T be the operator of a symmetry transformation that leaves the Hamiltonian H invariant. Then H and T commute, i.e. $[H, T] = 0$. This means that according to a general theorem of quantum mechanics [7] the matrix elements $\langle\psi_i|H|\psi_j\rangle$ vanish, if $|\psi_i\rangle$ and $|\psi_j\rangle$ are eigenfunctions of T belonging to different eigenvalues $T_i \neq T_j$.

This property of the eigenfunctions can help us enormously in solving the Schrödinger equation. Imagine we want to determine the eigenvalues of a Hamiltonian by expanding the wave function in an appropriate basis set. Then we only need to expand the wave function within a certain class of functions having all the same eigenvalue with respect to a commuting symmetry operator. Functions having another symmetry will belong to a different eigenvalue. Since the numerical effort to solve the Schrödinger equation can scale very unfavorably with the number n of basis functions (up to n^7 for very accurate quantum chemical methods), any reduction in this number can mean a huge reduction in computer memory and time.

The mathematical tool to deal with symmetries is group theory. It is beyond the scope of this book to provide an introduction into group theory. There are many text books that can be used as a reference, for example [6, 8, 9]. I will rather describe the symmetries present at surfaces, which has also the important aspect of defining the terminology commonly used to specify surface structures. To set the stage, we will first start with ideal three-dimensional crystal structures.

A three-dimensional periodic crystal is given by an infinite array of identical cells. These cells are arranged according to the so-called Bravais lattice. It is given by all the position vectors of the form

$$\boldsymbol{R} = n_1\boldsymbol{a}_1 + n_2\boldsymbol{a}_2 + n_3\boldsymbol{a}_3 \,. \tag{2.13}$$

The $\boldsymbol{a}_i$ are the three non-collinear unit vectors of the lattice, the n_i are integer numbers. The lattice vectors $\boldsymbol{R}$ are not necessarily identical with atomic positions of the crystal, but in most cases they are indeed identified with atomic positions. In addition to the translational symmetry, there are also so-called point operations that transform the crystal into itself. Operations such as rotation, reflection and inversion belong to this point group. Furthermore, translations through a vector not belonging to the Bravais lattice and point operations can be combined to give additional distinct symmetry operations, such as screw axes or glide planes.

There are 14 different types of Bravais lattices in three dimensions. Now there can be more than one atom per unit cell of the Bravais lattice. Then the crystal structure is given as a Bravais lattice with a basis which corresponds to the positions of the additional atoms in the unit cell. If the lattice has a basis, the symmetry of the corresponding crystal will usually be reduced compared to a crystal with just spherical symmetric atoms at the Bravais

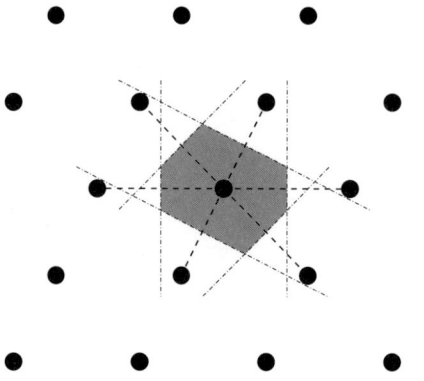

Fig. 2.1. Wigner–Seitz cell for a two-dimensional Bravais lattice. The six sides of the cell bisect the lines joining the central point to its six nearest neighbors. (After [10])

lattice sites. This enhances the number of symmetrically distinct lattices to in total 230. Details of these structure can be found in any text book of solid-state physics such as in [10, 11].

There is no way to uniquely choose the primitive unit cell of a Bravais lattice. Any cell that, when translated through all the Bravais lattice vectors, fills all space can serve as a unit cell. However, it is convenient to select a unit cell that has the full symmetry of the Bravais lattice. The so-called *Wigner–Seitz cell* has this property. It is defined as the region of space around a lattice point that is closer to that point than any other lattice point [10]. The construction of the Wigner–Seitz cell is demonstrated in Fig. 2.1 for a two-dimensional Bravais lattice. Select a lattice point and draw lines to the nearest-neighbors. Then bisect each connection with a line and take the smallest polyeder that contains the points bounded by these lines. Note that in two dimensions the Wigner–Seitz cell is always a hexagon unless the lattice is rectangular (see Exercise 4.3).

The periodicity of a crystal lattice leads to the existence of a dual space that mathematically reflects the translational symmetry of a lattice. The dual space to the real space for periodic structure is called reciprocal space. The basis vectors are obtained from the basis vectors of the real space $\boldsymbol{a}_i$ via

$$\boldsymbol{b}_1 = 2\pi \frac{\boldsymbol{a}_2 \times \boldsymbol{a}_3}{|\boldsymbol{a}_1 \cdot (\boldsymbol{a}_2 \times \boldsymbol{a}_3)|}. \tag{2.14}$$

The other two basis vectors of the reciprocal lattice $\boldsymbol{b}_2$ and $\boldsymbol{b}_3$ are obtained by a cyclic permutation of the indices in (2.14). By construction, the lattice vectors of the real space and the reciprocal lattice obey the relation

$$\boldsymbol{a}_i \cdot \boldsymbol{b}_j = 2\pi \delta_{ij}, \tag{2.15}$$

where δ_{ij} is the Kronecker symbol.

The reciprocal space is often called k-space since plane wave characterized by their wave vector $\boldsymbol{k}$ are represented by single points in the reciprocal space. The eigenenergies of the electronic wave functions in periodic lattices are usually plotted as a function of their $\boldsymbol{k}$-vector in the *first Brillouin zone*

2.3 Structure of the Hamiltonian

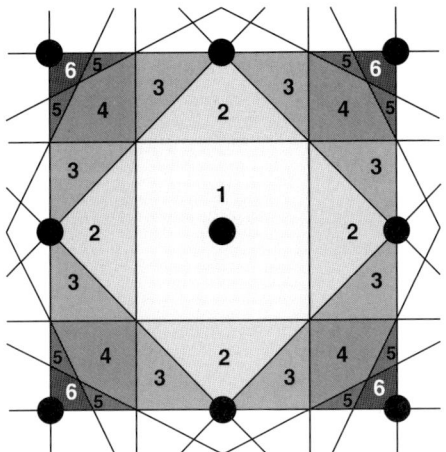

Fig. 2.2. Illustration of the definition of the Brillouin zones for a two-dimensional square reciprocal lattice. Note that only the first three Brillouin zones are entirely within the shaded areas. (After [10])

which is defined as the Wigner–Seitz cell of the reciprocal lattice. As the name *first Brillouin zone* suggests, there are also higher-order Brillouin zones. Their construction is illustrated in Fig. 2.2 in two dimensions for a square reciprocal lattice. The n-th Brillouin zone is defined as the set of points that can be reached from the origin by crossing the $n-1$ nearest bisecting planes. Note that each Brillouin zone is also a primitive unit cell of the reciprocal lattice. In fact, by translating the different sections of the higher-order Brillouin zones by appropriate reciprocal lattice vectors they can be rearranged to cover the first Brillouin zone. This can be easily checked for the second and third Brillouin zone in Fig. 2.2. In the periodic electronic structure theory this is called *backfolding*.

Reciprocal lattice vectors are used to denote lattice planes of the real-space lattice. Lattice planes of a Bravais lattice are described by the shortest reciprocal lattice vector $h\boldsymbol{b}_1 + k\boldsymbol{b}_2 + h\boldsymbol{b}_3$ that is perpendicular to this plane. The integer coefficients hkl are called *Miller indices*. Lattice planes are specified by the Miller indices in parentheses: (hkl). Family of lattice planes, i.e. lattice planes that are equivalent by symmetry, are denoted by $\{hkl\}$. Finally, indices in square brackets $[hkl]$ indicate directions. For face-centered cubic (fcc) and body-centered cubic (bcc) crystals the Miller indices are usually related to the underlying simple cubic lattice, i.e., fcc and bcc crystals are described as simple cubic lattices with a basis.

For hcp crystals such as Ru, Co, Zn or Ti, the Miller index notation used to describe the orientation of lattice planes is slightly more complex since no standard cartesian set of axes can be used. Instead the notation is based upon three axes at 120 degrees in the close-packed plane, and one axis (the c-axis) perpendicular to these planes. This leads to a four-digit index structure. However, since the first three axes are coplanar, the first three indices have to add up to zero. Hence the third index is redundant; in fact, it is sometimes omitted. For example, both Ru(001) and Ru(0001) describe the close-packed hexagonal plane of the hcp metal Ru.

The power of group theory to derive rigorous results can be nicely illustrated for periodic structures. The solution Ψ of the electronic Schrödinger equation (2.9) is a many-body wave function that incorporates the electron-electron interaction. However, as we will see below, there are many schemes to solve the electronic Schrödinger equation that involve the solution of effective one-particle Schrödinger equations of the form

$$\left\{-\frac{\hbar^2}{2m}\nabla^2 + v_{\text{eff}}(\boldsymbol{r})\right\}\psi_i(\boldsymbol{r}) = \varepsilon_i \psi_i(\boldsymbol{r}), \tag{2.16}$$

where the effective one-particle potential $v_{\text{eff}}(\boldsymbol{r})$ satisfies translational symmetry:

$$v_{\text{eff}}(\boldsymbol{r}) = v_{\text{eff}}(\boldsymbol{r} + \boldsymbol{R}), \tag{2.17}$$

with $\boldsymbol{R}$ being any Bravais lattice vector.

The translational operations $T_{\boldsymbol{R}}$ form an Abelian group since the order of translations does not matter for the result of applying two succesive translations. As mentioned above, the solutions of the Hamiltonian can be classified according to their symmetry properties. In group theory one says that solutions of different symmetries belong to so-called different representations of the symmetry group. Now there is an important theorem that the representations of an Abelian group are one-dimensional [6], which means that the eigenfunctions of the translational group can be written as

$$T_{\boldsymbol{R}}\psi_i(\boldsymbol{r}) = \psi_i(\boldsymbol{r} + \boldsymbol{R}) = c_i(\boldsymbol{R})\psi_i(\boldsymbol{r}). \tag{2.18}$$

The eigenvalues $c_i(\boldsymbol{R})$ are complex numbers of modulus unity that have to satisfy

$$c_i(\boldsymbol{R})c_i(\boldsymbol{R}') = c_i(\boldsymbol{R} + \boldsymbol{R}'), \tag{2.19}$$

which can be derived by applying two successive translation. From this relation it follows that the eigenvalues $c_i(\boldsymbol{R})$ can be expressed in an exponential form

$$c_i(\boldsymbol{R}) = \mathrm{e}^{\mathrm{i}\boldsymbol{k}\cdot\boldsymbol{R}}. \tag{2.20}$$

The eigenfunction $\psi_i(\boldsymbol{r})$ is thus characterized by the *crystal-momentum* $\boldsymbol{k}$ that acts as a quantum number. Equation (2.18) can now be reformulated to state that the eigenstates of a periodic Hamiltonian can be written in the form

$$\psi_{\boldsymbol{k}}(\boldsymbol{r}) = \mathrm{e}^{\mathrm{i}\boldsymbol{k}\cdot\boldsymbol{r}} u_{\boldsymbol{k}}(\boldsymbol{r}) \tag{2.21}$$

with the periodic function

$$u_{\boldsymbol{k}}(\boldsymbol{r}) = u_{\boldsymbol{k}}(\boldsymbol{r} + \boldsymbol{R}) \tag{2.22}$$

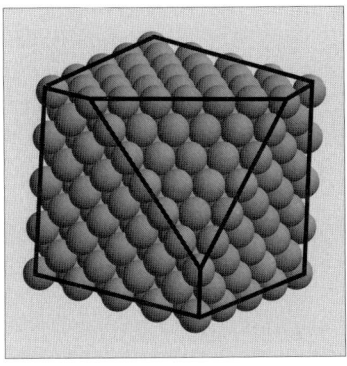 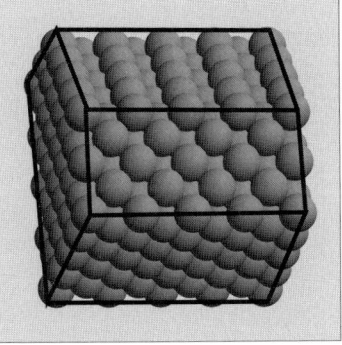

Fig. 2.3. Left panel: fcc crystal with 100 faces and one 111 face, right panel: fcc crystal with 100 faces and one 110 face

for all Bravais lattice vectors $\boldsymbol{R}$. This is the famous *Bloch theorem* which is an *exact* result since it is purely based on symmetry properties. Functions that obey the relation (2.21) are usually called *Bloch functions*.

A surface can be thought to be created by just cleaving an infinite crystal along one surface plane. A bulk-terminated surface, i.e. a surface whose configuration has not changed after cleavage, is called an *ideal* surface. Such ideal surfaces are shown in Fig. 2.3 where two fcc crystals are plotted that are terminated by the square {100} faces. In addition, the cubes are further cleaved to indicate the other low-index faces of a face-centered cubic crystal. In the left panel of Fig. 2.3, the (111) face is shown which is perpendicular to the diagonal of the cubic unit cell. This (111) face with its hexagonal structure is the closest-packed fcc surface. In the right panel a (110) face is shown that is perpendicular to the diagonal of one of the square faces. The (110) surface has already a rather open structure with troughs running along the [1$\bar{1}$0] direction.

A semi-infinite solid with an ideal surface no longer has the three-dimensional periodicity of the crystal. Still there is a two-dimensional periodicity present parallel to the surface. Equivalently to the three-dimensional case, in two dimensions Bravais lattices can be defined. Furthermore, there is also a two-dimensional Bloch theorem for a crystal having a periodic structure parallel to the surface which says that the electronic single-particle wave functions can be written as

$$\psi_{\mathbf{k}_\parallel}(\boldsymbol{r}) = e^{i\mathbf{k}_\parallel \cdot \boldsymbol{r}} u_{\mathbf{k}_\parallel}(\boldsymbol{r}), \tag{2.23}$$

where $u_{\mathbf{k}_\parallel}(\boldsymbol{r})$ has the two-dimensional periodicity of the surface.

There are five two-dimensional Bravais lattices which are sketched in Table 2.1. In fact, the centered rectangular lattice is just a special case of an oblique lattice, but it is usually listed separately. Examples of low-index planes of fcc and bcc crystals with the corresponding symmetry are also plotted. The square (100) surface has a fourfold symmetry axis. The hexagonal

14   2. The Hamiltonian

Table 2.1. The five two-dimensional Bravais lattices. In addition, examples of low-index planes of fcc and bcc crystal with the corresponding symmetry are plotted

2D lattice	Schematic sketch	Examples
Square $|a_1| = |a_2|$ $\varphi = 90°$		fcc(100)
Hexagonal $|a_1| = |a_2|$ $\varphi = 120°$		fcc(111)
Rectangular $|a_1| \neq |a_2|$ $\varphi = 90°$		fcc(110)
Centered rectangular $|a_1| \neq |a_2|$ $\varphi = 90°$		bcc(110)
Oblique $|a_1| \neq |a_2|$ $\varphi \neq 90°$		fcc(210)

Table 2.2. Illustration of relaxation and reconstruction of the fcc(110) surface. In the relaxed structure just the distance between the top and the second layer is changed leaving the surface symmetry unchanged while in the (2 × 1) missing-row reconstruction every second row on the surface is missing

Structure	Ideal 1×1	Relaxed 1×1	Reconstructed 2×1
Top view			
Side view			

(111) surface with its sixfold symmetry axis corresponding to the closest-packed surface is usually the most stable surface. Rectangular surfaces such as the (110) surface have already a more open structure. In fact, the low-index (100), (111) and (110) faces are the most often studied surfaces in surface science. Oblique surfaces are usually rather complex. Often they correspond to stepped surface like the example of the (210) surface that is shown in Table 2.1.

The planes plotted in Table 2.1 correspond to ideal surfaces where the interatomic distances are the same as in the bulk. However, at a real surface the fact that the bonding situation is entirely different compared to the bulk situation will cause a rearrangement of the atoms at and close to the surface. If this rearrangement preserves the symmetry of the bulk plane of termination, it is called relaxation. The corresponding surface structure is refered to as a (1 × 1) structure. However, if a significant restructuring of the surface occurs that changes the periodicity and symmetry of the surface, it is termed *reconstruction*.

Such a structure is labeled with respect to the ideal termination of the corresponding surface plane. If the new surface unit cell is spanned by new vectors $\boldsymbol{a}_1^s = m\boldsymbol{a}_1$ and $\boldsymbol{a}_2^s = n\boldsymbol{a}_2$, the surface is labeled by $(hkl)(m \times n)$.

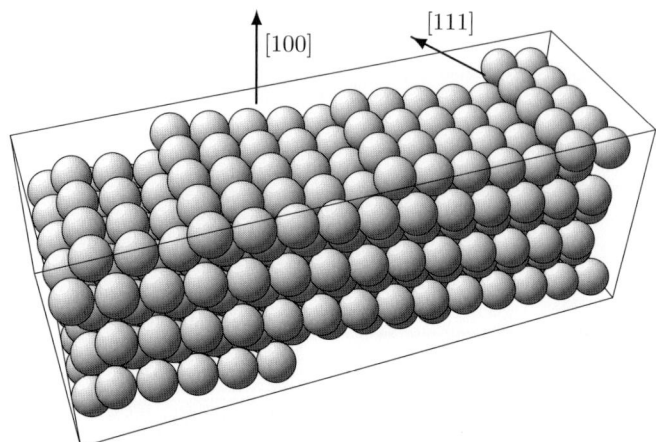

Fig. 2.4. A stepped (911) = 5(100) × (111) vicinal surface. Steps with ledges of (111) orientations separate (100) terraces that are 5 atom rows wide

Sometimes $(hkl)p(m \times n)$ is written, where p stands for *primitive*. Frequently, surface structures are observed with two atoms in the unit cell where the second atom occupies the centre of the unit cell. Such a situation is then labeled by $(hkl)c(m \times n)$, where c stands for *centered* [12].

The difference between relaxation and reconstruction is illustrated in Table 2.2 using the fcc(110) surface as an example. In the relaxed geometry just the distance between the top and the second layer is decreased with respect to the ideal surface. The top view of the relaxed structure indicates that the lateral symmetry of the surface remains unchanged. The last column of Table 2.2 presents a very prominent example for surface reconstructions, namely the so-called *missing-row reconstruction* which occurs for a number of materials such as Au(110) [13] or Pd(110). Every second row of the (110) surface is missing. The surface unit cell becomes twice as large resulting in a 2×1 structure.

Semiconductor surfaces often show much more complex reconstruction patterns than metals. This is caused by the covalent nature of bonding in semiconductors where creating a surface strongly perturbs the bonding situation. The most famous example is the 7×7 reconstruction of the Si(111) surface. But also compound semiconductor such as GaAs exhibit extremely complex reconstruction patterns, as will be demonstrated in Sect. 4.3.

The periodicity of a surface can also be perturbed by the presence of adsorbates. For sufficiently strong adsorbate-substrate interactions commensurate adlayers will be created that result in larger surface unit cells as, e.g., for the $O(2 \times 2)/Pt(111)$ structure, where one fourth of the surface three-fold hollow sites are occupied by oxygen atoms. If the adsorbate-substrate interaction is weaker than the adsorbate-adsorbate coupling strength, as is often

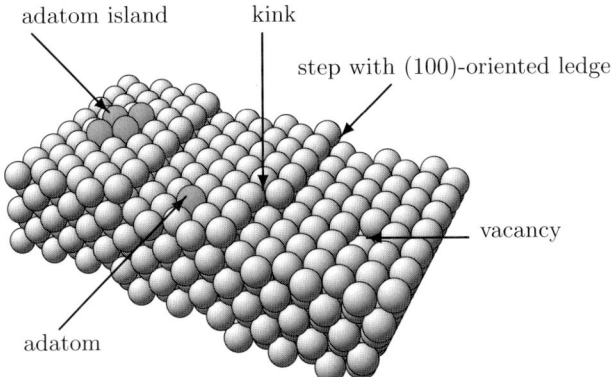

Fig. 2.5. Illustration of defects on surfaces such as steps, kinks, adatoms, adatom islands, and vacancies

the case for organic adlayers, the adsorbate adlayer is not nessessarily in registry with the surface resulting in an incommensurate adlayer for which no longer any surface periodicity can be expressed. This makes any theoretical treatment rather cumbersome.

A surface that is only slightly misaligned from a low index plane is called a *vicinal surface*. A vicinal surface is structured as a periodic array of terraces of a low-index orientation separated by monoatomic steps. In Fig. 2.4 a (911) surface is shown illustrating the structure of a vicinal surface. The high-index (911) surface consists of 5 atomic rows of (100) orientation separated by a step with a (111) ledge, i.e., the ledge represents (111) microfacets. In fact, in order to make the structure of a vicinal surfaces immediately obvious, they are often denoted by $n(hkl) \times (h'k'l')$ where (hkl) and $(h'k'l')$ are the Miller indices of the terraces and of the ledges, and n gives the width of the terraces in number of atomic rows parallel to the ledges. By studying vicinal surfaces the influence of steps on, e.g., adsorption properties or reactions on surfaces can be studied in a well-defined way. Further defects that can exist on surfaces are kinks, adatoms, vacancies and adatom islands. These defects are illustrated in Fig. 2.5. In fact, the plotted surface corresponds to a defected $(755) = 5(111) \times (100)$ surface where the steps are made of (100)-oriented microfacets. The creation of defects is usually associated with a cost of energy. Yet, at non-zero surface temperature there will always be a certain amount of defects present because of entropic reasons. This is a particular problem for experimentalists who have to check whether their observed results on nominally flat surfaces might be dominated by minority defect sites.

On the other hand, the study of defects is important because often the defects are considered to be the active sites for surface reactions. This is relevant for the understanding of, for example, real catalysts which usually exhibit a very defect-rich structure. Furthermore, the defects depicted in Fig. 2.5 all appear during growth processes on surfaces. Thus the properties are relevant for a true understanding of growth, as will be shown in Sect. 7.5.

Exercises

2.1 Born–Oppenheimer Approximation

Expand the eigenfunctions of the total Hamiltonian

$$H = T_{\text{nucl}} + T_{\text{el}} + V_{\text{nucl-nucl}} + V_{\text{nucl-el}} + V_{\text{el-el}} \tag{2.24}$$

according to

$$\Phi(\boldsymbol{R},\boldsymbol{r}) = \sum_{\mu} \Lambda_\mu(\boldsymbol{R})\, \Psi_\mu(\boldsymbol{r},\boldsymbol{R}), \tag{2.25}$$

where the $\Psi_\mu(\boldsymbol{r},\boldsymbol{R})$ are the eigenfunctions of the electronic Hamiltonian

$$H_{\text{el}}(\{\boldsymbol{R}\}) = T_{\text{el}} + V_{\text{nucl-nucl}} + V_{\text{nucl-el}} + V_{\text{el-el}}. \tag{2.26}$$

By multiplying the many-body Schrödinger equation (2.7) by $\langle \Psi_\nu |$, a set of coupled differential equations for the nuclear wave functions $\Lambda_\mu(\boldsymbol{R})$ can be obtained.

a) Write down the coupled equations for the nuclear wave functions $\Lambda_\mu(\boldsymbol{R})$. Which terms are neglected in the Born–Oppenheimer approximation (compare with (2.10))?

b) Discuss the meaning of the neglected terms. Give an estimate for the terms that are diagonal in the electronic wave functions.

2.2 Surface Structures

a) Determine the structure and the reciprocal lattice of the (100), (110) and (111) unreconstructed surfaces of bcc and fcc crystals. Give the basis vectors of the corresponding unit cells in units of the bulk cubic lattice constant a.

b) Find the surface first Brillouin zone for each surface.

2.3 Wigner–Seitz Cell in Two Dimensions

Consider a Bravais lattice in two dimensions.

a) Prove that the Wigner–Seitz cell is a primitive unit cell.

b) Show that the Wigner–Seitz cell for any two-dimensional Bravais lattice is either a hexagon or a rectangle.

2.4 Reciprocal Lattice

a) Show that the reciprocal lattice belongs to the same symmetry group as the underlying Bravais lattice in real space.

b) Determine the reciprocal lattices and the first Brillouin zones of all the two-dimensional Bravais lattices shown in Table 2.1.

2.5 Lattice Spacing and Vicinal Surfaces

Consider a lattice plane in a three-dimensional crystal described by the Miller indices (hkl).

a) Show that the reciprocal lattice vector $\boldsymbol{G} = h\boldsymbol{b}_1 + k\boldsymbol{b}_2 + l\boldsymbol{b}_3$ is perpendicular to this plane.

b) Show that the distance between two adjacent (hkl) planes is given by

$$d_{hkl} = \frac{2\pi}{|\boldsymbol{G}|}$$

c) Consider the unreconstructed $(11n)$ surface of a fcc crystal with cubic lattice constant a and n an odd number. This surface consists of (001) terraces terminated by steps of $[\bar{1}10]$ orientation. Show that the terraces have a width of $a \times n/\sqrt{8}$ and that the interlayer distance is given by $a/\sqrt{n^2+2}$. Prove that the miscut angle between $(11n)$ and (001) surfaces is $\arctan(\sqrt{2}/n)$.

d) The unreconstructed $(10n)$ surface of a fcc crystal with cubic lattice constant a also consists of (001) terraces terminated by monoatomic steps. Determine the terrace width, the lattice distance between adjacent $(10n)$ planes and the miscut angle to the (001) plane.

3. Electronic Structure Methods and Total Energies

In this chapter we discuss electronic structure methods to determine the total energies of particular systems. The determination of the total energies is a prerequisite for a theoretical treatment of any property or process at surfaces. There are two main techniques, wave-function and electron-density based methods that originate from quantum chemistry and solid-state physics, respectively. Both types of methods will be introduced and discussed in some detail. Special attention will be paid to the discussion of electronic exchange and correlation effects.

Electronic structure calculations for surface science problems are dominated by density functional theory (DFT) methods. Therefore we will have a closer look at some specific implementations of electronic structure algorithms based on DFT. This chapter will be concluded by an introduction to the tight-binding method which is well-suited for a qualitative discussion of band structure effects.

3.1 Hartree–Fock Theory

We start the sections about electronic structure methods with the so-called Hartree and Hartree-Fock methods. This does not only follow the historical development [14, 15], but it also allows to introduce important concepts such as self-consistency or electron exchange and correlation.[1] In this whole chapter we are concerned with ways of solving the time-independent electronic Schrödinger equation

$$H_{el}\Psi(r) = E_{el}\Psi(r). \qquad (3.1)$$

Here we have omitted the parametric dependence on the nuclear coordinates (c.f. (2.9)) for the sake of convenience. As already stated, except for the simplest cases there is no way to solve (3.1) in a close analytical form. Hence we have to come up with some feasible numerical scheme to solve (3.1). Mathematically, it corresponds to a second order partial differential equation. There

---
[1] I thank Matthias Scheffler for providing me with his lecture notes on his theoretical solid-state physics class held at the Technical University Berlin on which the sections about Hartree-Fock and density functional theory are based to a large extent.

are methods to directly integrate partial differential equations (see, e.g., [16]). However, if N is the number of electrons in the system, then we have to deal with a partial differential equation in $3N$ unknowns with N commonly larger than 100. This is completely intractable to solve. The way out is to expand the electronic wave function in some suitable, but necessarily finite basis set whose matrix elements derived from (3.1) can be conveniently determined. This will then convert the partial differential equation into a set of algebraic equations that are much easier to handle. Of course, we have to be aware that by using a finite basis set we will only find an approximative solution to the true many-body wave function. However, by increasing the size of the basis set we have a means to check whether our results are converged with respect to the basis set. Hence this corresponds to a controlled approximation because the accuracy of the calculations can be improved in a systematic way.

Furthermore, for the moment we are mainly interested in the electronic ground-state energy E_0. There is an important quantum mechanical principle – the *Rayleigh–Ritz variational principle* [17] – that provides a route to find approximative solutions for the ground state energy. It states that the expectation value of the Hamiltonian in any state $|\Psi\rangle$ is always larger than or equal to the ground state energy E_0, i.e.

$$E_0 \leq \frac{\langle \Psi|H|\Psi\rangle}{\langle \Psi|\Psi\rangle} \, . \tag{3.2}$$

Hence we can just pick some suitable guess for $|\Psi\rangle$. Then we know that $\langle \Psi|H|\Psi\rangle/\langle \Psi|\Psi\rangle$ will always be an upper bound for the true ground state energy. By improving our guesses for $|\Psi\rangle$, preferentially in a systematic way, we will come closer to the true ground state energy.

Before we proceed, we note that the potential term $V_{\text{nucl-el}}$ (2.5) acts as an effective external one-particle potential for the electrons. Hence we define the external potential for the electrons as

$$v_{ext}(\boldsymbol{r}) = -\sum_I \frac{Z_I e^2}{|\boldsymbol{r} - \boldsymbol{R}_I|} \, . \tag{3.3}$$

Now let us assume that the number of electrons in our system is N and that we have already determined the N lowest eigenfunctions $|\psi_i\rangle$ of the one-particle Schrödinger equation

$$\left\{-\frac{\hbar^2}{2m}\nabla^2 + v_{ext}(\boldsymbol{r})\right\}\psi_i(\boldsymbol{r}) = \varepsilon_i^o \psi_i(\boldsymbol{r}) \, . \tag{3.4}$$

Here we have completely neglected the electron-electron interaction. Still, we might simply consider the product wave function

$$\Psi_{\text{H}}(\boldsymbol{r}_1, \ldots, \boldsymbol{r}_N) = \psi_1(\boldsymbol{r}_1) \cdot \ldots \cdot \psi_N(\boldsymbol{r}_N) \, , \tag{3.5}$$

in which every one-particle state is only occupied once, as a first crude guess for the true many-particle wave function. Then we can determine the expection value of the electronic Hamiltonian (2.8) using the wave function (3.5). Thus we obtain

$$\langle \Psi_H | H | \Psi_H \rangle = \sum_{i=1}^{N} \int d^3r \psi_i^*(\bm{r}) \left(-\frac{\hbar^2}{2m} \nabla^2 + v_{ext}(\bm{r}) \right) \psi_i(\bm{r})$$

$$+ \frac{1}{2} \sum_{i,j=1}^{N} \int d^3r d^3r' \frac{e^2}{|\bm{r}-\bm{r}'|} |\psi_i(\bm{r})|^2 |\psi_j(\bm{r}')|^2 + V_{\text{nucl-nucl}} . \tag{3.6}$$

Now we would like to minimize the expectation value (3.6) with respect to more suitable single-particle functions $\psi_i(\bm{r})$ under the constraint that the wave functions are normalized. This is a typical variational problem with the constraint taken into account via Lagrange multipliers. If we consider the wave functions $\psi_i(\bm{r})$ and $\psi_i^*(\bm{r})$ as independent, we can minimize (3.6) with respect to the ψ_i^* under the constraint of normalization via

$$\frac{\delta}{\delta \psi_i^*} \left[\langle \Psi_H | H | \Psi_H \rangle - \sum_{i=1}^{N} \{ \varepsilon_i (1 - \langle \psi_i | \psi_i \rangle) \} \right] = 0 . \tag{3.7}$$

The ε_i act as Lagrange multipliers ensuring the normalisation of the eigen functions. This minimization leads to the so-called *Hartree equations* [14]:

$$\left\{ -\frac{\hbar^2}{2m} \nabla^2 + v_{ext}(\bm{r}) + \sum_{j=1}^{N} \int d^3r' \frac{e^2}{|\bm{r}-\bm{r}'|} |\psi_j(\bm{r}')|^2 \right\} \psi_i(\bm{r}) = \varepsilon_i \psi_i(\bm{r}) . \tag{3.8}$$

The Hartree equations correspond to a mean-field approximation. Equation (3.8) shows that an effective one-particle Schrödinger equation is solved for an electron embedded in the electrostatic field of *all electrons including the particular electron itself*. This causes the so-called *self interaction* which is erroneously contained in the Hartree equations.

Using the electron density

$$n(\bm{r}) = \sum_{i=1}^{N} |\psi_i(\bm{r})|^2 , \tag{3.9}$$

the *Hartree potential* v_H can be defined:

$$v_H(\bm{r}) = \int d^3r' n(\bm{r}') \frac{e^2}{|\bm{r}-\bm{r}'|} . \tag{3.10}$$

It corresponds to the electrostatic potential of all electrons. With this definition the Hartree equations can be written in a more compact form as

$$\left\{ -\frac{\hbar^2}{2m} \nabla^2 + v_{ext}(\bm{r}) + v_H(\bm{r}) \right\} \psi_i(\bm{r}) = \varepsilon_i \psi_i(\bm{r}) . \tag{3.11}$$

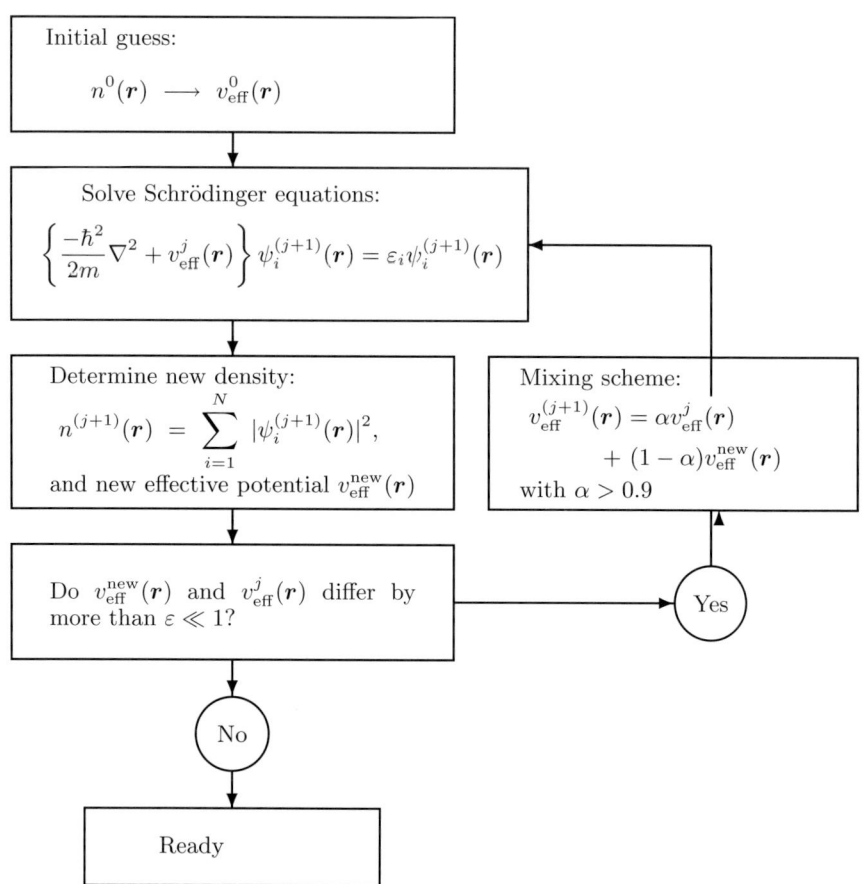

Fig. 3.1. Flow-chart diagram of a self-consistent scheme to solve the Hartree equations

The Hartree equations have the form of one-particle Schrödinger equations. However, the solutions $\psi_i(r)$ of the Hartree equations entering the effective one-particle Hamiltionan are in fact used to obtain the solutions. Consequently, the Hartree equations can only be solved in an iterative fashion: One starts with some initial guess for the wave functions which enter the effective one-particle Hamiltonian. The Hartree equations are then solved and a new set of solutions is determined. This cycle is repeated so often until the iterations no longer modify the solutions, i.e. until self-consistency is reached. Approximative methods such as the Hartree approximation that include a self-consistency cycle are also known as *self-consistent field* (SCF) approximations.

Such a self-consistent scheme is illustrated in a flow-chart diagram in Fig. 3.1 where we have combined the external and the Hartree potential to an effective potential $v_{\text{eff}}(\boldsymbol{r}) = v_{\text{ext}}(\boldsymbol{r}) + v_{\text{H}}(\boldsymbol{r})$. Furthermore, we have included a mixing scheme between the new effective potential and the one of the previous step for the construction of the effective potential entering the next iteration cycle. Usually a mixing scheme speeds up the convergence of the iteration scheme significantly; sometimes convergence can even not be reached without a mixing scheme. Note that the general self-consistency cycle depicted in Fig. 3.1 is not only appropriate for the solution of the Hartree scheme but for any method that requires a self-consistent solution of one-particle equations.

The expectation value of the total energy in the Hartree approximation E_{H} can be written as

$$\langle \Psi_{\text{H}} | H | \Psi_{\text{H}} \rangle = \sum_{i=1}^{N} \varepsilon_i - \frac{1}{2} \int d^3r\, d^3r' \frac{e^2 n(\boldsymbol{r}) n(\boldsymbol{r}')}{|\boldsymbol{r} - \boldsymbol{r}'|} + V_{\text{nucl-nucl}}$$

$$= \sum_{i=1}^{N} \varepsilon_i - V_{\text{H}} + V_{\text{nucl-nucl}} = E_{\text{H}} \quad (3.12)$$

The integral in (3.12) is the so-called *Hartree energy* V_{H}. It corresponds to the classical (or mean-field) electrostatic energy of the electronic charge distribution. It is contained twice in the Hartree eigenvalue; in order to correct for this double-counting it has to be subtracted in (3.12). In fact, the total energy in (3.12) would only be a sum over single-particle energies if the particles were non-interacting (except for the term $V_{\text{nucl-nucl}}$, which in this context for fixed nuclei just acts as an energy renormalization constant). If we evaluate the total energy for interacting particles by self-consistently solving a set of effective single-particle equations, the total energy is not just a sum over single-particle energies, but there will always be correction terms reflecting the interaction between the particles.

The Hartree ansatz obeys the Pauli principle only to some extent by populating each electronic state once. However, it does not take into account the anti-symmetry of the wave function. The Pauli principle requires that the sign of $|\Psi\rangle$ changes when two electrons are exchanged. The simplest ansatz to take account of this antisymmetry requirement is to replace the product wave function (3.5) by a single Slater determinant. In order to correctly incorporate the Pauli principle we have to take the spin degree of freedom into account in the following, i.e. we write the single-particle wave functions as $\psi(\boldsymbol{r}\sigma)$, where σ denotes the spin. The Slater determinant is then constructed from the single-particle functions by

$$\Psi_{\text{HF}}(\boldsymbol{r}_1\sigma_1, \ldots, \boldsymbol{r}_N\sigma_N) = \frac{1}{\sqrt{N!}} \begin{vmatrix} \psi_1(\boldsymbol{r}_1\sigma_1) & \psi_1(\boldsymbol{r}_2\sigma_2) & \ldots & \psi_1(\boldsymbol{r}_N\sigma_N) \\ \psi_2(\boldsymbol{r}_1\sigma_1) & \psi_2(\boldsymbol{r}_2\sigma_2) & \ldots & \psi_2(\boldsymbol{r}_N\sigma_N) \\ \vdots & \vdots & \ddots & \vdots \\ \psi_N(\boldsymbol{r}_1\sigma_1) & \psi_N(\boldsymbol{r}_2\sigma_2) & \ldots & \psi_N(\boldsymbol{r}_N\sigma_N) \end{vmatrix}. \quad (3.13)$$

Now we follow the same procedure as for the Hartree ansatz: first we write down the expectation value of the total energy:

$$\langle\Psi_{\text{HF}}|H|\Psi_{\text{HF}}\rangle = \sum_{i=1}^{N} \int d^3r \psi_i^*(\bm{r}) \left(-\frac{\hbar^2}{2m}\nabla^2 + v_{ext}(\bm{r})\right) \psi_i(\bm{r})$$
$$+\frac{1}{2}\sum_{i,j=1}^{N} \int d^3r d^3r' \frac{e^2}{|\bm{r}-\bm{r}'|} |\psi_i(\bm{r})|^2 |\psi_j(\bm{r}')|^2 + V_{\text{nucl-nucl}}$$
$$-\frac{1}{2}\sum_{i,j=1}^{N} \int d^3r d^3r' \frac{e^2}{|\bm{r}-\bm{r}'|} \delta_{\sigma_i \sigma_j} \psi_i^*(\bm{r}) \psi_i(\bm{r}') \psi_j^*(\bm{r}') \psi_j(\bm{r}) . \quad (3.14)$$

There is now an additional negative term for electrons with the same spin. This extra term is called the exchange energy E_x. Note that the total-energy expression (3.14) is *self-interaction free* because the diagonal terms of E_x with $i=j$ exactly cancel the corresponding terms in the Hartree energy E_H.

Again, we minimize the expression (3.14) with respect to the ψ_i^* under the constraint of normalization. This yields the *Hartree–Fock equations* [15]:

$$\left\{-\frac{\hbar^2}{2m}\nabla^2 + v_{ext}(\bm{r}) + v_{\text{H}}(\bm{r})\right\} \psi_i(\bm{r})$$
$$-\sum_{j=1}^{N} \int d^3r' \frac{e^2}{|\bm{r}-\bm{r}'|} \psi_j^*(\bm{r}')\psi_i(\bm{r}')\psi_j(\bm{r}) \delta_{\sigma_i \sigma_j} = \varepsilon_i \psi_i(\bm{r}). \quad (3.15)$$

The additional term, called the *exchange term*, introduces quite some complexity to the equations. It is of the form $\int V(\bm{r},\bm{r}')\psi(\bm{r}')d^3r'$, i.e., it is an integral operator. In more compact form, the expectation value of the total energy in the Hartree-Fock approximation is given by

$$\langle\Psi_{\text{HF}}|H|\Psi_{\text{HF}}\rangle = E_{\text{HF}} = \sum_{i=1}^{N} \varepsilon_i - V_{\text{H}} - E_x + V_{\text{nucl-nucl}} . \quad (3.16)$$

Analogously to (3.12), the Hartree energy V_H and the exchange energy E_x have to be substracted since they enter the Hartree–Fock eigenvalues twice. In order to understand the physical nature and the consequences of the exchange term, we will focus on a case where the exchange can be exactly determined: the *homogeneous electron gas*.

In this case the resulting electron density $n(\bm{r})$ will be just uniform. Let us assume that the electrostatic potential of the electrons is compensated by a positive charge background. This is also called the "jellium model". Then the external potential and the Hartree potential exactly cancel: $v_{\text{ext}}(\bm{r}) + v_{\text{H}}(\bm{r}) = 0$. Let us further apply the other mathematical methods usually employed in order to deal with infinite systems: we normalize the wave functions with respect to a cube of volume V and assume periodic boundary conditions at the faces of the cube. Within the cube we place N electrons. The eigenfunctions of the Hartree equations for the homogeneous electron gas are just plane waves:

$$\psi_i(\boldsymbol{r}) = \frac{1}{\sqrt{V}} e^{i\boldsymbol{k}_i \cdot \boldsymbol{r}} . \tag{3.17}$$

Furthermore, for a positive charge background the Hartree term V_H entering the expression for the total energy exactly cancels with $V_{\text{nucl-nucl}}$ in (3.12). The total energy is then simply given by the sum over the kinetic energy of the electrons:

$$E_H = \sum_{i=1}^{N} \varepsilon(\boldsymbol{k}_i) = 2 \sum_{|\boldsymbol{k}|<k_F} \frac{\hbar^2 k^2}{2m} . \tag{3.18}$$

Here we have introduced the Fermi vector k_F which gives the wave vector of the occupied one-electron levels of highest energy. It can be related to the electron density via [10] (see Exercise 3.1)

$$k_F = \left(3\pi^2 \frac{N}{V}\right)^{1/3} = (3\pi^2 n)^{1/3} \tag{3.19}$$

The factor of 2 in (3.18) is due to the spin.

Now we also consider the exchange term. It can be shown that the solutions of the Hartree-Fock equations for the homogeneous electron gas are still plane waves, but an extra term due to the exchange appears in the expression for the one-particle energies. If we set $|\boldsymbol{k}| = k$, we obtain (see Exercise 3.2)

$$\varepsilon(\boldsymbol{k}) = \frac{\hbar^2 k^2}{2m} - \frac{2e^2}{\pi} k_F F\left(\frac{k}{k_F}\right) , \tag{3.20}$$

where

$$F(x) = \frac{1}{2} + \frac{1-x^2}{4x} \ln\left|\frac{1+x}{1-x}\right| . \tag{3.21}$$

Note that the kinetic energy remains unchanged if the anti-symmetry of the wave-functions is taken into account.

In Fig. 3.2 the one-particle energies of the Hartree and the Hartree–Fock approximation for the homogeneous electron gas are compared. They are plotted in units of the free-electron Fermi energy $\varepsilon_F = (\hbar^2 k_F^2)/(2m)$ as a function of the wave vector normalized to k_F. It is apparent that taking into account the exchange leads to a strong decrease in the one-particle energies that stabilizes the homogeneous electron gas. Furthermore, the band width of the occupied electron states in the Hartree–Fock approximation is increased dramatically with respect to the Hartree approximation. In fact, this large increase of the band width is not supported by photoemission experiments [10]. There is another alarming feature of the Hartree–Fock single-particle energies (3.20). The derivative $\partial \varepsilon / \partial k$ becomes logarithmically infinite at $k = k_F$ which results in a vanishing density of states at the Fermi level.

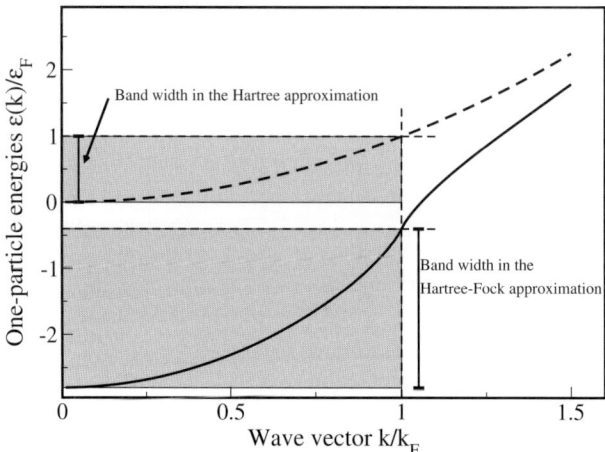

Fig. 3.2. One-particle energies of the homogeneous electron gas in the Hartree and the Hartree–Fock approximation

The total energy of the homogeneous electron gas in the Hartree–Fock approximation is given by

$$E_{\text{HF}} = 2 \sum_{k<k_F} \frac{\hbar^2 k^2}{2m} - \sum_{k<k_F} \frac{e^2}{\pi} k_F \, F\left(\frac{k}{k_F}\right)$$
$$= N \left(\frac{3}{5} \varepsilon_F - \frac{3}{4} \frac{e^2}{\pi} k_F \right). \quad (3.22)$$

The *exchange energy* ε_x per electron in the homogeneous electron gas can be expressed as

$$\varepsilon_x = -\frac{3}{4} \frac{e^2}{\pi} (3\pi^2 n)^{1/3}$$
$$= -\frac{3}{4} \frac{e^2}{\pi} \left(\frac{9\pi}{4}\right)^{1/3} \frac{1}{r_s}, \quad (3.23)$$

where we have used the *Wigner–Seitz radius* r_s

$$r_s = \left(\frac{3}{4\pi n}\right)^{1/3}, \quad (3.24)$$

which corresponds to the radius of the sphere whose volume $V/N = 1/n$ equals the volume per electron in the homogeneous electron gas.

Taking into account exchange leads to a strong decrease in the one-particle energies and thus also in the total energy of the homogeneous electron gas. However, it is important to note that the exchange energy is not a new kind of energy form. Both the Hartree and the Hartree–Fock wave function describe the homogeneous electron gas. But while in the Hartree approximation the

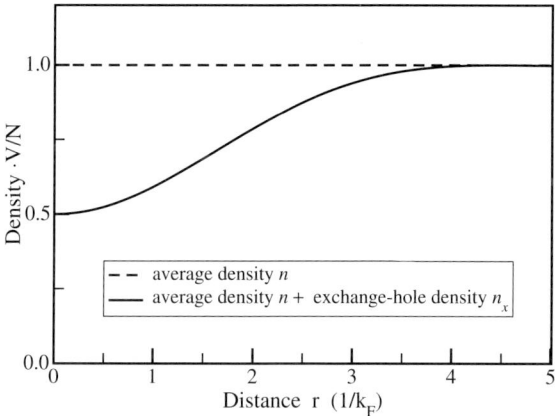

Fig. 3.3. Normalized mean electron density $n = N/V$ and exchange-hole density $n_x(\bar{r}, 0)$ in the homogeneous electron gas as a function of the distance in units of the inverse Fermi vector k_F

positions of the electrons are not correlated, the antisymmetry of the electronic wave function in the Hartree–Fock ansatz results in the so-called *exchange hole*: electrons of the same spin can not be at the same position. This can be illustrated by introducing the concept of the exchange-hole density $n_x^i(r, r')$ of a electron in state i. It can be defined as

$$n_x^i(r, r') = -\sum_{j=1}^{N} \frac{\psi_j^*(r')\psi_i(r')\psi_j(r)}{\psi_i(r)} \delta_{\sigma_i \sigma_j}. \tag{3.25}$$

The exchange-hole density satisfies

$$\int d^3r'\, n_x^j(r, r') = -1. \tag{3.26}$$

With this exchange-hole density the exchange term in the Hartree–Fock equations (3.15) can be written as

$$-\sum_{j=1}^{N} \int d^3r'\, \frac{e^2}{|r-r'|}\, \psi_j^*(r')\psi_i(r')\psi_j(r)\delta_{\sigma_i \sigma_j}$$

$$= \int d^3r'\, \frac{e^2}{|r-r'|}\, n_x^i(r, r')\psi_i(r). \tag{3.27}$$

The exchange-hole density $n_x^i(r, r')$ describes a nonlocal exchange hole of the electron in state i. Due to

$$n_x^i(r, r) = -\sum_{j=1}^{N} |\psi_j(r)|^2\, \delta_{\sigma_i \sigma_j}, \tag{3.28}$$

the exchange-hole density $n_x^i(r, r')$ represents the depletion of the remaining electrons with the same spin from the location of the i-th electron.

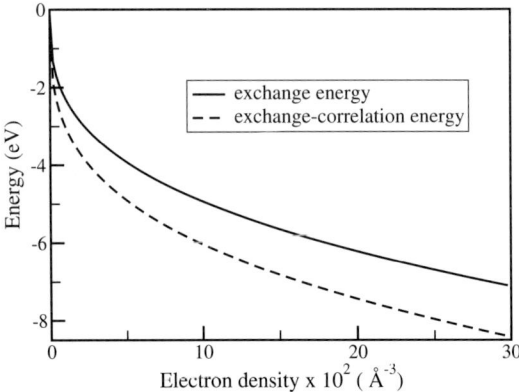

Fig. 3.4. Exchange and exchange-correlation energy per particle in the homogeneous electron gas. Typical valence electron densities of metals are in the range of $5-10 \times 10^2$ (Å^{-3})

Usually the local exchange density is defined as

$$n_x(\mathbf{r}, \mathbf{r}') = g(\mathbf{r}, \mathbf{r}') - n(\mathbf{r}'), \tag{3.29}$$

where $g(\mathbf{r}, \mathbf{r}')$ is the conditional density to find an electron at $\mathbf{r}'$ if there is already an electron at $\mathbf{r}$. It can be expressed as the average of $n_x^i(\mathbf{r}, \mathbf{r}')$ over all electrons

$$n_x(\mathbf{r}, \mathbf{r}') = \sum_{i=1}^{N} \frac{\psi_i^*(\mathbf{r}') n_x^i(\mathbf{r}, \mathbf{r}') \psi_i(\mathbf{r})}{n(\mathbf{r}')} \delta_{\sigma_i \sigma_j}. \tag{3.30}$$

For the homogeneous electron gas $n_x(\mathbf{r}, \mathbf{r}')$ can be determined analytically. It only depends on the distance $|\mathbf{r} - \mathbf{r}'| = \bar{r}$ between two electrons and is given by (see Exercise 3.3)

$$n_x(\bar{r}) = \frac{9}{2} \frac{N}{V} \left(\frac{k_F \bar{r} \cos(k_F \bar{r}) - \sin(k_F \bar{r})}{(k_F \bar{r})^3} \right)^2. \tag{3.31}$$

We have plotted the sum of the average density $n = N/V$ and the normalized exchange-hole density $n_x(\mathbf{r}, 0)$ as a function of the distance from the origin for a electron at $\mathbf{r}' = 0$ for the homogeneous electron gas and compared it to n in Fig. 3.3. Since taking into account the exchange leads to a depletion of the remaining electron density close to the position of the electron, the exchange energy just corresponds to the repulsive Coulomb interaction between electrons which is saved when the antisymmetry of electrons with the same spin is taken into account. The denser the homogeneous electron gas, the larger the reduction caused by the exchange. This is demonstrated in Fig. 3.4 where the mean exchange energy per electron in the homogeneous electron gas is plotted. The net energy gain due to exchange increases monotonously with increasing electron density.

Since the Hartree ansatz does not take into account the antisymmetry of the electronic many-body wave function, it does not yield a proper solution of the many-body Schrödinger equation. The Hartree–Fock method incorporates the antisymmetry requirement and leads to a reduction of the total energy. Hence it should be a more appropriate solution for the true ground-state already on the basis of the Rayleigh-Ritz variational principle. However, in the Hartree–Fock ansatz, electrons of opposite spin are still not correlated. If these electrons are also avoiding each other, the energy can be further reduced. This additional effect is called *electron correlation*. The *electron correlation energy* is defined as the difference between the exact energy of the system and the Hartree–Fock energy. The distinction between electron correlation and exchange is somehow artificial because the Hartree–Fock exchange is in principle also an electron correlation effect. The correlation energy in the homogeneous electron gas has been determined by quantum Monte Carlo calculations [18]. In Fig. 3.4, the exchange-correlation energy which is the sum of the exchange and the correlation energy is also plotted. The additional reduction due to the consideration of electron correlation in the homogeneous electron gas is obvious.

3.2 Quantum Chemistry Methods

Many chemical reactions are enormously accelerated through the presence of a catalyst. Therefore, historically chemists were the first to be interested in the theoretical description of surfaces and processes at surfaces. The theoretical tools used by chemists are designed to describe finite systems such as molecules. In the quantum chemistry approach, surfaces are regarded as big molecules and modelled by an finite cluster. This ansatz is guided by the idea that bonding on surfaces is a local process. It depends on the localisation of the electronic orbitals which can be estimated to be inversely proportional to the width of the band gap. Hence the cluster approach is most appropriate for wide-gap insulators, but it has also been used extensively for metal surfaces for which this approach is more questionable.

One typical example of a cluster used to describe surfaces is shown in Fig. 3.5. A Si_9H_{12} cluster is plotted which is used as a model for the Si(100) surface. At an ideal surface there are two broken Si-Si bonds per surface atom. In order to reduce the number of these *dangling bonds* that correspond to unsaturated orbitals, the Si(100) surface reconstructs by creating Si dimers at the surface. The Si_9H_{12} cluster just contains one of these surface dimers. The hydrogen atoms are added to the Si cluster in order to saturate the silicon dangling bonds that would be bonded to other silicon atoms.

Once the surface is modelled by a finite cluster, it can be treated by *quantum chemistry methods* [19]. Usually Hartree–Fock theory is a good starting point for the theoretical description of molecules and clusters. The exact total energy of a molecule is often reproduced to up to 99% [20]. Unfortunately,

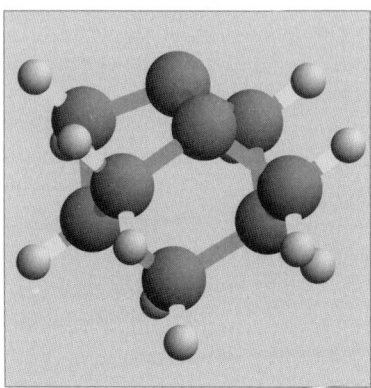

Fig. 3.5. Si_9H_{12} cluster used to model a Si(100) substrate. The silicon atoms are plotted as the darker large balls while the hydrogen atoms are presented by the lighter small balls

the missing part, namely the electron correlation energy, is rather important for a reliable description of chemical bond formation. As Fig. 3.4 shows, the correlation energy per electron can easily be more than 1 eV. Chemists often demand an accuracy of 1 kcal/mol ≈ 0.04 eV ("chemical accuracy") for energy differences in order to consider the calculations useful. If the uncertainty of the total energies is already above several eV, then only by a fortuitous cancellation of errors chemical accuracy can be achieved.

Quantum chemists have developed a whole machinery of theoretical methods that treat the electron correlation at various levels of sophistication [21, 22]. These methods can be devided into to two categories, the so-called single-reference and the multiple-reference methods.

In the single-reference methods, one starts with the Slater determinant that is the solution of the Hartree–Fock equations. This means that the electron correlation between electrons with opposite spin is neglected. One way to introduce correlation effects is by considering virtually excited states. These can be generated by replacing occupied orbitals in the Slater determinant by unoccupied ones, i.e., by states that do not correspond to the N lowest Hartree–Fock eigenvalues. By replacing one, two, three, four or more states, single, double, triple, quadruple or higher excitations can be created. In the Møller–Plesset theory [23], these excitations are treated perturbatively by regarding the Hartree–Fock Hamiltonian as the unperturbed Hamiltonian and the difference to the true many-body Hamiltonian as the perturbation.

To derive the Møller-Plesset theory, we first rewrite the Hartree–Fock equations (3.15) as

$$h_i^{\mathrm{HF}} \, \psi_i(\boldsymbol{r}) = \varepsilon_i \, \psi_i(\boldsymbol{r})\,, \tag{3.32}$$

where the effective one-electron operator h_i^{HF} acts on the i-th electron. If we neglect the nuclear-nuclear interaction $V_{\text{nucl-nucl}}$ for a moment, the Hartree–Fock Hamiltonian H_{H} can be defined as

$$H_{\text{HF}} = \sum_i h_i^{\text{HF}} - V_{\text{H}} - E_{\text{x}}. \quad (3.33)$$

Here V_{H} and E_{x} are just treated as constants determined with the true Hartree–Fock ground state wave-function. Note that H_{HF} does not correspond to the correct many-body electronic Hamiltonian H_{el} (2.8) since the correlation effects between electron with opposite spins are not included. The difference between H_{el} and H_{HF} is now treated as the perturbation H':

$$H' = H_{\text{el}} - H_{\text{HF}}. \quad (3.34)$$

The expression for the ground-state energy in second-order perturbation theory becomes

$$E^{(2)} = E_{\text{HF}} + \langle \Psi_0 | H' | \Psi_0 \rangle + \sum_{l \neq 0} \frac{|\langle \Psi_l | H' | \Psi_0 \rangle|^2}{E_0 - E_l} \text{ts}. \quad (3.35)$$

The sum over states l other than the ground state corresponds to Slater determinants with single, double, triple, etc. excitations. In fact it can be easily shown [23] that the Hartree–Fock theory is correct to first order, i.e., the first-order correction $\langle \Psi_0 | H' | \Psi_0 \rangle$ vanishes. If we now introduce the following notation for the Coulomb integral

$$\int d^3r d^3r' \psi_l^*(r) \psi_m^*(r') \frac{e^2}{|r-r'|} \psi_p(r') \psi_q(r) = \langle lm | \frac{e^2}{|r-r'|} | qp \rangle, \quad (3.36)$$

the second-order expression is given by

$$E^{(2)} = E_{\text{HF}} - \sum_{l<m}^{\text{occ}} \sum_{p<q}^{\text{unocc}} \frac{|\langle lm | \frac{e^2}{|r-r'|} | qp \rangle - \langle lm | \frac{e^2}{|r-r'|} | pq \rangle|^2}{\varepsilon_l + \varepsilon_m - \varepsilon_p - \varepsilon_q}, \quad (3.37)$$

where the sum is performed over all occupied and unoccupied (virtual) orbitals. Second-order Møller–Plesset theory is usually denoted by MP2. If higher-order corrections are included, the methods are named MP3, MP4 and so forth.

MP2 is a very popular method due to its conceptual simplicity. However, due to the perturbative treatment of electron correlation its applicability is still limited. Instead of perturbatively treating single and double excitations one might just directly express the wave function as a sum of the Hartree–Fock determinant plus other determinants obtained by replacing one or two orbitals:

$$\Psi_{\text{CISD}} = \Psi_{\text{HF}} + \sum c_i^{(1)} \Psi_i^{(1)} + \sum c_i^{(2)} \Psi_i^{(2)}. \quad (3.38)$$

The optimum wave function is then found by varying the coefficients $c_i^{(j)}$. Since different configurations (or Slater determinants) are included in (3.38), this method is called *configuration interaction* (CI). If only single (S) and double (D) excitations are included in the sum, one refers to the method as CISD. This approach does not obey one important desirable property of electronic structure methods, namely the so-called *size consistency* or *size extensivity*. In principle, with size extensitivity the linear scaling of the energy with the number of electrons is meant. For infinitely separated systems, this comes down to additive energies of the separated components. This property is not only important for large systems, but even for small molecules [20]. The CISD method does not fulfil size extensivity because the product of two fragment CISD wave functions contains triple and quadruple excitations and is therefore no CISD function. One elegant way to recover size extensivity is to exponentiate the single and double excitations operator:

$$\Psi_{\text{CCSD}} = \exp(T_1 + T_2)\,\Psi_{\text{HF}}\,. \tag{3.39}$$

This approach is called coupled cluster (CC) theory, the limitation to single and double excitations is denoted by CCSD. If also triple excitations are included, the method is called CCSDT. However, the computational effort of this method has an eighth-power dependence on the size of the system and is therefore rather impractical. The scaling is less unfavorable if the triple excitations are incorporated perturbatively. This CCSD(T) method is still very accurate.

The single-reference methods can be very reliable in the vicinity of equilibrium configurations, but they are often no longer adequate to describe a bond-breaking process. One Slater determinant plus excited states derived from this determinant are not sufficient because the dissociation products should be described by a linear combination of two Slater determinants taking into account the proper spin state. Any many-particle electronic wave function can in principle be represented by a sum over Slater determinants. Hence by considering more and more configurations in the calculations, the accuracy can be systematically improved. If all possible determinants are included in an electronic structure calculation, the method is called full configuration interaction (FCI). Because of the large computational effort required, FCI calculations are limited to rather small systems. In particular, the treatment of larger clusters necessary to model surface science problems is not possible with FCI. Hence approximate multi-reference methods are needed.

In the multiconfigurational self-consistent field (MCSCF) approach, a relatively small number of configurations is selected and both the orbitals and the configuration interactions coefficients are determined variationally. The selection of the configurations included in a MCSCF calculation cannot be done automatically; usually chemical insight is needed [20] which, however, might introduce a certain bias in the calculations. This can be avoided by the complete active space (CAS) approach. In a CASSCF calculations a set of active orbitals is identified and all excitations within this active space are included. This method is again computationally very costly. After all, if the active space is increased to include all electrons and orbitals, we end up with a FCI calculation.

For a proper treatment of the electronic correlation, not only the appropriate method has to been chosen, but also the basis set has to be sufficiently large enough. Quantum chemical methods usually describe the electrons by a localized basis set derived from atomic orbitals. The preferred basis functions are Gaussian functions because they allow the analytical evaluation of the matrix elements necessary to perform an electronic structure calculations. In fact, the most popular electronic structure code used in quantum chemistry, the GAUSSIAN program [22], is named after the type of basis functions employed in the program.

Quantum chemists have developed a particular nomenclature to describe the quality of a basis set. It is beyond the scope of this book to give a full introduction into the terminology. I will only give a short overview. The simplest choice of just one atomic orbital per valence state is called "minimal basis set" or "single zeta". If two or more orbitals are included, the basis set is called "double zeta"(DZ), "triple zeta" (TZ) and so on. Often polarization functions are added which correspond to one or more sets of d functions on first row atoms. Then a "P" is added to the acronym of the basis set resulting in, e.g., DZ2P. These polarization functions describe small displacements of the atomic orbitals from the nuclear centers. For rather delocalized states such as anionic or Rydberg excited states, further diffuse functions are added.

In quantum chemical methods the accuracy of the treatment of electron correlation can be improved in a systematic way by either choosing a more refined method or by increasing the basis set. This is a very attractive property of these wave function based methods. Unfortunately, the accuracy is paid by an immense computational effort so that accurate calculations are limited to a rather small number of atoms, typically about 10–20. This is often not sufficient to model an extended substrate. The accuracy is worthless if the evaluated system does not correspond to the "real" system it is supposed to model. There has been a very lively and sometimes heated discussion about the value of cluster calculations to model surfaces. Adsorption geometries are usually well-reproduced by cluster calculations, but adsorption energies often exhibit an strong dependence on the cluster size and can be seriously in error, in particular for metal surfaces [24]. Cluster calculations can still yield qualitative trends, but they are "often best used for explanatory rather than predictive purposes" [25]. In recent years one method has become more and more popular that is not based on a representation of the many-body wave function, but on the electron density: density functional theory [26–30].

3.3 Density Functional Theory

There is a simple predecessor of density functional theory, the *Thomas–Fermi theory* [31]. In the homogeneous electron-gas, in the presence of a constant external potential v_{ext} the chemical potential μ can be expressed as

$$\mu = \frac{\hbar^2}{2m}(3\pi^2 n)^{2/3} + v_{\text{ext}}, \tag{3.40}$$

which follows for example from (3.19). Assuming that (3.40) also holds for weakly varying effective potentials $v_{\text{eff}}(\boldsymbol{r})$, the *Thomas–Fermi equation* is obtained

$$\frac{\hbar^2}{2m}(3\pi^2 n(\boldsymbol{r}))^{2/3} + v_{\text{eff}}(\boldsymbol{r}) = \mu, \tag{3.41}$$

which can also been derived from a variational principle (see Exercise 3.4). However, it was not clear whether there is a strict relation between the electron density appearing in the Thomas–Fermi equation and the corresponding many-body wave function [30].

This connection is in fact provided by the Hohenberg–Kohn theorem [26] density functional theory (DFT) is based upon. This theorem states that the ground-state density $n(\boldsymbol{r})$ of a system of interacting electrons in an external potential uniquely determines this potential. The proof for this theorem which is rather simple will be presented here in order to demonstrate that also theories that are based on simple ideas can lead to a Nobel prize (Walter Kohn, Nobel prize for chemistry 1998). However, the formulation of its rigorous mathematical foundations was only completed several years after the first presentation of the Hohenberg–Kohn theorem (see, e.g., [28]).

Let us assume for the sake of simplicity that the system of interacting electrons has a nondegenerate ground state (the extension to degenerate cases is straightforward). Let the wave function Ψ_1 be the nondegenerate ground state of the Hamiltonian H_1 with external potential $v_1(\boldsymbol{r})$ and the corresponding ground-state density $n(\boldsymbol{r})$. The ground-state energy E_1 is then given by

$$\begin{aligned} E_1 &= \langle\, \Psi_1 \mid H_1 \mid \Psi_1 \,\rangle \\ &= \int v_1(\boldsymbol{r}) n(\boldsymbol{r})\, d^3\boldsymbol{r} + \langle\, \Psi_1 \mid (T+U) \mid \Psi_1 \,\rangle. \end{aligned} \tag{3.42}$$

Here T and U are the operators of the kinetic and the interaction energy. Now let us assume that there is a second potential $v_2(\boldsymbol{r})$ which differs from $v_1(\boldsymbol{r})$ by not just a constant, i.e. $v_2(\boldsymbol{r}) \neq v_1(\boldsymbol{r}) + \text{const.}$, but leads to the same electron density $n(\boldsymbol{r})$. The corresponding ground state energy is

$$\begin{aligned} E_2 &= \langle\, \Psi_2 \mid H_2 \mid \Psi_2 \,\rangle \\ &= \int v_2(\boldsymbol{r}) n(\boldsymbol{r})\, d^3\boldsymbol{r} + \langle\, \Psi_2 \mid (T+U) \mid \Psi_2 \,\rangle. \end{aligned} \tag{3.43}$$

Now we can apply the Rayleigh-Ritz variational principle. Since the ground state Ψ_1 is assumed to be nondegenerate, we obtain the true inequality

$$E_1 < \langle \Psi_2 | H_1 | \Psi_2 \rangle$$
$$= \int v_1(\boldsymbol{r}) n(\boldsymbol{r}) \, d^3\boldsymbol{r} + \langle \Psi_2 | (T+U) | \Psi_2 \rangle$$
$$= E_2 + \int (v_1(\boldsymbol{r}) - v_2(\boldsymbol{r})) \, n(\boldsymbol{r}) \, d^3\boldsymbol{r} \,. \tag{3.44}$$

Equivalently, we can use the Rayleigh–Ritz variational principle for H_2. We have not explicitly assumed that Ψ_2 is nondegenerate, hence we obtain

$$E_2 \leq \langle \Psi_1 | H_2 | \Psi_1 \rangle$$
$$= E_1 + \int (v_2(\boldsymbol{r}) - v_1(\boldsymbol{r})) \, n(\boldsymbol{r}) \, d^3\boldsymbol{r} \,. \tag{3.45}$$

If we add (3.44) and (3.45), we end up with the contradiction

$$E_1 + E_2 < E_1 + E_2 \,. \tag{3.46}$$

Hence the initial assumption that two different external potential can lead to the same electron density is wrong. This concludes the proof of the Hohenberg–Kohn theorem.

Since the density $n(\boldsymbol{r})$ is uniquely related to the external potential and the number N of electrons via $N = \int n(\boldsymbol{r}) d^3\boldsymbol{r}$, it determines the full Hamiltonian. Thus in principle it determines all quantities that can be derived from the Hamiltonian such as, e.g., the electronic excitation spectrum. However, unfortunately this has no practical consequences since the dependence is only implicit.

In the derivation of the Hartree and the Hartree–Fock methods we have used the Rayleigh-Ritz variational principle. This demonstrated the importance of variational principles. In fact, there is also a variational principle for the energy functional, namely that the exact ground state density and energy can be determined by the minimisation of the energy functional $E[n]$:

$$E_{\text{tot}} = \min_{n(\boldsymbol{r})} E[n] = \min_{n(\boldsymbol{r})} (T[n] + V_{\text{ext}}[n] + V_{\text{H}}[n] + E_{\text{xc}}[n]) \,. \tag{3.47}$$

$V_{\text{ext}}[n]$ and $V_H[n]$ are the functionals of the external potential and of the classical electrostatic interaction energy that corresponds to the Hartree energy, while $T[n]$ is the kinetic energy functional for non-interacting electrons, i.e. the kinetic energy functional of a non-interacting reference system that is exposed to the same external potential as the true interacting system. All quantum mechanical many-body effects are contained in the so-called exchange-correlation functional $E^{\text{xc}}[n]$. Yet, this non-local functional is not known; probably it is even impossible to determine it exactly in a closed form. However, it has the important property that it is a well-defined universal functional of the electron density, i.e., it does not depend on any specific system or element. Instead of using the many-body quantum wave function which depends on $3N$ coordinates now only a function of three coordinates has to be varied. In practice, however, no direct variation of the density is performed. One of the reasons is that the kinetic energy functional $T[n]$ is not well-known either.

The density is rather expressed as a sum over single-particle states

$$n(\boldsymbol{r}) = \sum_{i=1}^{N} |\psi_i(\boldsymbol{r})|^2 . \qquad (3.48)$$

Now we make use of the variational principle for the energy functional and minimize $E[n]$ with respect to the single particle states under the constraint of normalization. This procedure is entirely equivalent to the derivation of the Hartree and the Hartree–Fock equations (3.8) and (3.15), respectively. Thus we obtain the so-called Kohn–Sham equations [27]

$$\left\{ -\frac{\hbar^2}{2m}\nabla^2 + v_{ext}(\boldsymbol{r}) + v_{\mathrm{H}}(\boldsymbol{r}) + v_{\mathrm{xc}}(\boldsymbol{r}) \right\} \psi_i(\boldsymbol{r}) = \varepsilon_i \, \psi_i(\boldsymbol{r}) . \qquad (3.49)$$

Thus the effective one-electron potential acting on the electrons is given in the Kohn–Sham formalism by

$$v_{\mathrm{eff}}(\boldsymbol{r}) = v_{\mathrm{ext}}(\boldsymbol{r}) + v_{\mathrm{H}}(\boldsymbol{r}) + v_{\mathrm{xc}}(\boldsymbol{r}) . \qquad (3.50)$$

The exchange-correlation potential $v_{\mathrm{xc}}(\boldsymbol{r})$ is the functional derivative of the exchange-correlation functional $E_{\mathrm{xc}}[n]$

$$v_{\mathrm{xc}}(\boldsymbol{r}) = \frac{\delta E_{\mathrm{xc}}[n]}{\delta n} . \qquad (3.51)$$

The ground state energy can now be expressed as

$$E = \sum_{i=1}^{N} \varepsilon_i + E_{\mathrm{xc}}[n] - \int v_{\mathrm{xc}}(\boldsymbol{r}) n(\boldsymbol{r}) \, d^3\boldsymbol{r} - V_{\mathrm{H}} + V_{\mathrm{nucl-nucl}} . \qquad (3.52)$$

Here we have added the term $V_{\mathrm{nucl-nucl}}$ in order to have the correct total energy of the electronic Hamiltonian (2.8). In solid-state applications, the sum over the single-particle energies in (3.52) is often called the band-structure energy. However, it is important to keep in mind that the "single-particle energies" ε_i enter the formalism just as Lagrange multipliers ensuring the normalisation of the wave functions. The Kohn–Sham states correspond to quasiparticles with no specific physical meaning except for the highest occupied state [28]. Still it is almost always taken for granted that the Kohn–Sham eigenenergies can be interpreted, apart from a rigid shift, as the correct electronic one-particle energies. This is justified by the success since the Kohn–Sham eigenenergy spectrum indeed very often gives meaningful physical results, as will be shown in the next chapters.

Note that if the exchange-correlation terms E_{xc} and v_{xc} are neglected in (3.49)–(3.52), we recover the Hartree formulation of the electronic many-body problem. Hence the Kohn–Sham theory may be regarded as a formal extention of the Hartree theory. In contrast to the total energy expression in the Hartree and the Hartree–Fock approximation, the ground-state energy

3.3 Density Functional Theory

(3.52) is in principle exact. The reliability of any practical implementation of density functional theory depends crucially on the accuracy of the expression for the exchange-correlation functional.

The exchange-correlation functional $E_{\text{xc}}[n]$ can be written as

$$E_{\text{xc}}[n] = \int d^3r \, n(\boldsymbol{r}) \, \varepsilon_{\text{xc}}[n](\boldsymbol{r}), \tag{3.53}$$

where $\varepsilon_{\text{xc}}[n](\boldsymbol{r})$ is the exchange-correlation energy per particle at the point $\boldsymbol{r}$, but depends on the whole electron density distribution $n(\boldsymbol{r})$. In order to discuss the properties of $E_{\text{xc}}[n]$, it is helpful to introduce the exchange-correlation hole distribution

$$n_{\text{xc}}(\boldsymbol{r}, \boldsymbol{r}') = g(\boldsymbol{r}, \boldsymbol{r}') - n(\boldsymbol{r}'), \tag{3.54}$$

where $g(\boldsymbol{r}, \boldsymbol{r}')$ is the conditional density to find an electron at $\boldsymbol{r}'$ if there is already an electron at $\boldsymbol{r}$. Every electron creates a hole corresponding to exactly one electron out of the average density $n(\boldsymbol{r})$. This is expressed through the sum rule

$$\int d^3r' \, n_{\text{xc}}(\boldsymbol{r}, \boldsymbol{r}') = -1. \tag{3.55}$$

Furthermore, the exchange-correlation hole vanishes for large distances:

$$n_{\text{xc}}(\boldsymbol{r}, \boldsymbol{r}') \xrightarrow[|\boldsymbol{r}-\boldsymbol{r}'|\to\infty]{} 0, \tag{3.56}$$

and there is an asymptotical result for the integral

$$\int d^3r' \, \frac{n_{\text{xc}}(\boldsymbol{r}, \boldsymbol{r}')}{|\boldsymbol{r} - \boldsymbol{r}'|} \xrightarrow[|\boldsymbol{r}|\to\infty]{} -\frac{1}{|\boldsymbol{r}|}. \tag{3.57}$$

Since the exchange-correlation functional $E_{\text{xc}}[n]$ is not known in general, the exchange-correlation energy $\varepsilon_{\text{xc}}[n](\boldsymbol{r})$ cannot be exactly derived either. What is known is the exchange-correlation energy for the homogeneous electron gas, i.e. for a system with a constant electron density [18]. This energy is plotted in Fig. 3.4. In the so-called Local Density Approximation (LDA), the exchange-correlation energy for the homogeneous electron gas is also used for non-homogeneous situations,

$$E_{\text{xc}}^{\text{LDA}}[n] = \int d^3r \, n(\boldsymbol{r}) \, \varepsilon_{\text{xc}}^{\text{LDA}}(n(\boldsymbol{r})), \tag{3.58}$$

As (3.58) shows, at any point in space the local exchange-correlation energy $\varepsilon_{\text{xc}}^{\text{LDA}}(n(\boldsymbol{r}))$ of the homogeneous electron gas is used for the corresponding density, ignoring the non-locality of the true exchange-correlation energy $\varepsilon_{\text{xc}}[n]$.

In a wide range of bulk and surface problems the LDA has been surprisingly successful [29]. This is still not fully understood but probably due to a cancellation of opposing errors in the exchange and the correlation expression in the LDA. Furthermore, the LDA satisfies the sum rule (3.55) which is apparingly also very important. For chemical reactions in the gas phase and at

Table 3.1. O_2 binding energy obtained by DFT calculations using LDA and different GGA exchange-correlation functionals [36]

functional	LDA	PW91	PBE	RPBE	Exp.
O_2 binding energy (eV)	7.30	6.06	5.99	5.59	5.23

surfaces, however, the LDA results are not sufficiently accurate. Usually LDA shows *over-binding*, i.e. binding and cohesive energies turn out to be too large compared to experiment. This overbinding also leads to lattice constants and bond lengths that are smaller than the experimental values. These shortcominigs of LDA were the reason why many theoretical chemists were rather reluctant to use DFT for a long time. There had been attempts to formulate a Taylor expansion of the exchange-correlation energy $\varepsilon_{xc}[n]$, but these first attempts had not been successful because by a straightforward gradient expansion (3.55) is violated. Only with the advent of exchange-correlation functionals in the Generalized Gradient Approximation (GGA) [32–36] this situation has changed. In the GGA the gradient of the density is also included in the exchange-correlation energy,

$$E_{xc}^{GGA}[n] = \int d^3\boldsymbol{r}\; n(\boldsymbol{r})\; \varepsilon_{xc}^{GGA}(n(\boldsymbol{r}), |\nabla n(\boldsymbol{r})|)\,, \qquad (3.59)$$

but the dependence on the gradient is modified in such a way as to satisfy the sum rule (3.55). In addition, general scaling properties and the asymptotic behavior of effective potentials are taken into account in the construction of the GGA. DFT calculations in the GGA achieve chemical accuracy (error $\leq 0.1\,\text{eV}$) for many chemical reactions. This improvement in the accuracy of DFT calculations finally opened the way for Walter Kohn to be honored with the Nobel prize in chemistry in 1998 for the development of DFT which is somewhat paradox because DFT was accepted in the physics community much earlier than in the chemistry community.

Still there are important exceptions where the GGA also does not yield sufficient accuracy. In Table 3.1 DFT results for the O_2 binding energy obtained using LDA and different GGA exchange-correlation functionals [36] are compared to the experimental value. The LDA result shows the typical large overbinding. The GGA functional by Perdew and Wang (PW91) [34] and by Perdew, Burke and Ernzerhof (PBE) [35] have been constructed to give similar results. The revised PBE functional (RPBE) [36] follows the same construction scheme as the PBE functional, just a different interpolation that is not specified by the construction scheme is used. This leads to a difference of almost half an eV for the O_2 binding energy. This is a rather unsatisfactorily result because this means that there is an intrinsic uncertainty of up to half an eV for energy differences obtained within the generalized gradient approximation. And still the theoretical O_2 binding energies are much larger than measured in experiment.

The binding energy of O_2 is not the only case where DFT calculations are rather inaccurate. A list of the failures of DFT with present-day functionals includes: (i) van der Waals forces are not properly described, (ii) negative ions are usually not bound, i.e. electron affinities are too small, (iii) the Kohn–Sham potential falls off exponentially for large distances instead of $\propto 1/r$, (iv) band gaps are underestimated in both LDA and GGA by approximately 50%, (v) cohesive energies are overestimated in LDA and underestimated in GGA, (vi) strongly correlated solids such as NiO and FeO are predicted as metals and not as antiferromagnetic insulators.

The problem in the development of a more accurate exchange-correlation functional is the reliable representation of the non-locality of this functional. One could say that all present formulations of the exchange-correlation functional in principle still represent an uncontrolled approximation. There is no systematic way of improving the functionals since there is no expansion in some controllable parameter.

Still the development of more accurate exchange-correlation function is a very active research field. One route is to include to some extent "exact exchange" in addition to a standard functional. Another *ansatz* is the development of so-called *meta-GGA*'s that include higher-order powers of the gradient or the local kinetic energy density. Very accurate results for small molecules can be obtained by methods based on *orbital functionals* such as the optimized potential method (OPM) or the optimized effective potential (OEP) method [37]. In this approach, the exchange-correlational functional does not explicitly depend on the density but on the individual orbitals. Thus the self-interaction can be avoided. It is still true that all improved functionals mentioned above require a significant increase in the computational effort. Therefore they have not been used yet in standard applications of DFT calculations for surface science problems.

In any practical implementation of DFT the computational effort of course increases significantly with the number of electrons that have to be taken into account. However, most chemical and solid-state properties are determined almost entirely by the valence electrons while the influence of the core electrons on these properties is negligible. Indeed there is a way to replace the effect of the core electrons by an effective potential so that they do not have to be taken into account explicitly, namely by constructing so called *pseudopotentials*. Since this significantly reduces the number of electrons that have to be taken into account, the use of pseudopotentials leads to an enormous saving of computer time.

3.4 Pseudopotentials

The concept of pseudopotentials is based on the observation that the chemical properties of most atoms are determined by their valence electrons. Core electrons hardly participate in any chemical interaction. The starting point for the generation of pseudopotentials is an all-electron calculation for the isolated atom. We rewrite the one-particle wave function of a valence electron as [38, 39]

$$|\psi_v\rangle = |\psi_{\text{ps}}\rangle - \sum_i |\psi_{c_i}\rangle \langle \psi_{c_i} | \psi_{\text{ps}} \rangle. \quad (3.60)$$

The $|\psi_{c_i}\rangle$ are core states with one-particle energies ε_{c_i}. The valence wave function is assumed to be orthogonal to the core states. Indeed, in the form (3.60) the matrix element $\langle \psi_v|\psi_{c_i}\rangle$ of the valence wave-function with any core state will vanish by construction. However, the pseudo-wave function $|\psi_{\text{ps}}\rangle$ is not uniquely defined by (3.60) since the coefficients $\langle \psi_{c_i}|\psi_{\text{ps}}\rangle$ can be chosen arbitrarily. Substituting the wave function (3.60) into an effective one-particle Schrödinger equation such as the Kohn–Sham equations (3.49) for the isolated atom

$$\left\{-\frac{\hbar^2}{2m}\nabla^2 + v_{\text{eff}}\right\} |\psi_v\rangle = \varepsilon_v |\psi_v\rangle \quad (3.61)$$

leads to

$$\left\{-\frac{\hbar^2}{2m}\nabla^2 + v_{\text{ps}}\right\} |\psi_{\text{ps}}\rangle = \varepsilon_v |\psi_{\text{ps}}\rangle, \quad (3.62)$$

with the pseudopotential v_{ps} defined by given by

$$v_{\text{ps}} = v_{\text{eff}}(\boldsymbol{r}) + \sum_i (\varepsilon_v - \varepsilon_{c_i}) |\psi_{c_i}\rangle \langle \psi_{c_i}|. \quad (3.63)$$

It is a non-local potential since it does not simply operate on a wave function by a multiplication of a $\boldsymbol{r}$-dependent function. In addition, it is energy-dependent. This constitutes the difference to a "true" potential and is the reason why it is called a "pseudo-potential". Since $\varepsilon_v > \varepsilon_{c_i}$ and the core states are localised, the sum in (3.63) acts as a short-range repulsiv potential. Note that the wave function $\psi_{\text{ps}}(\boldsymbol{r})$ has the same one-particle energy ε_v as the true valence wave function $\psi_v(\boldsymbol{r})$. Furthermore, $\psi_{\text{ps}}(\boldsymbol{r})$ does not need to be orthogonal to the core states. Consequently, it does not have to have a nodal structure in the core region. Therefore one has the freedom to choose a smooth pseudo-wave function which is rather advantageous when the wave function is expanded in some set of basis functions.

The derivation (3.60)–(3.63) captures the essentials of the pseudopotential generation. Still this simple formulation has some drawbacks. For example, the pseudo-wave function $|\psi_{\text{ps}}\rangle$ entering (3.60) is not normalized. This can easily be checked by taking the norm on both sides of (3.60) under the assumption that $|\psi_v\rangle$ is normalized. The deviation of the norm from unity is given by

$$1 - \langle \psi_{\text{ps}} | \psi_{\text{ps}} \rangle = \sum_i |\langle \psi_{c_i} | \psi_{\text{ps}} \rangle|^2, \quad (3.64)$$

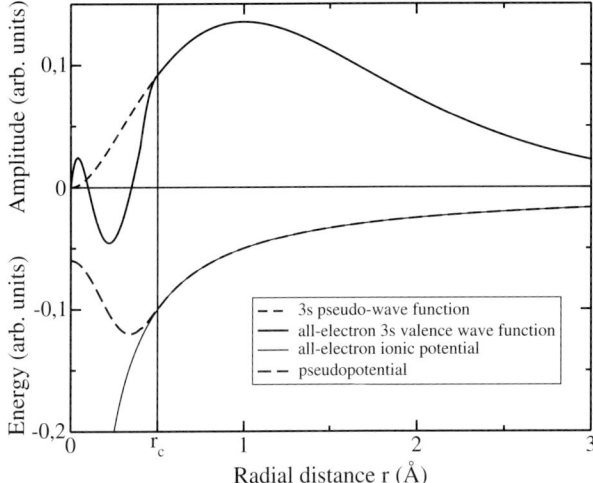

Fig. 3.6. Schematic illustration of the difference between the all-electron (*solid line*) and pseudo 3s wave function (*dashed line*) and their corresponding potentials. The core radius r_c is indicated by the vertical line

which is typically of the order of 0.1. One can of course explicitly normalize the pseudo-wave function but this leads to an incorrect distribution of the valence charge. This problem can be avoided by the construction of so-called *norm-conserving pseudopotentials* [40,41]. Their construction is guided by the following requirements. Asymptotically a pseudopotential should describe the long-range interaction of the core. Outside of a core radius r_c the pseudo-wavefunction should coincide with the full wavefunction. Inside of this radius both the pseudopotential and the wavefunction should be as smooth as possible to reduce the computational effort. This property is also referred to as the *softness* of the pseudopotential. The requirement that the norm of the pseudo-wave function is conserved then automatically ensures that it has the same one-particle energy ε_v as the true valence wave function.

The general form of norm-conserving pseudopotentials is given by

$$v_{\mathrm{ps}}(\boldsymbol{r}) = \sum_{lm} |Y_{lm}\rangle \, v_l(\boldsymbol{r}) \, \langle Y_{lm}| \,. \tag{3.65}$$

This form is called semi-local because it is local in the radial part and non-local in the angular part. The requirements for the construction of norm-conserving pseudopotentials still leaves a lot of freedom for the specific generation. A further important property of pseudopotentials which should be considered in their construction is the transferability. A pseudopotential should give reliable results independent of the particular environment in which it is used.

The most common pseudopotential generation schemes have been developed by Bachelet, Hamann and Schlüter [41] and by Troullier and Martins [42,43]. The Troullier–Martins pseudopotentials are constructed in such a way as to give particularly soft potentials. A comparison of a typical pseudo-wave function ψ_{ps} and the pseudopotential v_{ps} with the corresponding all-electron-results is illustrated in Fig. 3.6 for a $3s$ state.

A further significant improvement has been the development of the *Vanderbilt* or *ultra-soft* pseudopotentials [44]. In the generation of this pseudopotentials the norm-conserving constraint has been removed. They are rather constructed by a generalized orthonormality condition. In order to recover the correct charge density, augmentation charges are introduced in the core region. The electron density can thus be subdivided in a delocalized smooth part and a localized hard part in the core regions. As the name already suggest, by this procedure very soft pseudopotentials can be created that enable a dramatic reduction in the necessary size of the basis set.

3.5 Implementations of Density Functional Theory

Density functional theory was first accepted as a very valuable method in solid-state physics where one deals with periodic structures. Many bulk properties of materials were accurately reproduced by DFT calculations in the local density approximation. The natural basis to describe periodic structure is made of plane waves

$$\psi_{\boldsymbol{k}}^{\boldsymbol{G}}(\boldsymbol{r}) = \frac{1}{\sqrt{V}}\, e^{i(\boldsymbol{k}+\boldsymbol{G})\cdot\boldsymbol{r}}, \tag{3.66}$$

because plane waves are already of the form (2.21) required by the Bloch theorem. In (3.66), $\boldsymbol{G}$ is a reciprocal lattice vector and $\boldsymbol{k}$ is supposed to lie within the first Brillouin zone. Due to the symmetry properties of an infinite crystal, plane wave with wave vectors that do not differ by a reciprocal lattice vector do not couple, i.e.

$$\langle \psi_{\boldsymbol{k}} | h | \psi_{\boldsymbol{k}'} \rangle = 0 \quad \text{for} \quad \boldsymbol{k} \neq \boldsymbol{k}' + \boldsymbol{G}, \tag{3.67}$$

where h is an effective one-particle Hamiltonian, for example the one entering the Kohn–Sham equations. Hence the expansion of any wave function solving the Kohn–Sham equations only contains plane waves that differ by reciprocal lattice vectors. To determine the total energy of a crystal, still a summation over the lowest eigenvalues has to be performed. For infinite periodic systems, this band structure energy in the total energy expression has to be replaced by an integral over the first Brillouin zone

$$\sum_{i} \varepsilon_i \; \longrightarrow \; \sum_{\text{bands } j} \frac{V}{(2\pi)^3} \int_{BZ} d^3k \; \varepsilon_j(\boldsymbol{k}), \tag{3.68}$$

where over all occupied energy bands has to be summed. Fortunately this integral can be approximated rather accurately by a sum over a finite set

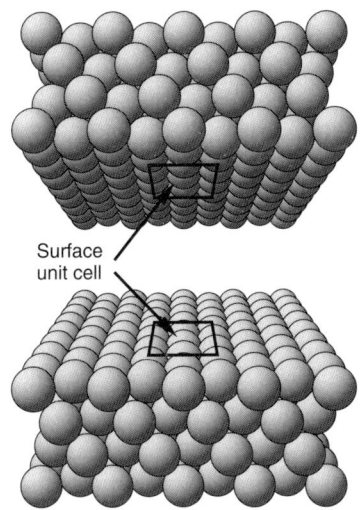

Fig. 3.7. Illustration of the supercell approach, in which surfaces are represented through an infinite array of slabs

of k-points, either by using equally spaced k-points within the first Brillouin zone or by using so-called *special k-points* [45]. In practice this means that one performs a number of calculations for different k-points. The electron density $n(r)$ in real space that enters the self-consistency cycle is also constructed by a Fourier transform over these k-points. Finally the eigenenergies $\varepsilon_j(k)$ at the different k-points are summed up for the band structure term.

The expansion of the electronic wave functions in plane waves is computationally very efficient, in particular due to Fast-Fourier-Transformation (FFT) techniques [16]. However, this expansion usually requires a three-dimensional periodicity of the considered system. If one wants to use plane-wave codes for surface science problems, the surface has to be cast into a three-dimensional periodicity. This is achieved in the so-called supercell approach, in which a surface is represented through an infinite array of slabs. The supercell approach for surfaces is illustrated in Fig. 3.7. In order to give a reliable descriptions of surfaces, firstly the vacuum layer between the slabs has to be sufficiently wide to avoid any interaction between the slabs, and secondly the slabs have to be thick enough to be a reasonable model for a surface of a semi-infinite substrate. Both these properties can be checked by convergence tests of any calculated property with respect to the width of the vacuum layer and the thickness of the slab.

In the discussion of the pseudopotentials we already mentioned that their introduction leads to a significant reduction in the computational cost because of the smaller number of electrons that have to be taken into account explicitly. In plane wave calculations, however, the softness of a pseudopotential plays an essential role. The smoother, i.e, the softer a pseudopotential is, the smaller the number of plane waves necessary in the expansion of the wave function. So to say, less Fourier coefficients are needed in order to resolve smooth structures. This is the reason why many of the modern plane wave DFT codes [46] use the ultrasoft pseudopotentials [44].

Even within the pseudopotential concept the solution of the single-particle equations would still require the diagonalization of a rather large matrix. The size of this matrix is determined by the number of plane waves in the expansion of wave functions. As a parameter characterizing the size of the plane wave basis set usually the kinetic energy of the highest Fourier component, the so-called cutoff energy

$$E_{\text{cutoff}} = \max_{\boldsymbol{G}} \frac{\hbar^2 (\boldsymbol{k} + \boldsymbol{G})^2}{2m} \tag{3.69}$$

is used. Employing norm-conserving Troullier–Martins pseudopotentials, this cutoff energy is typically about 10 Ryd for semiconductors and more than 50 Ryd for transition metals. Using ultrasoft pseudopotentials, the cutoff energy for transition metals can be reduced to about 20 Ryd. Depending on the number of atoms in the supercell, the number of plane waves in the expansion can easily be larger than 10,000. The diagonalization of a $10,000 \times 10,000$ matrix is computationally still very demanding. A full diagonalization is anyways not required because often only the lowest 100 to 1000 eigenvalues are needed. Hence in almost all modern DFT algorithms the diagonalization is avoided by using the fact that the diagonalization can be regarded as a minimization problem for which many efficient algorithms exist.

The use of pseudopotentials still represents an approximation. For some elements, there is a significant interaction between core and valence electrons. Hence all-electron calculations are desirable for systems containing these elements. There are indeed electronic structure methods for extended periodic systems that do take into account the core electrons. The main idea, proposed by Slater in 1937 [47], is to expand the electronic wave function in the core region in a different basis set than in the interstitial region. In the first implementations, the effective potential was approximated by the *muffin-tin potential*

$$v_{\text{eff}}(\boldsymbol{r}) = \begin{cases} 0 & \text{interstitial region} \\ v_{\text{MT}}(|\boldsymbol{r} - \boldsymbol{R}_i|) & |\boldsymbol{r} - \boldsymbol{R}_i| < r_{\text{MT}} \end{cases}. \tag{3.70}$$

This means that the effective potential is approximated by a constant potential in the interstitial region and by a radial symmetric potential within the muffin-tin radius r_{MT} in the core region. As the basis set for the expansion of the correct solution of the effective one-particle equations within the muffin-tin approximation (3.70) *augmented plane waves* (APW) are used which are defined as

$$\phi_{\boldsymbol{k},\varepsilon}(\boldsymbol{r}) = \begin{cases} \frac{1}{\sqrt{V}} e^{i\boldsymbol{k}\cdot\boldsymbol{r}} & \text{interstitial region} \\ \sum_{lm} a_{lm}(\boldsymbol{k}, \varepsilon) \, u_l(r_i, \varepsilon) \, Y_{lm}(\theta, \phi) & |\boldsymbol{r} - \boldsymbol{R}_i| < r_{\text{MT}} \end{cases}, \tag{3.71}$$

where we have set $r_i = |\mathbf{r} - \mathbf{R}_i|$ and where θ and ϕ are related to the origin at $\mathbf{R}_i$. The plane waves are so-to-say augmented by spherical harmonics times a radial function in the core region. The augmented plane waves are defined to be continuous at the boundary between core and interstitial region. The functions $u_l(r, \varepsilon)$ are solutions of the radial Schrödinger equation

$$\left\{ -\frac{\hbar^2}{2m} \frac{d^2}{dr^2} + \frac{\hbar^2}{2m} \frac{l(l+1)}{r^2} + v_{\mathrm{MT}}(r) - \varepsilon \right\} r u_l(r, \varepsilon) = 0. \qquad (3.72)$$

Note that one can define augmented plane waves for any wave vector $\mathbf{k}$ and any energy ε; there is no constraint relating the two quantities. The continuity requirement at the boundary between core and interstitial region then uniquely determines the expansion coefficients $a_{lm}(\mathbf{k}, \varepsilon)$ for any given combination of $\mathbf{k}$ and ε, but the derivative is discontinuous at the boundary. The correct solution of the one-particle Schrödinger equation is written as a superposition of augmented plane waves, all with the same energy:

$$\psi_{\mathbf{k}}(\mathbf{r}) = \sum_{\mathbf{G}} c_{\mathbf{G}}\, \phi_{\mathbf{k}+\mathbf{G}, \varepsilon}(\mathbf{r}), \qquad (3.73)$$

where the sum is over reciprocal lattice vectors. The APW method for a given muffin-tin potential can in principle be exact, but it is computationally rather costly. This is so because the basis functions, the augmented plane waves, depend on the energy. Therefore the basis set cannot be used for the whole energy spectrum.

This problem is avoided by the concept of the *linearized augmented plane waves* (LAPW) that was proposed by Andersen in 1975 [48]. First, the basis functions *and* their first derivative are required to be continuous at the boundary between core and interstitial region. This makes only an approximative solution of the Schrödinger equation possible, but the associated error is rather small [48]. Secondly and more importantly, the radial functions $u_l(r, \varepsilon)$ are expanded around a fixed energy ε_l, i.e. they are written as

$$u_l(r, \varepsilon) = u_l(r, \varepsilon_l) + \dot{u}_l(r, \varepsilon_l)\, (\varepsilon - \varepsilon_l) + \ldots, \qquad (3.74)$$

where $\dot{u}_l(r, \varepsilon_l)$ is the energy derivative

$$\dot{u}_l(r, \varepsilon_l) = \left. \frac{du_l(r, \varepsilon)}{d\varepsilon} \right|_{\varepsilon = \varepsilon_l}. \qquad (3.75)$$

The fixed energy ε_l should be in the middle of the corresponding energy band with l character.

The LAPW basis functions are then given by

$$\phi_{\mathbf{k}}(\mathbf{r}) = \begin{cases} \frac{1}{\sqrt{V}} e^{i\mathbf{k}\cdot\mathbf{r}} & \text{interstitial region} \\ \sum_{lm} [a_{lm}(\mathbf{k})\, u_l(r_i, \varepsilon_l) + \\ \quad b_{lm}(\mathbf{k})\, \dot{u}_l(r_i, \varepsilon_l)]\, Y_{lm}(\theta, \phi) & |\mathbf{r} - \mathbf{R}_i| < r_{\mathrm{MT}} \end{cases}. \qquad (3.76)$$

The augmentation parameters $a_{lm}(\bm{k})$ and $b_{lm}(\bm{k})$ are determined through the continuity requirement of both the wave function and its first derivative. Through the linearization (3.74) the basis functions become energy independent and the corresponding radial functions $u_l(r,\varepsilon_l)$ only have to be determined once. This leads to an enormous increase in the efficiency of the method compared to the original APW method.

The restriction to spherical symmetric potentials in the core region and constant potentials in the interstitial region can also be lifted in the LAPW method. The potential is expanded corresponding to the wave function as

$$v_{\text{eff}}(\bm{r}) = \begin{cases} \sum_{\bm{G}} v_{\text{eff}}(\bm{G})\, e^{i\bm{G}\cdot\bm{r}} & \text{interstitial region} \\ \sum_{lm} v_{\text{eff}}^{lm}(r_i)\, Y_{lm}(\theta,\phi) & |\bm{r}-\bm{R}_i| < r_{\text{MT}} \end{cases}. \quad (3.77)$$

Since now the "full potential" (FP) can be taken into account, this method is called FP-LAPW method.

One problem arises for the determination of the atomic forces. The LAPW basis functions (3.76) depend on the atomic position and move according to the dynamics of the atoms. For such a basis set the atomic forces are not simply given by the Hellmann–Feynman theorem (2.12), but in addition basis set corrections have to taken into account, the so-called *Pulay forces*. This makes the evaluation of the atomic forces more complex, but they are now implemented in standard FP-LAPW packages [49].

All-electron DFT calculations using the FP-LAPW method are considered to give the most accurate results apart from the errors associated with the exchange-correlation functional. This requires a large computational effort. Therefore FP-LAPW calculations are usually more expensive than plane-wave calculations using ultrasoft pseudopotentials.

An all-electron method that only requires the computational effort of ultrasoft pseudopotential calculations is based on the so-called *projected augmented waves* (PAW) [50, 51]. The augmentation procedure differs from the LAPW method in that partial-wave expansions are not determined through the continuity requirement of both the wave function and its first derivative at the muffin-tin radius, but rather by the overlap with localized projector functions [50]. In fact, there is a formal relationship between the ultrasoft pseudopotential method and the PAW method. It can be shown that they only differ by one-center terms [51]. This makes the PAW method to a computationally very efficient all-electron method.

One of the main goals of total-energy calculations is to find equilibrium structures of a particular system. These structures corresponds to the minima in the Born–Oppenheimer energy surface. Usually the determination of the minima requires many calculations of the energies and the gradients of the Born–Oppenheimer surface. In fact, it is not necessary to perform many calculations on the Born–Oppenheimer surface in order to find the minima. Instead, taking advantage of the variational principle of density functional theory, both nuclear and electronic degrees of freedom can be relaxed simultaneously. This approach which is now known as the *Car–Parrinello method* [52] was first proposed by Bendt and Zunger [53]. For a satisfactory convergence, however, this method reqires the existence of an electronic band gap. Therefore, for metallic systems the Born–Oppenheimer approach is recommended [54]. Still the Car–Parrinello has been used extensively, in particular for biologically relevant systems [55].

3.6 Further Many-Electron Methods

Although density functional theory has been remarkably successful, there are severe shortcomings, as listed on p. 40, that need to be overcome. For weakly correlated solids, very accurate ground-state expectation values can be obtained by *quantum Monte Carlo* (QMC) methods [56]. In general, the term *Monte Carlo* denotes computational techniques that are based on random sampling. There are indeed also several versions of quantum Monte Carlo methods. In the simplest of these, variational Monte Carlo (VMC), the expectation values with respect to a chosen trial many-body wave function are evaluated by a Monte Carlo integration scheme. This requires a good initial guess for the trial wave function. The more sophisticated diffusion quantum Monte Carlo avoids this limitation. It corresponds to a projector technique in which a stochastic imaginary-time evolution is used to suppress the higher-states components of the trial wave function.

Quantum Monte Carlo calculations can provide very accurate results. For example, the parametrization of the exchange-correlation energy in the local density approximation is based on quantum Monte Carlo [18]. On the other hand, QMC simulations are computationally very demanding so that their applications are still limited to systems with a rather small number of atoms. Furthermore, while probabilistic methods such as DMC usually require positive distributions, many-fermion wave functions change sign due to their antisymmetry. This leads to the fermion sign problem in quantum Monte Carlo. Therefore there are almost no approximation-free QMC algorithms treating fermion systems. Usually the fixed-node approximation is employed in which the nodal structure of the wave-functions is determined by the trial wave function and kept fixed in the simulation. In spite of these problems, remarkable progress has been made in the development of efficient and accurate quantum Monte Carlo algorithms, and first applications to surface science problems have been carried out.

As an intermediate approach that is computationally less demanding than quantum chemistry or quantum Monte Carlo methods but more accurate than standard DFT methods, *natural orbital functional theory* [57] has been proposed. Natural orbits are the eigenfunctions of the one-particle density matrix with the occupation numbers n_i as their eigenvalues. In this approach, the true many-body kinetic energy can be expressed in terms of the natural orbitals so that the still unknown exchange-correlation functional only includes potential energy contributions whereas the exchange-correlation functional in the Kohn–Sham formulation contains both potential and kinetic energy contributions. This method has not matured yet, but it can produce atomic energies as good as, and atomic densities that are better than those obtained from the most accurate DFT implementations [57].

3.7 Tight-Binding Method

Total energy calculations using density functional theory are still computationally rather expensive. Consequently, the systems treated are often limited to well below 100 atoms within the supercell. There is a computational scheme of the early days of computational physics [58] that is still rather popular: the tight-binding method [59] in which the essentials of quantum mechanics are retained. Still it is computationally much more effective than a true self-consistent electronic structure calculation. Compared to ab initio methods, tight-binding is about two to three orders of magnitude faster, depending on the particular system. This allows the treatment of large systems with more than 1000 atoms, in particular in combination with so-called order(N) methods [60] in massively parallel calculations. Tight-binding is also two to three orders of magnitude slower than empirical methods which, however, hardly reproduce the quantum mechanical nature of bonding.

In short, the tight-binding method can be characterized by saying that it assumes an expansion of the eigenstates of the effective one-particle Hamiltonian in an atomic-like basis set and replaces the exact many-body Hamiltonian with parametrized Hamiltonian matrix elements [59]. The atomic-like basis functions are usually not considered explicitly, but the matrix elements are assumed to have the same symmetry properties as matrix elements between atomic states. Tight-binding is very similar to the Hückel and extended Hückel methods used in quantum chemistry.

To be specific, in tight-binding one formally starts with a basis of atomic functions

$$\phi_{i\alpha}(\boldsymbol{r}) = \phi_\alpha(\boldsymbol{r} - \boldsymbol{R}_i), \tag{3.78}$$

where i labels the lattice site and α the type of the atomic orbital such as s, p, d etc. From the atomic orbitals (3.78) periodic functions can be constructed by forming Bloch sums

$$\phi_{\alpha\boldsymbol{k}}(\boldsymbol{r}) = \frac{1}{\sqrt{N}} \sum_i e^{i\boldsymbol{k}\cdot\boldsymbol{R}_i} \phi_\alpha(\boldsymbol{r} - \boldsymbol{R}_i), \tag{3.79}$$

where N is the number of unit cells within the periodic boundary conditions. For simplicity we have assumed here that there is only one atom per unit cell. The extension to a lattice with a basis is straightforward. The sums (3.79) are then used to determine the matrix elements

$$\begin{aligned}
h_{\alpha\beta}(\boldsymbol{k}) &= \langle\phi_{\alpha\boldsymbol{k}}|h|\phi_{\beta\boldsymbol{k}}\rangle \\
&= \frac{1}{N}\sum_{ij} e^{i\boldsymbol{k}\cdot(\boldsymbol{R}_j-\boldsymbol{R}_i)} \int \phi_\alpha^*(\boldsymbol{r}-\boldsymbol{R}_i)\, h\, \phi_\beta(\boldsymbol{r}-\boldsymbol{R}_j)\, d^3r \\
&= \sum_j e^{i\boldsymbol{k}\cdot\boldsymbol{R}_j} \int \phi_\alpha^*(\boldsymbol{r})\, h\, \phi_\beta(\boldsymbol{r}-\boldsymbol{R}_j)\, d^3r \\
&= \sum_j e^{i\boldsymbol{k}\cdot\boldsymbol{R}_j} \langle\phi_{0\alpha}|h|\phi_{j\beta}\rangle = \sum_j e^{i\boldsymbol{k}\cdot\boldsymbol{R}_j}\, h_{\alpha\beta}(\boldsymbol{R}_j), \quad (3.80)
\end{aligned}$$

where we have used the translational invariance of the Bravais lattice. The key idea in tight-binding which was formulated by Slater and Koster [58] is to replace the explicit determination of the integrals $h_{\alpha\beta}(\boldsymbol{R})$ by a parametrized function depending on the interatomic distances $\boldsymbol{R}$. This requires the so-called *two-center approximation*. The evaluation of the matrix elements $h_{\alpha\beta}(\boldsymbol{R})$ usually involves integration over two atomic orbitals and the potential part of the Hamiltonian. This potential is due to all atoms in the system. This leads to three-center integrals in the evaluation of $h_{\alpha\beta}(\boldsymbol{R})$ which are neglected in the two-center approximation. In this approximation $h_{\alpha\beta}(\boldsymbol{R})$ becomes a function of the square modulus $R = |\boldsymbol{R}|$ and of the direction cosines k, l, m of $\boldsymbol{R}$. These direction cosines have been tabulated [58]. Often the atomic orbitals and matrix elements are further expanded as a sum over functions with well-defined angular momentum with respect to the axis between the two atoms, i.e., in σ, π, δ bonds etc.

Any eigenfunction $\chi_{\boldsymbol{k}}$ of the one-particle Hamiltonian can be written as

$$\chi_{\boldsymbol{k}}(\boldsymbol{r}) = \sum_\alpha c_\alpha(\boldsymbol{k})\, \phi_{\alpha\boldsymbol{k}}(\boldsymbol{r}). \quad (3.81)$$

The band energies $\varepsilon(\boldsymbol{k})$ can then be evaluated as the eigenvalues of

$$h(\boldsymbol{k})\, c(\boldsymbol{k}) = S(\boldsymbol{k})\, \varepsilon(\boldsymbol{k})\, c(\boldsymbol{k}), \quad (3.82)$$

where $S(\boldsymbol{k})$ is the overlap matrix given by

$$S_{\alpha\beta}(\boldsymbol{k}) = \sum_j e^{i\boldsymbol{k}\cdot\boldsymbol{R}_j} \langle\phi_{0\alpha}|\phi_{j\beta}\rangle = \sum_j e^{i\boldsymbol{k}\cdot\boldsymbol{R}_j}\, S_{\alpha\beta}(\boldsymbol{R}_j). \quad (3.83)$$

This means that the dispersion curves $\varepsilon(\boldsymbol{k})$ can be obtained by the diagonalization of $S(\boldsymbol{k})^{-1}h(\boldsymbol{k})$. This method is called non-orthogonal tight-binding. A further simplification results if one assumes that the atomic orbitals are already orthogonalized according to the Löwdin scheme [61]

$$\psi_{i\alpha}(\boldsymbol{r}) = \sum_{j\beta} S^{-1/2}_{i\alpha j\beta}\, \phi_{j\beta}(\boldsymbol{r}). \quad (3.84)$$

In this orthogonal tight-binding the eigenenergies are just obtained by the diagonalization of the Hamilton matrix $h(\boldsymbol{k})$. If there are N_c atoms in the unit cell and l atomic orbitals per atoms, then a $N_c l \times N_c l$ matrix has to be diagonalized.

In (3.82), often only the valence electrons are taken into account, i.e. it is assumed that pseudopotentials are used in the effective one-particle Hamiltonian h. As already mentioned, the atomic basis functions do not explicitly appear in the tight-binding formulation, but only implicitly via the matrix elements $h_{\alpha\beta}(\mathbf{R})$ and $S_{\alpha\beta}(\mathbf{R})$ which are both written in a parametrized form. Hence it is the parametrization scheme that determines the reliability and accuracy of any tight-binding calculation.

In order to obtain total energies rather than just band energies, usually a repulsive term written as a sum of pair terms is added

$$E_{\text{tot}} = E_{\text{band}} + E_{\text{rep}}$$
$$= \sum_{k} \varepsilon(\mathbf{k}) + \sum_{ij} U_{ij} . \tag{3.85}$$

Such a form looks similar to the total energy in DFT (3.52) under the assumption that the exchange-correlation energy and the Hartree energy can be combined as a pairwise repulsive interaction. In fact, the validity of (3.85) has been derived from DFT considerations [62]. There is, however, also a tight-binding scheme in which the repulsive term is contained in the band energies [63].

Tight-binding supplies a very useful scheme to understand qualitative trends of band structures. For example, if one assumes in orthogonal tight-binding that only nearest neighbors (nn) integrals contribute significantly, then the dispersion for a s-band in a metal is given by

$$\varepsilon(\mathbf{k}) = \beta + \sum_{nn} \gamma \, \cos \mathbf{k} \cdot \mathbf{R}, \tag{3.86}$$

where $\beta = \langle \phi_{0s} | h | \phi_{0s} \rangle$ is the so-called *on-site* term and $\gamma = \langle \phi_{0s} | h | \phi_{(nn)s} \rangle$ is the hopping matrix element connecting nearest neighbors. For a fcc crystal the s-band width turns out to be 12γ. Hence it is the magnitude of the overlap integral to the neighboring atoms that determines the band width: the larger the overlap, the broader the band. This qualitative picture will be important for the so-called d-band model (see Sect. 5.6).

Exercises

3.1 Fermi Energy of Free Electrons

Consider a system of N free electrons in a volume $V = L^3$ that corresponds to a cube with sides L. Assume that there are periodic boundary conditions at the faces of the cube.

a) How does the ground state of this system look like in $\mathbf{k}$-space?

b) Show that in the ground state the maximum energy of the electrons is given by

$$\varepsilon_F = \frac{\hbar^2 k_F^2}{2m} \tag{3.87}$$

with

$$k_F = |\mathbf{k}_F| = (3\pi^2 n)^{1/3} . \tag{3.88}$$

c) Determine k_F for a two-dimensional electron gas.

3.2 Hartree–Fock Theory for Free Electrons

Assume that the electrostatic potential of the electrons is compensated by a uniform positive charge background.

a) Show that then the Hartree–Fock equations are

$$-\frac{\hbar^2}{2m} \nabla^2 \psi_i(\mathbf{r}) - e^2 \sum_j \int d^3 r' \frac{\psi_j^*(\mathbf{r}')\psi_i(\mathbf{r}')}{|\mathbf{r}-\mathbf{r}'|} \psi_j(\mathbf{r}) = \varepsilon_i \, \psi_i(\mathbf{r}). \tag{3.89}$$

b) Show that the one-particle eigenenergies are given by

$$\varepsilon(\mathbf{k}) = \frac{\hbar^2 k^2}{2m} - 4\pi e^2 \int_{k'<k_F} \frac{d^3 k'}{(2\pi)^3} \frac{1}{|\mathbf{k}-\mathbf{k}'|^2}$$

$$= \frac{\hbar^2 k^2}{2m} - \frac{2e^2}{\pi} k_F \, F\!\left(\frac{k}{k_F}\right) \tag{3.90}$$

where

$$F(x) = \frac{1}{2} + \frac{1-x^2}{4x} \ln\left|\frac{1+x}{1-x}\right| . \tag{3.91}$$

Hint: Replace $\frac{1}{|\mathbf{r}-\mathbf{r}'|}$ by $4\pi \int \frac{d^3 q}{(2\pi)^3} \frac{1}{q^2} \exp(i\,\mathbf{q}\cdot(\mathbf{r}-\mathbf{r}'))$.

c) Explain briefly the lowering of the energy.

d) Show that for small k the Hartree–Fock one-particle energies can be approximated by

$$\varepsilon(\mathbf{k}) \approx \frac{\hbar^2 k^2}{2m^*} - \frac{2e^2}{\pi} k_F , \tag{3.92}$$

where the effective mass of the electrons m^* is given by

$$\frac{m^*}{m} = \left(1 + \frac{4e^2}{3\pi} \frac{m}{\hbar^2 k_F}\right)^{-1} . \tag{3.93}$$

3.3 Exchange Hole

Verify that the exchange hole in the homogeneous electron gas is given by

$$n_x(\bar{r}) = \frac{9}{2}\frac{N}{V}\left(\frac{k_F\bar{r}\cos(k_F\bar{r}) - \sin(k_F\bar{r})}{(k_F\bar{r})^3}\right)^2. \quad (3.94)$$

Hint: Express the sums in (3.30) as the appropriate integrals.

3.4 Thomas–Fermi Theory

Show that the Thomas–Fermi equation in the Hartree approximation

$$\frac{\hbar^2}{2m}(3\pi^2 n(r))^{2/3} + v_{\text{ext}}(r) + \int d^3r' n(r')\frac{e^2}{|r-r'|} = \mu, \quad (3.95)$$

can be derived from a variational principle.

Hint: Express the expectation value of the total energy in the Hartree approximation (3.12) in terms of the density $n(r)$ and assume that the kinetic energy term

$$T[n] = \int d^3r \, n(r) \, t[n(r)] \quad (3.96)$$

can be evaluated using the kinetic energy density $t(n)$ of the homogeneous electron gas (see (3.22)). Then mimimize the energy expression with respect to the electron density under the constraint of particle conservation.

3.5 Density Functional Theory

Consider an inhomogeneous electron gas of N electrons in the volume V_g subject to the external potential $v_{\text{ext}}(r)$.

a) Show that a system of noninteracting electrons exists that has the same ground state density as the inhomogeneous electron gas. Derive that the functional of the kinetic energy of the noninteracting electrons

$$T_s[n] = \sum_{i=1}^{N}\int \varphi_i^*(r)\frac{-\hbar^2}{2m}\nabla_r^2\,\varphi_i(r)\,d^3r \text{ with density } n(r) = \sum_{i=1}^{N}|\varphi_i(r)|^2$$

can be written as

$$T_s[n] = \sum_{i=1}^{N}\epsilon_i - \int v_{\text{eff}}(r)n(r)d^3r \quad (3.97)$$

where the effective potential is given by (3.50) and ϵ_i and φ_i are determined by the Kohn–Sham equation

$$\left[-\frac{\hbar^2}{2m}\nabla_r^2 + v_{\text{eff}}(r)\right]\varphi_i(r) = \epsilon_i\varphi_i(r). \quad (3.98)$$

b) Show that the total energy

$$E_{\text{tot}} = T[n_0] + V_{\text{ext}}[n_0] + V_{\text{H}}[n_0] + E_{\text{xc}}[n_0] + V_{\text{nucl-nucl}} \quad (3.99)$$

can be expressed as (3.52)

$$E = \sum_{i=1}^{N} \varepsilon_i + E_{\text{xc}}[n_0] - \int v_{\text{xc}}(\boldsymbol{r}) n_0(\boldsymbol{r}) \, d^3r - V_{\text{H}} + V_{\text{nucl-nucl}}, \quad (3.100)$$

where n_0 is the ground state density of the inhomogeneous electron gas.

c) We assume now that the external potential depends parametrically on the position of the nuclei: $v(\boldsymbol{r}) \to v(\boldsymbol{r}, \{\boldsymbol{R}_I\})$. Show that the gradient of the ground state energy is given by

$$\frac{\partial E}{\partial \boldsymbol{R}_J} = \int \frac{\partial v_{\text{ext}}(\boldsymbol{r}, \{\boldsymbol{R}_I\})}{\partial \boldsymbol{R}_J} n_0(\boldsymbol{r}, \{\boldsymbol{R}_I\}) d^3r + \frac{\partial V_{\text{nucl-nucl}}}{\partial \boldsymbol{R}_J}. \quad (3.101)$$

3.6 Surface States in the Tight-Binding Approximation

Consider an equidistant linear chain of identical atoms having only one s orbital $|\phi_l\rangle$ at each site l. Use the orthogonal tight-binding approximation with only nearest-neighbor interactions. The on-site and hopping matrix elements are given by

$$\langle \phi_l | h | \phi_m \rangle = \beta \delta_{l,m} + \gamma \delta_{l,m\pm 1}. \quad (3.102)$$

a) Determine the "bulk" band structure for an infinite chain of atoms.

b) Now we consider a semi-infinite chain of atoms with sites $n = 0, 1, \ldots$ as a model for a crystal with a surface. The on-site term for the surface atom differs from the bulk value, i.e.

$$\langle \phi_0 | h | \phi_0 \rangle = \beta' \neq \beta. \quad (3.103)$$

Show that for a strong perturbation of the on-site term at the surface $|\beta' - \beta| > |\gamma|$, new states above and below the bulk continuum appear. These states are called *Tamm surface states*.

c) Show that the Tamm states are surface states, i.e., that they are localized at the surface.

4. Structure and Energetics of Clean Surfaces

In the preceding chapter electronic structure methods were introduced which allow the evaluation of the total energy of a particular system. At zero temperature, the stable structure of a specific system is given by the structure with the minimal total energy. Therefore total-energy calculations are so important for the structural determination of surfaces. As far as finite temperature effects are concerned, the minimum of the free energy is the appropriate quantity. In the following sections, the electronic and geometric structure and the energetics of clean surfaces and their determination by first-principles calculations will be addressed. In addition, the underlying principles that lead to a particular structure will be thoroughly discussed. Since the surface vibrational modes are strongly related to the structure of the surface, surface phonons will also be addressed in this chapter.

4.1 Electronic Structure of Surfaces

Naturally, at the surface of a solid the electronic structure is strongly modified compared to the bulk electronic structure. The three-dimensional periodicity of an infinite crystal is broken so that the wave number k_z of the Bloch waves no longer is a good quantum number. Still the periodicity parallel to the surface is conserved. As we will see, this can lead to electronic bands localized at the surface. Here we will first introduce some basics about the electronic structure at surfaces.

Some fundamental properties of the electronic structure of metal surfaces, in particular simple metal surfaces, can be deduced from a very simple model in which the positive ion charges are replaced by a uniform charge background. In this jellium model, which has already been introduced on p. 26, the positive ion charges at a surface are simply represented by

$$n_+(\mathbf{r}) = \begin{cases} \bar{n}, & z \leq 0 \\ 0, & z > 0 \end{cases}. \quad (4.1)$$

Here z denotes, as usual, the direction perpendicular to the surface. The charge density in the jellium model is commonly specified by the corresponding Wigner–Seitz radius in atomic units, i.e., in multiples of the Bohr radius.

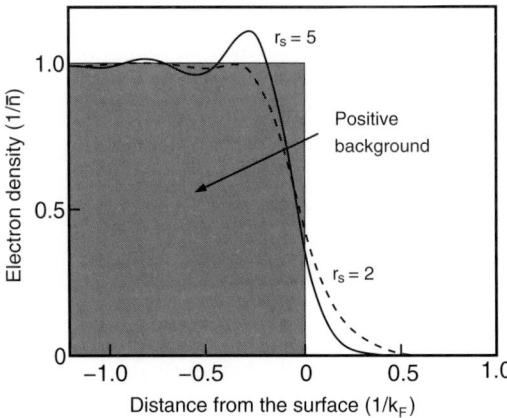

Fig. 4.1. Charge density as a function of the distance from the surface in Fermi wavelengths determined within the jellium model for two different background densities. (After [64])

The electronic charge distribution evaluated within the jellium model using density functional theory in the local density approximation [64] is plotted in Fig. 4.1. Two different background densities have been chosen corresponding to a high-density ($r_s = 2$) and a low-density metal ($r_s = 5$). The electron distribution does not follow the sharp edge of the positive background. Instead, it decreases smoothly and the electrons spill out into the vacuum. In fact this creates an electrostatic dipole layer at the surface because above the surface there is now an excess negative charge density while directly below the jellium edge there is an excess positive charge density. This dipole layer is sometimes also called double layer [10].

Furthermore, the charge density profile exhibits a damped oscillatory structure inside the jellium. These *Friedel oscillations* are a consequence of the sharp edge of the background density in the jellium model. The electrons try to screen out the positive background. Only electrons with wave vectors up to the Fermi wave vector k_f are available while in principle arbitrarily large wave vectors are needed. Thus the screening is incomplete and the Friedel oscillations with wavelength π/k_F result. For the high-density case ($r_s = 2$), however, these oscillations are already rather small.

The work function Φ is defined as the minimum work that must be done to remove an electron from a solid at 0 K. Consider a neutral slab representing the solid. Then the work function is given by

$$\Phi = \phi(\infty) + E_{N-1} - E_N. \qquad (4.2)$$

Here $\phi(\infty)$ is the total electrostatic potential far from the surface and E_M is the ground-state energy of the slab with M electrons but with an unchanged number of positive charges. As Fig. 4.1 indicates, the spilling out of the electrons at a surface creates a dipole layer. In order to carry the electron

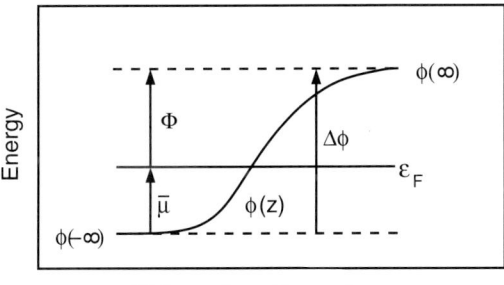

Fig. 4.2. Schematic representation of the electrostatic potential $\phi(z)$, the chemical potential $\bar{\mu}$, and the work function Φ

through the electric field of the dipole double layer at the surface, work has to be done. The work function can therefore be expressed as [65]

$$\Phi = \Delta\phi - \bar{\mu}$$
$$= \phi(\infty) - \varepsilon_F , \qquad (4.3)$$

where $\Delta\phi$ is the change in the electrostatic potential across the dipole layer, $\bar{\mu}$ is the intrinsic chemical potential of the electrons inside the bulk relative to the mean electrostatic potential there (see Fig. 4.2), and ε_F is the Fermi energy. It is important to note that there are two contributions to the work function: an intrinsic one due to the binding of the electrons and the effect of the dipole layer at the surface (see, e.g., the detailed discussion in [10]).

The jellium model has been used to evaluate the work function of simple and noble metals [65]. In order to estimate the variation of the work function from one crystal face to another, the ions have been modeled by pseudopotentials

$$v_{\text{ps}}(r) = \begin{cases} 0, & r \leq r_c \\ -\dfrac{Z}{r}, & r > r_c \end{cases} . \qquad (4.4)$$

These potentials have been added *a posteori* in this ion lattice model, i.e., the energies and work functions have been calculated using the electron distribution determined self-consistently within the jellium model without the pseudopotentials.

The calculated values of the work function are compared with experimental results for polycrystalline samples in Fig. 4.3. The plotted results of the ion lattice model correspond to the mean value of the work function for the (110), (100) and (111) surface for the cubic metals and to the (0001) surface for the hcp metals Zn and Mg. For the simple metals there is a rather good agreement between the jellium calculations and the experiment which is somewhat improved by taking into account the ion lattice contributions. The variations in the work function between the different surface orientations

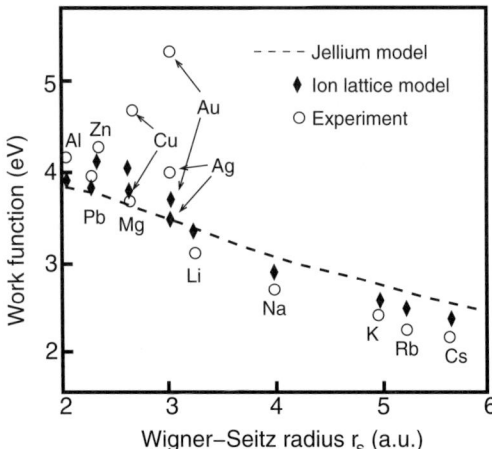

Fig. 4.3. Comparison of theoretical values of the work function obtained in the jellium model and the ion lattice model with the experiment. The experimental values were measured for polycrystalline samples, while the plotted results of the ion lattice gas model correspond to an average over the (110), (100) and (111) surface for the cubic metals and to the (0001) surface for the hcp metals Zn and Mg. (After [65])

is of the order of 10% of the mean work function [65]. In general, the lowest work function is found for the least densely packed surface considered which is the (110) face for the fcc metals and the (111) face for the bcc metals.

In addition to the simple metals, the work functions for the three noble metals Cu, Au and Ag have been calculated. As Fig. 4.3 demonstrates, there are already large quantitative differences between experiment and jellium calculations for these metals. Surprisingly, the rather crude jellium model is able to reproduce certain features of the *sp*-bonded simple metals with rather delocalized electron orbitals. However, the jellium approximation breaks down when it comes to metals with *d* electrons which are much more localized. For noble and transition metals, a more realistic theoretical description is needed.

Despite its shortcomings, the jellium model is well-suited to describe qualitative aspects of the change of the electron density in *real space* at a surface and related quantities such as the dipole layer. However, it neglects the lattice aspects in the description of the electronic structure at surfaces. These aspects related to the crystal structure can be best addressed qualitatively in the *nearly-free electron model* [10] in which the influence of the screened positive ion cores is approximated by a weak periodic pseudopotential.

Let us first focus on a simplified one-dimensional description. Within the simplest version of the nearly-free electron model, we describe an infinite solid as a chain of atoms creating an effective potential for the electrons given by

$$V(z) = V_0 + V_G \cos Gz, \tag{4.5}$$

where $G = 2\pi/a$ is the shortest reciprocal lattice vector of the chain. Using perturbation theory for degenerate states at the Brillouin zone edge, it is easy to show [11] that the periodic potential causes the opening up of a gap of $E_g = 2V_G$ at the zone boundary (see Exercise 4.1) In the gap, solutions of the one-particle Schrödinger equation with the potential (4.5) also exist, but they correspond to exponentially growing wave functions. Therefore they are physically unreasonable. However, at a surface the wave functions that increase exponentially towards the surface can be matched with wave functions that decay into the vacuum. This leads to the existence of localized states at the surface with energies in the gap that are called *Shockley surface states*.

If the local atomic orbitals are strongly perturbed at the surface, as for example in the case of semiconductor surfaces with broken bonds, additional surface states above and below the bulk continuum can appear. The existence of these so-called Tamm surface states can most easily be derived using a tight-binding description of the surface (see Exercise 3.6).

In the one-dimensional description, the surface state corresponds to a localized *bound* state. In three dimensions, crystal surfaces are still periodic in the lateral directions and can be characterized by a two-dimensional surface Brillouin zone. In other words, the wave vector $k_\parallel$ is still a good quantum number. This leads to a whole band of surface states. On the other hand, due to the broken symmetry in the z-direction, the discrete reciprocal lattice points along the surface normal are turned into rods which reflects that k_z is no longer a good quantum number. In order to analyse the surface band structure and to determine the nature of the electronic states at the surface, the presentation of the projected bulk band structure is rather helpful.

The construction of the projected bulk band structure is illustrated in Fig. 4.4. Two surface state bands are indicated by the solid lines in the band gaps of the projected bulk band structure. The chosen example corresponds to a hypothetical metal since for any energy ε there is at least one bulk state somewhere in the three-dimensional $\boldsymbol{k}$-space. A semiconductor or an insulator would have a band gap completely across the entire surface Brillouin zone (see, e.g., Fig. 4.13).

In the one-dimensional band structure for $\boldsymbol{k} = (0, 0, k_z)$, plotted in Fig. 4.4 two band gaps are present. The lower one is due to the interaction at the Brillouin zone boundary while the upper one results from an avoided crossing of two bands. In such a hybridization gap, also true surface states can exist, as is indicated by the upper surface band. The lower surface band, on the other hand, joins the projected bulk band structure and mixes with delocalized bulk states. By this mixing, a bulk state with a significantly enhanced amplitude at the surface is created. Such a state is called a surface resonance.

The analysis of the band structure is not always a convenient tool for the determination and discussion of the bonding situation a surfaces, in particular when it comes to the understanding of adsorption phenomena [66]. Here, in

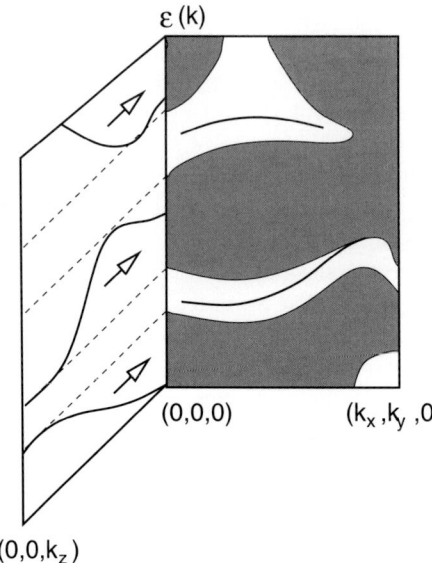

Fig. 4.4. Schematic illustration of the projected bulk band structure which is indicated by the grey-shaded areas. In addition, two surface state bands are included in the band gaps of the projected bulk band structure. While the upper band corresponds to true surface states, the lower surface band mixes with bulk states leading to surface resonances

particular the local density of states $n(\boldsymbol{r}, \varepsilon)$ (LDOS) can be rather useful. The LDOS is defined as

$$n(\boldsymbol{r}, \varepsilon) = \sum_i |\phi_i(\boldsymbol{r})|^2 \, \delta(\varepsilon - \varepsilon_i) \,. \tag{4.6}$$

Using (4.6) the global density of states and the electron density can be conveniently expressed through the following integrals

$$n(\varepsilon) = \int n(\boldsymbol{r}, \varepsilon) \, d^3 \boldsymbol{r} \,,$$

$$n(\boldsymbol{r}) = \int n(\boldsymbol{r}, \varepsilon) \, d\varepsilon \,. \tag{4.7}$$

Thus the band-structure energy in the total-energy expression (3.52) can also be written as an integral

$$\sum_{i=1}^{N} \varepsilon_i = \int n(\varepsilon) \, \varepsilon \, d\varepsilon \,. \tag{4.8}$$

Furthermore, the projected density of states (PDOS),

$$n_a(\varepsilon) = \sum_i |\langle \phi_i | \phi_a \rangle|^2 \, \delta(\varepsilon - \varepsilon_i) \,, \tag{4.9}$$

is also a useful tool since it allows to determine the nature and symmetry of chemical bonds. However, its analysis is already very illuminating even in the case of a clean surface. In Fig. 4.5, the layer-resolved local d-band density of states is plotted for the three uppermost surface layers of Pd(210). The

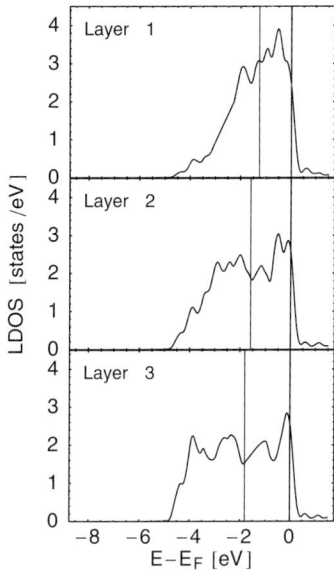

Fig. 4.5. Layer-resolved, local d-band density of states of Pd(210) determined by GGA-DFT calculations. The Fermi level and the center of the d-band are indicated by vertical lines. The third-layer PDOS is already very close to the bulk density of states of palladium. (After [67])

electronic structure had been determined by GGA-DFT calculations [67]. The (210) orientation corresponds to a rather open surface that can be regarded as a stepped surface with a high density of steps. Due to the lower coordination of the surface atoms, the d-band width is significantly reduced at the surface. Since the number of d-electrons remains the same, the local narrowing of the d-band leads to an upshift of the d-band center which is indicated by the vertical lines in Fig. 4.5. Otherwise the entire d-band would be located below the Fermi energy resulting in an increased occupation of the d-band. As we will see in the next chapter, the upshift of the d-band center leads to a higher reactivity of the surface (see p. 126).

The local d-band of the second layer is still somewhat narrower that the Pd bulk d-band, but already the third-layer d-band is practically indistinguishable from the bulk band. This is a consequence of the good screening properties of metals which lead to a rapid recovery of bulk properties in the vicinity of imperfections, which also includes surfaces.

4.2 Metal Surfaces

After introducing the basic concepts relevant for the discussion of the electronic structure of surfaces, I will first focus on metal surfaces, in particular on first-principles studies of noble and transition metals. Low-index metal surfaces do usually not reconstruct. The electron density of a Cu(100) surface determined by GGA-DFT calculations using ultrasoft pseudopotentials is shown in Fig. 4.6. Recall that the jellium model turned out not to be ap-

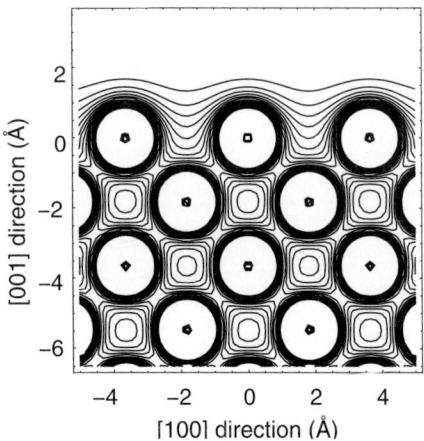

Fig. 4.6. Electron density of a Cu(001) surface along a (010) plane demonstrating electron smoothing at a metal surface. The electronic density has been calculated by DFT-GGA calculations using ultrasoft pseudopotentials

propriate for Cu surfaces (see Fig. 4.3). Figure 4.6 confirms that the electron density of Cu(100) is indeed rather inhomogeneous. Directly at the surface, however, the electron distribution is much smoother than in the bulk. At the surface, the electrons are free to lower their kinetic energy by becoming more uniformly distributed which results in the so-called Smoluchowski smoothing [68].

Let us now turn from the electronic structure in real space to the electronic structure in reciprocal space. The calculated band structure of Cu(111) is shown in Fig. 4.7. The projected bulk d-band is indicated by the darker shaded areas whereas the lighter shaded areas corresponds to states of the sp-bands. The sp states exhibit an almost free electron behaviour which can be infered from the parabolic shape of the lower and upper band edge. There is a pair of surface states in the upper band gap which also shows nearly-free-electron features. From the shape of the parabola around $\bar{\Gamma}$, an effective mass of $m^* = 0.37 m_e$ is derived [69]. These surface states correspond to Shockley states in the sp-band gap.

There is another surface state just above the d-band which is located approximately 1.5 eV below the Fermi level. This is a Tamm state which is pushed out of the top of the d-band. Although it lies mostly in the sp-continuum along $\bar{\Gamma}\bar{M}$, it is still a true surface state since it has a different symmetry than the sp-states and is therefore orthogonal to the sp-continuum. There are also surface resonances present at the Cu(111) surface plotted as dashed lines in Fig. 4.7. For example, focus on the Tamm surface state band. This band emanating from $\bar{M}$ bends down and would enter the d-band continuum were it not repelled. At the point where the surface state is repelled, a surface resonance splits off and enters the d-band. Another surface resonance originates at the $\bar{\Gamma}$ point where it is degenerate with the Tamm surface state.

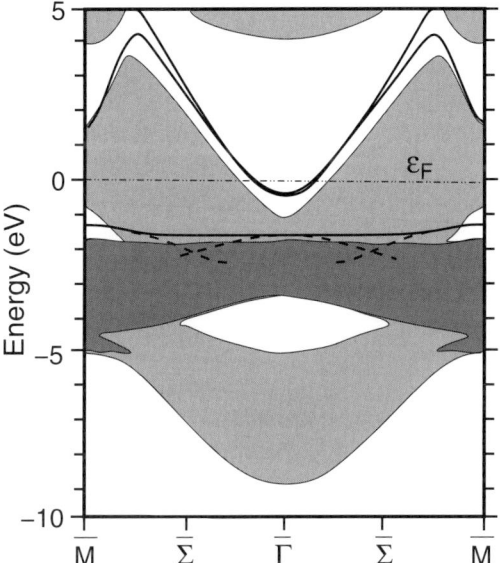

Fig. 4.7. Projected bulk band structure (*shaded areas*), surface states (*solid lines*) and surface resonances (*dashed lines*) at a Cu(111) surface determined by LDA-DFT calculations. The darker shaded areas correspond to states of *d*-character. (After [69])

The band structure of Cu(111) is shown up to the vacuum level 5 eV above the Fermi level. The calculated LDA-DFT workfunction of 5 eV is in good agreement with experimental values (see Fig. 4.3). A striking feature of the band structure is the large band gap along the surface normal at $\bar{\Gamma}$. For Cu(100), the projected bulk band structure looks qualitatively rather similar [70], although there is no Shockley surface state in the *sp*-band gap. However, the projected *sp*-band gap along the surface normal even extends above the vacuum level. This has the consequence that an electron with energy below the vacuum energy can be trapped in the potential well formed by the attractive image potential $V_{\rm im} = -e^2/4Z$ (see p. 98) and the repulsive surface barrier [71]. The resulting quantized image-potential states form a Rydberg series with energies

$$\varepsilon_n = \varepsilon_{\rm vac} - \frac{0.85 {\rm eV}}{(n+a)^2}, n = 1, 2, \ldots . \quad (4.10)$$

The constant a in the denominator is called the quantum defect which approximately takes into account the fact that the surface potential is no hard repulsive wall. These image-potential states which are sketched in Fig. 4.8 have indeed been detected in experiment and even the dynamical evolution of a coherent superposition of several quantum states has been observed by time-resolved photoelectron spectroscopy [72]. Still there is an overlap of the wave functions with the bulk electronic states which leads to a finite lifetime τ

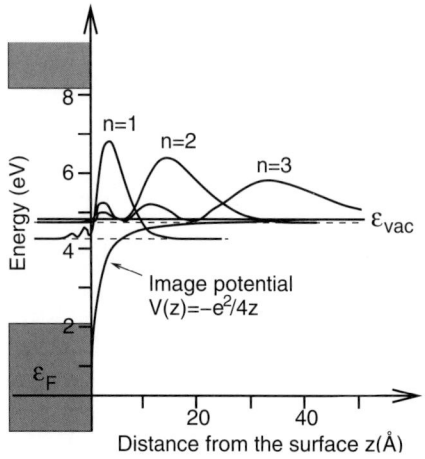

Fig. 4.8. Image potential and the lowest three image-potential states of an electron in front of a Cu(100) surface. The surface sp-band gap is indicated by the shaded areas. (Adapted from [72])

of the image-potential states. This lifetime increases from $\tau_1 = 40$ fs for the $n = 1$ state to $\tau_3 = 300$ fs for the $n = 3$ state [72] since the higher lying states are located further away from the surface and thus have a smaller overlap with the bulk states.

We will now turn from the electronic to the geometric structure of metal surfaces, still using Cu surfaces as our exemplary system. The stability of a particular surface structure is given by its surface energy. In Sect. 2.3, we demonstrated that a surface can be assumed to be created by just cleaving an infinite solid. This will cost energy because otherwise the crystal would cleave spontaneously. Put in other words, creating surfaces by cleaving a solid is energetically hindered because the bonds between the atoms of the cleavage planes have to be broken. The *surface energy* γ is defined as the surface excess free energy per unit area of a particular crystal face. The pure jellium model has some severe shortcomings as far as the evaluation of surface energies is concerned. Calculated surface energies are found to be negative for high densities ($r_s \leq 2.5$) [64] which means that the crystal would not be stable. Only if the lattice structure is included non-self-consistently by representing the ion cores by simple pseudopotentials (4.4), there is semi-quantitative agreement with the experiment [64].

For a quantitative determination of the surface energy, a realistic self-consistent electronic structure calculation using the slab model is required. At zero temperature the surface energy of a monoatomic crystal may be derived from a N-layer slab calculation for a 1×1 surface unit by

$$\gamma = \frac{1}{2A} \left(E_{\text{slab}} - N \cdot E_{\text{bulk}} \right) \tag{4.11}$$

where E_{slab} is the total energy of the slab per supercell, E_{bulk} is the bulk cohesive energy per atom, and A is the surface area in the supercell. The factor of 2 in the denominator takes account of the fact that in the slab

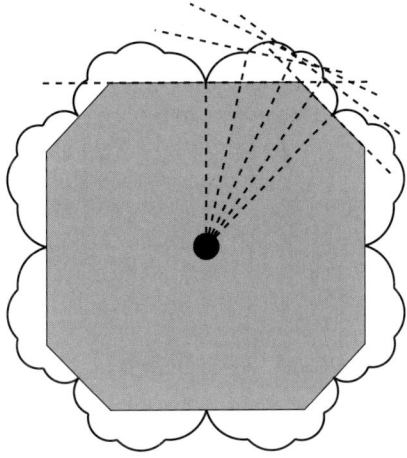

Fig. 4.9. Polar plot of the surface energy γ and the Wulff construction to determine the crystal equilibrium shape

we have *two* surfaces, one at the bottom and one at the top of the slab. In the limit of large N (4.11) should give the correct surface energy. For closed-packed metal surfaces often 4-6 layers are already sufficient to obtain converged results due to the good screening properties of metals. If the two surfaces are not equivalent as may happen in the case of binary compounds, then the value of γ evaluated according to (4.11) corresponds to the average of the two surface energies.

The equilibrium shape of finite mesoscopic crystals can be directly derived from the surface energies by the so-called *Wulff construction* which is based on the concept that the crystal seeks to minimize its total surface energy subject to the constraint of fixed volume (see Exercise 4.3). The Wulff construction is illustrated in Fig. 4.9. Draw radius vectors from the origin of the polar plot of the surface energies. At the points of the intersections, construct a plane perpendicular to the corresponding radius vector. These planes are known as Wulff planes. The equilibrium shape of the crystal is given by the interior envelope of all such possible Wulff planes. This is a convex figure where the distance of each face is proportional to its surface energy.

Besides its importance for the equilibrium shape, the determination of surface energies is also important for an understanding of crystal growth phenomena. Yet it is experimentally not trivial to determine surface energies. It is hard to directly measure absolute values. Often they are only determined relative to other surface energies. In fact, many reported surface energies are derived from surface tension measurements which are made in the liquid phase and extrapolated to zero temperature [73]. This does also mean that these surface energies are not related to any particular crystal face. Due to these problems associated with the measurements, the reliable theoretical determination of surface energies from first principles is of particular importance. For the low index surfaces of 60 metals databases of calculated surface energies using density functional theory are available [74, 75].

The more densely packed a certain lattice plane and the higher coordinated the atoms in that plane, the less bonds have to be broken upon cleavage. Hence the most densely packed surface should have the lowest surface energy. This is indeed the case for almost all $3d$, $4d$ and $5d$ transition metals. We have illustrated this trend in Table 4.1 where we have compiled the surface energies for some copper surfaces determined by DFT calculations as well as by experiment. The theoretical results have been obtained by a plane-wave DFT code [84] using ultrasoft pseudopentials [44] or the projector augmented-wave technique [50, 51], respectively, within the PW91-GGA functional [34].

For the low-index surfaces of Cu the trend $\gamma_{(111)} < \gamma_{(100)} < \gamma_{(110)}$ is obvious. Furthermore, we have included some stepped copper $((2n-1),1,1)$ surfaces. The $((2n-1),1,1)$ surfaces are in fact $n(100) \times (111)$ surfaces according to the notation introduced on page 17, i.e., they consist of (100) terraces separated by steps with (111)-oriented ledges. A $(911) = 5(100) \times (111)$ surface is plotted in Fig. 2.4. For large n the $((2n-1),1,1)$ surface energy will approach that of the (100) surface, therefore the surface energy decreases with increasing n. In fact, from the sequence of $((2n-1),1,1)$ surface energies the step formation energy can be derived.

We have included an experimentally determined surface energy for copper in Table 4.1. As already mentioned, the measured energy does not correspond to a particular crystal surface. The most important information gained by the comparison experiment–theory is thus the fact, that the measured surface energy has the same order of magnitude as the calculated ones.

Table 4.1. Surface energies γ and relaxations of the uppermost layer of various Cu surfaces. The relaxations are given in percent relative to the bulk layer spacing $d_0(hkl)$

Surface	Method	γ (J/m²)	Δd_{12}	Δd_{23}	Δd_{34}	$d_0(hkl)$ (Å)
Cu(111)	Theory[a]	1.30	-0.9	-0.3		2.10
Cu(111)	Exp.	~1.79[b]	-0.7[c]			
Cu(100)	Theory[a]	1.45	-2.6	1.5		1.821
Cu(100)	Exp.	~1.79[b]	-2.1[d]	0.4[d]	0.1[d]	1.807
Cu(110)	Theory[e]	1.53	-10.8	5.3	0.1	1.29
Cu(110)	Exp.	~1.79[b]	-8.5[f]	2.3[f]		
Cu(311)	Theory[g]	1.82	-15.0	4.0	-0.6	1.10
Cu(311)	Exp.[h]		-11.9	1.8		1.10
Cu(511)	Theory[g]	1.68	-11.1	-16.4	8.4	0.70
Cu(511)	Exp[i]		-14.2	-5.2	5.2	0.70

References: a) [76], b) [73], c) [77], d) [78], e) [79], f) [80], g) [81], h) [82], i) [83]

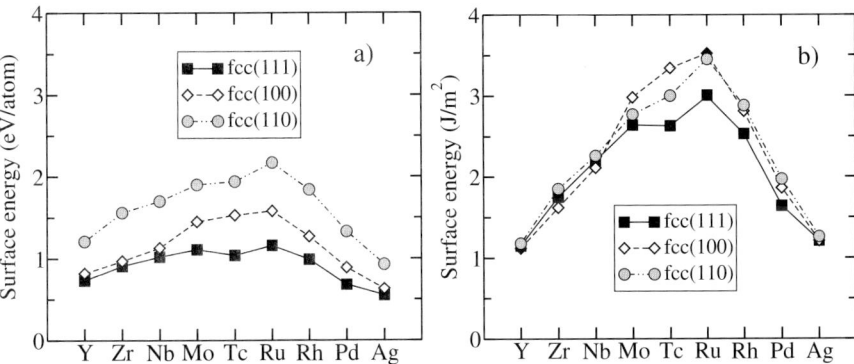

Fig. 4.10. First-principles surface energies for the 4d transition metals calculated using density functional theory within the local density approximation [74]. The energies have been determined for the (111), (100) and (110) surfaces in the fcc structure even for the hcp metals Y, Zr, Tc and Ru and for the bcc metals Nb and Mo. (**a**) surface energy in eV/atom, (**b**) surface energy in J/m^2

In order to analyse chemical trends within the transition metals, in Fig. 4.10 calculated surface energies in eV per surface atom and in J/m^2 [74] are plotted for the 4d transition metals. The results have been obtained using density functional theory within the local density approximation for the (111), (100) and (110) surfaces. To make the dependence on the d-band occupation more obvious, all surface energies have been calculated for the fcc structure, even for the hcp metals Y, Zr, Tc and Ru and for the bcc metals Nb and Mo.

The parabolic dependence of the surface energy on the d-band occupation is obvious. The surface energy is largest for a half-full band, while it is minimal for either an empty or a completely full d-band. The same trend is also observed for the 3d and the 5d transition metals [75] and is already well-known for the cohesive energies [10]. Using the so-called bond-cutting model, a more quantitative comparison between surface energies and cohesive energies can be made. A surface atom of a (111) surface is still nine-fold coordinated compared to the twelve-fold coordination in the bulk. Thus three out of twelve bonds have to be broken in order to create a surface; consequently, one would assume that the surface energy per atom is $3/12 = 0.25$ of the cohesive energy. However, this simple estimate gives surface energies that are about twice as large as the calculated ones [74] because it does not take into account the fact that the bond strength varies with the coordination number. For a low-coordinated atom the single bonds are stronger than for a high-coordinated atom. In a simple tight-binding picture, the band width is linearly related to the coordination number N_c which leads to an energy gain proportional to $(N_c)^{1/2}$ due to the down-shift of the occupied states (see Exercise 4.4). Thus the energy per bond can be assumed to scale with

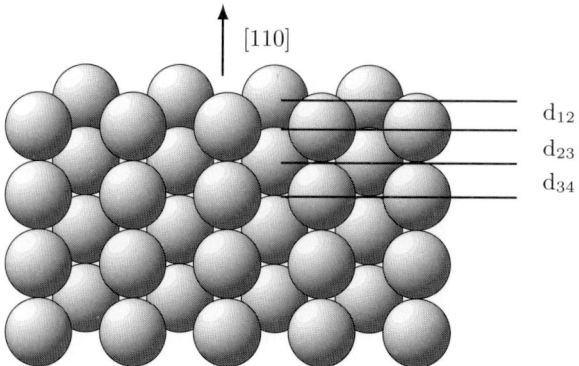

Fig. 4.11. Definition of the layer spacings illustrated for a (110) surface

$(N_c)^{1/2}$. If we denote the surface energy per atom by σ, we can estimate it by

$$\sigma = \frac{\sqrt{N_c^{bulk}} - \sqrt{N_c^{surf}}}{\sqrt{N_c^{bulk}}} E'_{coh}, \qquad (4.12)$$

where N_c^{bulk} and N_c^{surf} are the coordination number of the bulk and the surface, respectively, and E'_{coh} is the cohesive energy related to a *non-magnetic* atom for a non-magnetic surface. For a fcc(111) surface (4.12) yields a surface energy per atom of $\sigma = 0.134 E'_{coh}$ which gives results rather close to the ones plotted in Fig. 4.10a. Within this bond-cutting model also the increase in surface energy for the rougher, more open surfaces can be understood. The coordination of the surface atoms in the (100) surface is eight, while it is only six for the (110) surfaces. Thus the more bonds are broken to create a surface, the higher the surface energy per atom.

If we plot the surface energies in J/m² (Fig. 4.10b), the parabolic dependence on the d-band occupation is much more dramatic. This is due to the fact that the lattice constant becomes smaller with larger bond strength which increases the surface energy per unit area for the metals with a half-full band. On the other hand, the anisotropy in the surface energies between the (111), (100) and (110) surfaces becomes suppressed because the more open surfaces have a larger surface area per atom.

In Table 4.1, additionally the relaxations of the top crystal layers with respect to the bulk layer spacings are tabulated. The definition of the layer spacings is illustrated in Fig. 4.11. At metal surfaces, the smoothening of the electron density usually leads to a contractive relaxation of the first layer. For the densely packed (111) and (100) Cu surfaces, this contraction is rather small, for the more open surfaces, in particular the stepped surfaces, it can already be a rather significant effect. Furthermore, many metal substrates respond to the contraction of the first interlayer spacing by an expansion of the second interlayer, as for example the Cu(100), (110) and (311) surfaces.

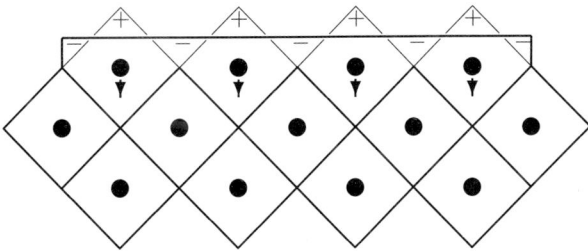

Fig. 4.12. Schematic sketch of the electron redistribution and the resulting first interlayer contraction at a metal surface

However, this oscillatory behavior does not necessarily occur, as the case of the Cu(511) surface demonstrates.

A simple model that explains the first interlayer contraction of metal surfaces was proposed by Finnis and Heine [85] based on the Smoluchowski smoothing [68] already demonstrated in Fig. 4.6. The electron smoothing is sketched schematically in Fig. 4.12. The plotted quadratic cells correspond to the Wigner–Seitz cells in this plane. In a first approximation the electron density is assumed to be uniform with the density falling off abruptly at a plane parallel to the surface. This causes a charge transfer from the areas denoted by + to the regions labelled by −. Thus the surface Wigner–Seitz cells become distorted.

Consequently, due to the modified charge distribution the surface ions are no longer in electrostatic equilibrium, and a net electrostatic force on the positively charged ions results. The ionic cores will rearrange to a new equilibrium structure which is determined by the requirement that the electric field of both the positive ions and negative electronic charge is small outside the distorted surface Wigner–Seitz cell. This electric field is determined by

$$\boldsymbol{E}(\boldsymbol{r}) = \int d^3r' \, \rho(\boldsymbol{r}') \, \frac{(\boldsymbol{r}-\boldsymbol{r}')}{|\boldsymbol{r}-\boldsymbol{r}'|^3} \tag{4.13}$$

Thus the atomic cores will relax to positions at the "electrostatic center" of the electron charge distribution that is given by

$$\int_{WS} d^3r \, n(\boldsymbol{r}) \, \frac{\boldsymbol{r}}{r^3} = 0 \, , \tag{4.14}$$

where the integral is performed over the surface Wigner–Seitz (WS) cell. This means that the atom cores will be located at positions where the electric field of the electron charge distribution of the Wigner–Seitz cells vanishes. This leads to an inward relaxation of the first atomic layer which is illustrated by the arrows in Fig. 4.12b. For fcc metals the surface Wigner–Seitz cells correspond to distorted rhombic dodecahedra. Performing the integrals for the (111), (100) and (110) surfaces of fcc metals yields inward relaxations of

−1.6%, −4.6% and −16%, respectively [85]. This inward relaxation is somewhat too large compared to experiment and ab initio calculations (see Table 4.1), but the trend and the relative differences are well-reproduced, in particular with respect to the fact that the assumptions of the simple model, uniform electron distribution in the Wigner–Seitz cells and abrupt electron density fall off at the surface, are crude and not realistic, as Fig. 4.6 demonstrates.

This simple model cannot explain, however, the outward relaxation of the second layer spacing, because no net effect occurs on the second layer atoms. In order to understand this mechanism, the change of the electron density upon the first interlayer contraction has to be taken into account. The inward relaxation causes a charge accumulation at the second layer. In order to reduce the density to a bulk-like value that is energetically more favorable, the second interlayer spacing expands.

4.3 Semiconductor Surfaces

While the surface energies of the transition metals can be reliably estimated by the modified bond-cutting model given by (4.12), the situation for the divalent fcc and bcc sp-metals is already more complicated. For metals such as Ca, Sr or Ba one finds that the surface energy of the second most close packed surface is consistently lower than that of the most closed packed surface [75]. Even more complex is the situation for semiconductor surfaces. Truly directional bonds between atoms will be broken upon cleavage. This creates an highly unstable state. The surface will try to minimize the number of unsaturated bonds, the so-called dangling bonds. A prominent example of the resulting surface reconstruction is provided by the Si(100) surface. In the ideal (1×1) surface termination, every silicon atom on the surface is only two-fold coordinated. One of the two surface dangling bonds per atom is used to bind to a neighboring atom. This dimerization creates a (2×1) surface structure. Still there is one dangling bond per surface atom left.

In Fig. 3.5 we have already shown a model of the Si(100)-(2×1) surface with a symmetric dimer. At each of the dimer atoms an equivalent dangling-bond orbital is located which are coupled by a π interaction. The π states are split into a bonding π band and an antibonding π^* band. In Fig. 4.13a, the LDA band structure of the Si(100) surface is plotted [86]. The upper panel shows the π and π^* band in the symmetric-dimer model (SDM). The interaction between neighboring dangling bonds is rather strong leading to a significant dispersion of the bands. In fact, for Si(100) these bands overlap and the surface becomes metallic.

Now it is energetically unfavorable for semiconductors to have metallic surfaces. The metallic state at the Si(100)-(2×1) can actually be avoided if there is charge transfer between the dangling bonds so that one of the dangling bonds per dimer is completely filled while the other one becomes

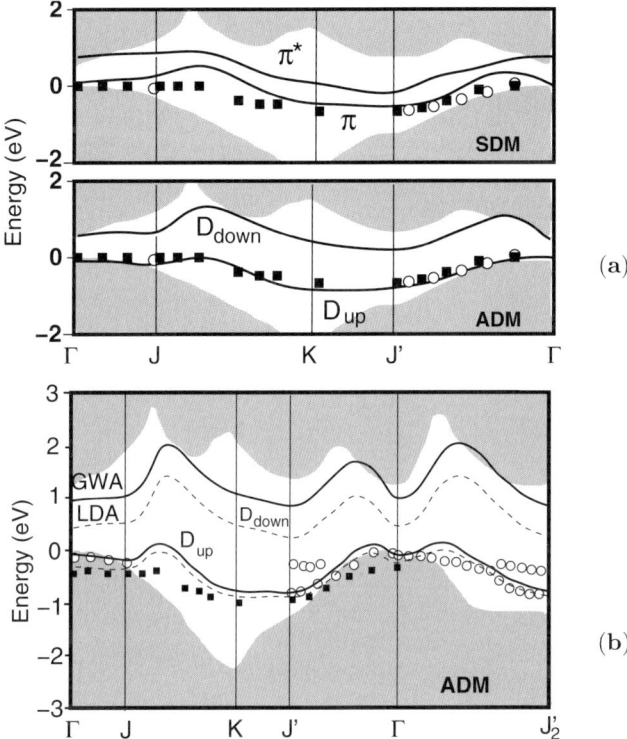

Fig. 4.13. Surface band structure of the Si(100)–(2 × 1) surface. (a) LDA calculations of the band structure in the symmetric dimer model (SDM) and the asymmetric dimer model (ADM) (after [86]). (b) Comparison of the LDA and GWA band structure for the asymmetric dimer model (after [87]). The grey-shaded areas correspond to the calculated projected bulk band structure. In all panels experimental results of [88] (*squares*) and [89] (*circles*) are included

empty. The effect is similar to the Jahn–Teller effect observed in solid-state physics. Through the interaction between the dangling bonds and the silicon lattice the symmetry of the surface will be reduced and thus the degeneracy of the two dangling bond states will be lifted. Geometrically this leads to asymmetric buckled dimers where the dangling bond at the atom closer to the surface is unoccupied. The asymmetrically buckled dimer is illustrated in Fig. 4.14.

Because of the asymmetry of the dimers, there is a pronounced splitting of the related energy bands D_{up} and D_{down} whose states are mainly located at the up or down atoms, respectively, of the surface dimer. According to the LDA calculations, there is a band gap of 0.1 eV between the two bands so that the surface is semiconducting, as it is found in the experiment. In Fig. 4.13 experimental photoemission data of the surface states [88, 89] are

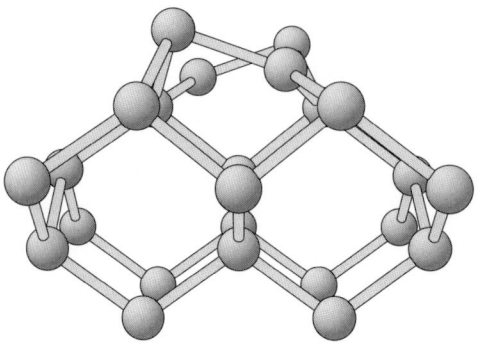

Fig. 4.14. Si(100) (2×2) surface structure with alternating asymmetric buckled dimers

also included, showing a good agreement for the occupied D_{up} band with the LDA calculations.

As far as the width of band gap is concerned, the LDA projected bulk band structure shown as the grey shaded areas in Fig. 4.13a only exhibits a band gap of approximately 0.5 eV. This is in fact much smaller than the experimental value of 1.17 eV [10]. Recall that the interpretation of the Kohn–Sham eigenenergies as electronic one-particle energies does not rest on a firm theoretical ground (see page 38). Still the underestimation of band gaps is a well-known shortcoming of LDA-DFT calculations which is also not corrected for within GGA. The local exchange-correlation potential $v_{\text{xc}}(\bm{r})$ does not adequately describe the dynamical correlations of the electrons in a strongly inhomogeneous environment. In order to obtain a correct treatment of the correlation effects, the potential $v_{\text{xc}}(\bm{r})$ has to be replaced by a nonlocal, energy-dependent self-energy operator $\Sigma(\bm{r},\bm{r}',\varepsilon)$ that enters the effective one-particle Schrödinger equation

$$\left\{ -\frac{\hbar^2}{2m}\nabla^2 + v_{\text{ext}}(\bm{r}) + v_{\text{H}}(\bm{r}) \right\} \psi_i(\bm{r})$$
$$+ \int d^3r'\, \Sigma(\bm{r},\bm{r}',\varepsilon_i)\, \psi_i(\bm{r}') = \varepsilon_i\, \psi_i(\bm{r}). \qquad (4.15)$$

An exact solution of the one-particle equations using the self-energy operator is not possible. In fact, several coupled integral and differential equations have to be solved. An approximate expression of the self-energy operator can be obtained by an expansion of the operator in a series containing the Green function G and the screened interaction W of the system [90, 91]. In the so-called *GW* approximation (GWA), only the first term of the expansion is retained so that the self-energy is written as

$$\Sigma(\bm{r},\bm{r}',\varepsilon) = \frac{\mathrm{i}}{2\pi} \int d\omega\, \mathrm{e}^{-\mathrm{i}\omega\delta}\, G(\bm{r},\bm{r}',\varepsilon-\omega)\, W(\bm{r},\bm{r}',\omega)\,. \qquad (4.16)$$

The single particle Green function can be expressed as

$$G(\mathbf{r},\mathbf{r}',\varepsilon) = \sum_i \frac{\psi_i(\mathbf{r})\psi_i^*(\mathbf{r}')}{\varepsilon - \varepsilon_i + i\delta\,\mathrm{sgn}(\varepsilon_i - \mu)}, \tag{4.17}$$

whereas the dynamically screened Coulomb interaction is defined as

$$W = \epsilon^{-1} v, \tag{4.18}$$

where ϵ is the dielectric function and v the unscreened Coulomb interaction. The dielectric function can be evaluated from the polarization within the random phase approximation. Still the evaluation is computationally rather complex and costly. Once the GW self-energy operator is defined, the effective one-particle equations (4.15) have to be solved self-consistently since the one-particle wave functions ψ_i enter the expression for the self-energy. However, although often the LDA energies are not reliable enough, the LDA wave functions ψ_i^{LDA} and consequently the LDA electron density are usually rather accurate. Hence the self-consistency cycle can be avoided in order to reduce the computational cost and the GW one-particle energies are determined from the LDA wave functions via

$$\varepsilon_i = \varepsilon_i^{\mathrm{LDA}} + Z_i \langle \psi_i^{\mathrm{LDA}} | \Sigma(\varepsilon_i^{\mathrm{LDA}}) - v_{\mathrm{xc}}^{\mathrm{LDA}} | \psi_i^{\mathrm{LDA}} \rangle. \tag{4.19}$$

The energy dependence of the self-energy is approximately taken into account by the renormalization constant Z_i. Interestingly enough, a self-consistent determination of the one-particle energies using the GW expression for the self-energy in general does not lead to improved results. It is important to realize that the GW formulation still yields an approximate expression for the self-energy. Apparently, there is an error cancellation if LDA wave functions are used in a non-self-consistent GW calculation. Note that no total energies can be derived from the GW approximation since it only yields the electronic band structure.

This band structure, however, is improved significantly within the GW approximation. Figure 4.13b shows a comparison between the LDA and GWA dangling bond bands of the buckled Si(100)-(2×1) surface. The shaded area corresponds to the projected bulk band structure obtained within the GW approximation For bulk silicon, the fundamental band gap is increased to 1.23 eV [87] in excellent agreement with experiment. Also the indirect fundamental surface band gap is increased from 0.2 eV (LDA) to 0.7 eV (GWA). Note that in Fig. 4.13b the top of the valence band has been defined as the energy zero for both the LDA and the GWA results. In fact, the GWA leads to a downshift of the D_{up} band by 0.20–0.35 eV while the D_{down} band is shifted up by 0.10–0.25 eV [87]. Although the shifts are slightly energy dependent, it is apparent from Fig. 4.13 that the improved description of the exchange-correlation effects with the GWA leads to a rather constant opening of the band gap. The shape and dispersion of the valence and conduction bands remain more or less unchanged.

Apart from the single-particle electronic excitations, in addition there are coupled electron-hole pair excitations. Because of the two-particle nature of these excitations, their theoretical description is much more demanding [92] because the so-called *Bethe-Salpeter equation* for the excited states has to be solved. This is not trivial, however, it can also be done from first principles. If the electron-hole pair is bound and localized, it is called an *exciton*. This requires that its energy is in a band gap because otherwise the electron-hole pair would couple to the continuum of delocalized states. For the Si(111)-(2×1) surface, a surface exciton at 0.43 eV above the valence band has been identified by an ab initio approch [93], in good agreement with the experiment [94]. This exciton is stabilized at the surface compared to the bulk because of the reduced screening at the surface which leads to a stronger Coulomb interaction between the hole and the electron.

Returning to the structure of the Si(100) surface, it turns out that the buckled dimer shown in Fig. 4.14 leads to a significant amount of mechanical stress at the surface. This surface stress can be partially released if the dimers are buckled in an alternating fashion [95, 96], thereby further reducing the surface energy. In fact, the structure plotted in Fig. 4.14 already corresponds to the Si(100)-(2×2) surface with alternating buckled dimers.

In spite of the unambiguous theoretical results, there has been a long debate about the microscopic structure of the dimers at the Si(100) surface. At room temperature, the dimers appear to be symmetric according to STM experiments [97]. It has been suggested that the symmetric images are caused by the thermal flipping motion of the dimers between the left- and right-tilted positions [95]. In fact, at low-temperatures asymmetric dimers have been observed in STM experiments [98].

The most famous example for a semiconductor surface reconstruction is the Si(111)-(7×7) structure [99] which can be found in almost all text books on surface science [100, 101]. The large (7×7) structure which is referred to as the DAS (Dimer-Adatom Stacking-fault) model contains twelve top-layer adatoms, six rest atoms, a stacking fault in one of the two triangular subunits of the second layer, nine dimers at the borders of the triangular subunit in the third layer and a deep corner hole at each apex of the surface unit cell. This reconstruction has already been addressed by first-principles total energy calculations [102]. Its driving force is again the minimization of the number of dangling bonds at the surface.

A compound semiconductor with rather complex surface reconstructions that has been investigated in great detail by first-principles electronic structure calculations is GaAs [103–105]. The evaluation of surface energies for compound materials is more complex than for elemental materials. Then the surface energy is not simply given by (4.11), it rather depends on the specific thermodynamic conditions, i.e., the reservoir with which the atoms are exchanged in a structural transition. Therefore the chemical potential of the constituents enters the surface energy. The most stable surface structure is determined by the minimum of the free energy which at zero temperature is given by

$$\gamma = \frac{1}{A}\left(E_{\text{surf}} - \sum_i \mu_i N_i\right). \tag{4.20}$$

Here E_{surf} is the total energy of the surface per unit cell which can be calculated in slab calculations according to

$$E_{\text{surf}} = \frac{1}{2}\left(E_{\text{slab}} - N \cdot E_{\text{bulk}}\right), \tag{4.21}$$

where N is now the number of bulk unit cells contained in the slab.

There is no complete freedom for the choice of the chemical potentials. Let us consider the case of GaAs in the following. Bulk GaAs has the zinc-blende-structure. For a GaAs crystal in thermal equilibrium with atomic reservoirs of Ga and As the sum of the chemical potentials of Ga and As must be equal to the chemical potential of bulk GaAs, i.e.

$$\mu_{\text{GaAs}} = \mu_{\text{Ga}} + \mu_{\text{As}} \tag{4.22}$$

Of course the atomic chemical potentials have to be the same in the bulk and on the surface, e.g. $\mu_{\text{As}}^{(\text{GaAs bulk})} = \mu_{\text{As}}^{(\text{GaAs surface})}$, otherwise we would have some macroscopic mass transport. Now the atomic chemical potentials can be varied between certain limits. They should be less than the chemical potential of the condensed phases of the respective elements, for example

$$\mu_{\text{As}} < \mu_{\text{As}}^{(\text{As bulk})} \tag{4.23}$$

because otherwise the elemental condensed phase would be formed. On the other hand, the GaAs chemical potential is related to the elemental bulk chemical potentials through the heat of formation ΔH_{GaAs} via

$$\mu_{\text{GaAs}} = \mu_{\text{Ga}}^{(\text{Ga bulk})} + \mu_{\text{As}}^{(\text{As bulk})} - \Delta H_{\text{GaAs}}. \tag{4.24}$$

Combining (4.22) and (4.24), we obtain a range for possible values of the As chemical potential

$$\mu_{\text{As}}^{(\text{As bulk})} - \Delta H_{\text{GaAs}} < \mu_{\text{As}} < \mu_{\text{As}}^{(\text{As bulk})}. \tag{4.25}$$

We can then write the surface energy of GaAs as a function of a single variable which we will take to be μ_{As}:

$$\gamma = \frac{1}{A}\left[E_{\text{surf}} - \mu_{\text{GaAs}} N_{\text{Ga}} - \mu_{\text{As}}(N_{\text{As}} - N_{\text{Ga}})\right]. \tag{4.26}$$

N_{As} and N_{Ga} are the number of As and Ga atoms, respectively, per 1×1 surface unit cell. The stoichiometry $\Delta N = N_{\text{As}} - N_{\text{Ga}}$ simply gives the slope of the surface energy with respect to the chemical potential. In Fig. 4.15 we have collected some of the possible reconstructions of the technologically most relevant GaAs surface, the (100) surface. Figure 4.15a shows a ball and stick representation of the As-terminated ideal (1×1) surface. This picture

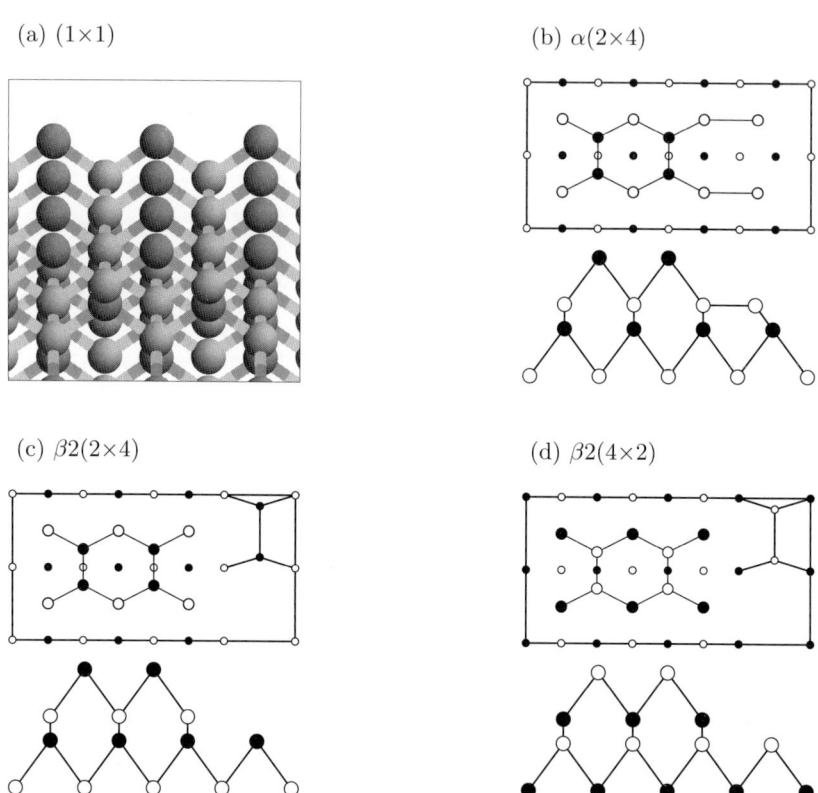

Fig. 4.15. Structural models for the GaAs(100) surface. Dark, filled circles and bright, empty circles represent As and Ga atoms, respectively. (**a**) Ball and stick model of the ideal (1 × 1) As-terminated GaAs(100) surface. (**b–d**) Schematic top and side view of the $\alpha(2\times 4)$, $\beta 2(2\times 4)$ and $\beta 2(4\times 2)$ reconstructions. Larger circles correspond to atoms closer to the surface

illustrates the problems that arise for a proper definition of the stoichiometry ΔN. The As and the Ga atoms are not equivalent at the ideal GaAs(100) surface. In fact, ΔN is defined in such a way that it is equal to $\frac{1}{2}$ for the (1×1) As-terminated surface and $-\frac{1}{2}$ for the (1×1) Ga-terminated surface. A simple way to understand this counting rule is to think of a symmetric slab with two identical, say, As-terminated surfaces. This slab has one As atom more than Ga across the slab, so that per (1×1) surface unit cell there is $\frac{1}{2}$ additional As atom. By this procedure one obtains stoichiometries of $\Delta N = 0$ for the $\alpha(2\times 4)$ GaAs(100) surface (Fig. 4.15b), $\Delta N = \frac{1}{4}$ per 1×1 unit cell for the $\beta 2(2\times 4)$ structure (Fig. 4.15c), and $\Delta N = -\frac{1}{4}$ per 1×1 unit cell for the $\beta 2(4\times 2)$ structure (Fig. 4.15d).

This simple counting argument cannot be applied if no symmetric slab can be constructed. This is the case for the (111) surfaces of zinc-blende-structures

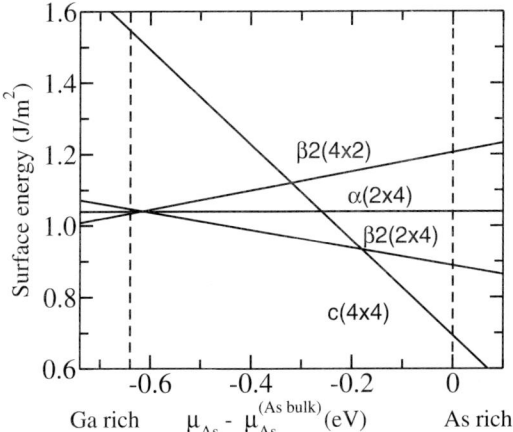

Fig. 4.16. Surface energies of different GaAs(100) reconstructions in J/m² as a function of the difference of the chemical potential of As and bulk As (after [105]). The perpendicular dashed lines indicate the range of possible As chemical potentials given by (4.25)

because there the (111) and the ($\bar{1}\bar{1}\bar{1}$) are inequivalent. Still a counting rule can be established based on bulk symmetries of the crystal [106].

The surface energies of the structures shown in Fig. 4.15 as a function of the As chemical potential are plotted in Fig. 4.16. These energies have been determined by DFT calculations in the local density approximation [105]. In addition, the surface energy of a completely As-terminated $c(4 \times 4)$ structure is plotted for which the surface stoichiometry is $\Delta N = \frac{5}{4}$. The perpendicular dashed lines indicate the range of possible As chemical potentials given by (4.25). Low As chemical potential corresponds to a Ga-rich environment while the As bulk chemical potential gives the As-rich limit. For a particular As chemical potential the surface with the lowest free energy corresponds to the thermodynamically stable one. As a function of increasing As coverage the order of the stable structures is given by $\beta 2(4 \times 2)$, $\alpha(2 \times 4)$, $\beta 2(2 \times 4)$ and $c(4 \times 4)$. The unreconstructed Ga and As terminated (1×1) surfaces have much higher surface energies in the order of 3 J/m².

It is not trivial to understand the reasons why a particular structure is the most stable one. First of all it is obvious that the GaAs surface tries to minimize the number of dangling bonds by dimerization. Still, at the dimer atoms of a (100) surface one dangling bond per atom remains. Secondly, also for the GaAs it is true that semiconducting surfaces have usually a lower energy than metallic structures [105]. Ga dangling bonds are energetically higher than As dangling bonds. The occupation of Ga dangling bonds would lead to a metallic surface which should be avoided. In general, polar semiconductors exhibit surface reconstructions with the anion dangling bonds filled and the cation dangling bonds empty. This is referred to as the *electron-counting principle*.

At the (100) surface of GaAs this causes an electron transfer from the Ga atoms to the As atoms which formally leads to negatively charged As and positively charged Ga atoms.

Both the $\beta 2(2\times 4)$ and the $\beta 2(4\times 2)$ structure minimize the electrostatic repulsion [107]. The $\beta 2(2\times 4)$ surface that is stable over a wide range of As chemical potentials is actually the As terminated counterpart of the $\beta 2(4\times 2)$ structure with As atoms exchanged by Ga atoms and *vice versa*. The negatively charged As atom at the surface with its completely filled dangling bond tends to form bonds with its three p orbitals. This leads to bond angle close to 90° which can only be achieved when the As atoms relaxes outward. The positively charged Ga atom at the surface, on the other hand, has lost an electron and prefers a sp^2-like hybridization. Therefore Ga favors a more planar configuration and relaxes inward. Apparently this is energetically more costly than the outward relaxation of the As atoms so that the Ga-terminated $\beta 2(4\times 2)$ surface is only stable under extrem Ga rich conditions.

4.4 Ionic Surfaces

In the case of the GaAs(100) surface we already realized the importance of electrostatic effects if there is a charge transfer between the two constituents of a compound semiconductor. These considerations are even more important in the case of ionic crystals where the bonding is entirely dominated by electrostatics. Depending on the difference in ionic radii, alkali halid solids crystallize in the sodium chloride or cesium chloride structure [10]. The sodium chloride structure is shown in Fig. 4.17. It corresponds to two fcc sublattices translated by $a/2(\mathbf{e}_x + \mathbf{e}_y + \mathbf{e}_z)$.

The electrostatic potential outside a slab structure can be derived by a Madelung summation [108]. Far outside the surface plane the potential approaches

$$\phi(z \to \infty) = \phi_G \exp\left(-\frac{2\pi}{z}a\right) + 2\pi\sigma_\perp , \quad (4.27)$$

where $\sigma_\perp$ is the dipole moment perpendicular to the plane per unit area. Hence, if there are normal dipoles in the slab, the asymptotic value of the potential is $2\pi\sigma_\perp$, but if there are no normal dipoles present, the potential vanishes exponentially. In fact, for a so-called non-polar surface without any normal dipole moment the electrostatic potential falls off very rapidly with a decay length in the order of the lattice constant. Due to cancellation effects the potential of a non-polar ionic crystal surfaces has thus a much shorter range than, e.g., that caused by the van der Waals interaction.

Since the formation of a dipole layer is energetically rather costly, polar surfaces are usually highly unstable. Hence alkali halid crystals in equilibrium are usually terminated by non-polar surfaces such as the {100} surfaces of

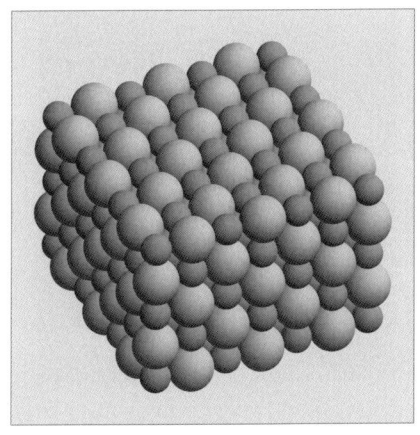

Fig. 4.17. Sodium chloride structure with non-polar {100} surfaces

the sodium chloride structure shown in Fig. 4.17. That is the reason why for example salt grains have an almost perfect cubic shape.

If one applies (4.27) to determine the potential inside the bulk, then it turns out that potentials at bulk sites only one lattice constant distant from the surface are practically indistinguishable from those in the bulk. This means that also the potential felt by the surface atoms is rather close to the bulk potential, and consequently no strong relaxations occur. Therefore alkali halids exhibit surface terminations that are almost ideal.

Surface structures are more complicated for insulating oxide materials where the bonds still have a covalent character although there is a significant charge transfer between the constituents. Here we focus on the (0001) surface of α-Al_2O_3 (corundum or sapphire) which has been studied extensively by both theory [109–112] and experiment [113, 114]. The α-Al_2O_3(0001) surface is an important substrate for very high frequency microelectronic devices due to its insulating character, but it is also of interest in the automobile industry, in atmospheric treatments and in catalytic reactions.

Oxides usually have rather complicated bulk structures. α-Al_2O_3 (sapphire) cystallizes in the corundum structure that can be described by a primitive rhombohedral unit cell with two Al_2O_3 formula units. More convenient is the hexagonal unit cell that contains 12 Al atoms and 18 O atoms. The side and the top view of the ideal Al-terminated α-Al_2O_3(0001) surface are shown in Fig. 4.18. The hexagonal cell corresponds to a layered structure with six oxygen planes associated with aluminum planes above and below it, forming stochiometric triple layers. There are three O atoms and just one Al atom in each layer per unit cell. The O atoms are stacked in a slightly distorted hcp structure, as can be seen in the top view of Fig. 4.18. The Al atoms occupy two-third of the octahedral holes in the oxygen sublattice. Due to this layered structure, there is no non-polar termination of the α-Al_2O_3(0001) surface.

Although Al_2O_3 crystallizes in a rather complex bulk structure, at room temperature no reconstructions of α-Al_2O_3(0001) are observed [113,115]. Cal-

1: Al →
2: O$_3$ →
3: Al →
4: Al →
5: O$_3$ →
6: Al →

Fig. 4.18. Side and top view of the ideal Al-terminated α-Al$_2$O$_3$(0001) surface. In addition the uppermost layers are labelled

culated surface energies as a function of the oxygen chemical potential for different (1×1) terminations are plotted in Fig. 4.19 [109, 112]. Over the entire range of oxygen chemical potentials the stoichiometric AlO$_3$Al-termination is by far the energetically most favorable one. This can be understood by simple electrostatic arguments because the triple AlO$_3$Al layer does not have a dipole moment while all other (1 × 1) surface terminations have one. Seemingly, at ionic oxide surfaces electrostatic considerations are more important for the determination of stable surface structures than bond-saturation arguments.

However, the situation is in fact not as simple as suggested above. In Table 4.2 calculated and measured interlayer relaxations of the Al-terminated α-Al$_2$O$_3$(0001) are listed. First of all it is remarkable that the theoretical values agree rather well with each other, independent of the functional that has been used. All calculations give a remarkable strong inward relaxation of the first Al-layer so that it practically becomes coplanar with the oxygen layer. In contrast, the interlayer spacing between the second and third layer is only slightly modified compared to the bulk spacing. This creates a surface dipole of the uppermost triple AlO$_3$Al layer and should thus be energetically unfavorable.

There are two seemingly conflicting explanations for the large inward relaxation. The inward relaxation could be viewed as driven by a rehybridization of the surface Al atom to an sp^2 orbital configuration which favors the planar configuration. By this rehybridization the Al $3p_z$ orbital perpendicular to the surface becomes empty so that no partially occupied dangling bonds are present at the surface. Therefore the surface becomes insulating with the lowest empty surface state 4.5 eV above the Fermi level [109]. In addition, the non-stoichiometric O$_3$-terminated surfaces which are already energetically unfavorable due to their large dipole moment are found to be metallic which should further increase their energy.

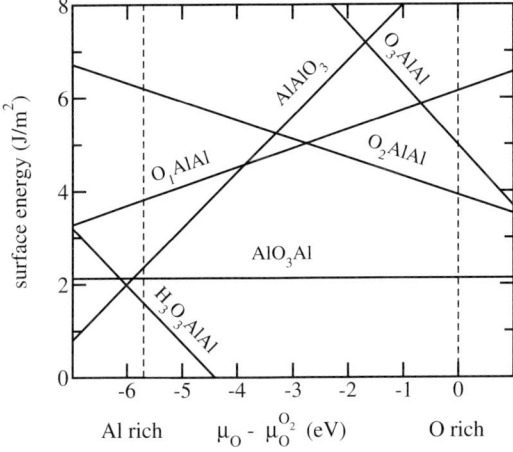

Fig. 4.19. Surface energies of different $Al_2O_3(0001)$ (1×1) structure in J/m^2 as a function of the difference of the oxygen chemical potential (after [109]). The uppermost layers of the corresponding structures are indicated in the figure. The perpendicular dashed lines indicate the range of possible oxygen chemical potentials

On the other hand, the strong inward relaxation of the surface Al layer spacing has also been explained by the reduction of the electrostatic dipole of just the first two layers [112]. This explanation is supported by the fact that early studies with just classical ionic interatomic potentials [117] that did not take into account any rehybridization effects have also found this large relaxation.

Restricting this electrostatic argumentation just to the first interlayer spacings, however, is obviously not appropriate with regard to the large relaxations that have been found for up to the fourth interlayer spacing (see

Table 4.2. Interlayer relaxations at the Al-terminated α-$Al_2O_3(0001)$ surface in percent of the corresponding bulk spacings

Interlayer		Theory[a] GGA	Theory[b] LDA	Theory[c] LDA	Theory[d] LDA	Exp.[e]	Exp.[f]	Exp.[g]
Al-O_3	1–2	−86	−87	−85	−77	−51	+30	−52.8
O_3-Al	2–3	+6	+3	+3	+11	+16	+6	+1.5
Al-Al	3–4	−49	−42	−45	−34	−29	−55	–
Al-O_3	4–5	+22	+19	+20	+19	+20	–	–
O_3-Al	5–6	+6	+6	–	+1	–	–	–

References: [a] Wang et al. [109], [b] Verdozzi et al. [110], [c] Di Felice et al. [111], [d] Batyrev et al. [112], [e] Guenard et al. [113], [f] Toofan et al. [114], [g] Soares et al. [116]

Table 4.2). These cannot be explained in terms of bond saturation effects, either, since the coordination is not changed for this inner layers. Hence the most reasonable mechanism for the strong relaxation effects is a mixture of hybridization effects at the uppermost layer and electrostatic effects that can be rather far-reaching into the bulk.

The agreement with the experiment is not too satisfactorily. At least there is a qualitative agreement with the results of Guenard et al. [113] and Soares et al. [116]. The agreement could even be made quantitative if one assumes that hydrogen atoms had been present in the experiment because this reduces the calculated relaxations [109]. The experimental results of Toofan et al. [114] do even not qualitatively agree with the calculations. Again, this could be explained by hydrogen-induced effects, in this case on the oxygen terminated surface. As Fig. 4.19 demonstrates, the oxygen terminated surfaces are energetically very unfavorable. However, the large surface dipole and the oxygen dangling bonds can be compensated by hydrogen adsorption. These hydrogen terminated surfaces have in fact the lowest surface energy in the range of physically realistic conditions if the energy is calculated with respect to the hydrogen chemical potential of H_2 (see Fig. 4.19). The result that the surface energies even become negative reflects the fact that the hydroxylated surface is lower in energy than bulk sapphire and H_2. The hydrogen termination actually also leads to an outward relaxation of the first layer thus giving a reasonable explanation for the experimental results of Toofan et al. [114].

However, in LEED experiments the hypothesis of the influence of hydrogen on the surface termination has been tested by processing the Al_2O_3 surface under hydrogen-rich, oxygen-rich and vacuum-like sample preparation conditions [116]. It turned out that the α-Al_2O_3(0001) surface structure is insensitive to the different processing methods: it is always terminated by a single Al layer. At the same time the experiments found unusually large vibrational amplitudes of the topmost Al layer at room temperature. This might suggest that the disagreement between theory and experiment is related to the fact that the experiments are performed at room temperature while the calculations correspond to zero temperature [116].

As for further temperature effects, the Al-terminated surface is in fact stable for temperature up to 1350 K. Above this temperature oxygen evaporates from the crystal leading to Al rich surfaces. Upon heating to more than 1600 K, a sequence of different surface reconstructions is observed that ends with a $\sqrt{31} \times \sqrt{31}$ structure [115]. This surface is terminated by Al atoms arranged in two fcc-like layers. While this structure is too large to be determined from first-principles, DFT calculations for 1×1 Al-terminated surfaces suggest that it might be terminated by a two-dimensional Al_3 fcc layer, i.e. a layer with three Al atoms per 1×1 surface unit cell, on top of a Al_2 layer [111]. For the $(3\sqrt{3} \times 3\sqrt{3})$r30° reconstructed α-Al_2O_3(0001) surface simulations using empirical potentials exist [118]. These calculations confirm the complex structure of these large reconstructions: the hexago-

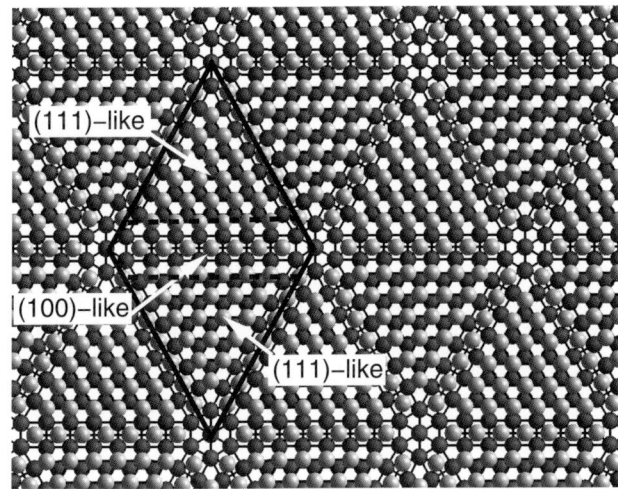

Fig. 4.20. Atomic configuration of the $(3\sqrt{3} \times 3\sqrt{3})\text{r}30°$ reconstructed α-$Al_2O_3(0001)$ surface obtained by simulations using empirical potentials [118] (Courtesy of F. Lançon). The surface unit cell and the fcc(111) and fcc(100)-like regions are indicated

nal reconstructed unit cell is composed of triangles where two layers of Al adatoms are ordered as fcc(111) whereas the stacking between the triangles is fcc(100)-like. This surface structure is illustrated in Fig. 4.20.

4.5 Interpretation of STM Images

The scanning tunneling microscope (STM) [13] has become one of the most valuable experimental tools for the microscopic determination of surface structures. Although we do not want to dwell on STM theory (for a comprehensive overview see, e.g., [119]), this chapter on the theoretical determination of the structure and energetics of surfaces would not be complete without some remarks about the use of ab initio electronic structure calculations for the interpretation of STM images.

It is important to note that in the STM the atoms are not directly imaged. Rather the tunneling current between surface and STM tip is monitored. In first-order perturbation theory, the current between two electrodes can be expressed as

$$I = \frac{2\pi e}{\hbar} \sum_{\mu\nu} f(E_\mu) \left[1 - f(E_\nu)\right] |M_{\mu\nu}|^2 \delta(E_\mu - (E_\nu + eV)) , \qquad (4.28)$$

where $f(E)$ is the Fermi function, V is the applied voltage and $M_{\mu\nu}$ is the tunneling matrix elements between tip states ψ_μ and surface states ψ_ν.

86   4. Structure and Energetics of Clean Surfaces

In the limit of small voltage and temperature, (4.28) can be simplified. Thus it takes the form

$$I = \frac{2\pi}{\hbar} e^2 V \sum_{\mu\nu} |M_{\mu\nu}|^2 \, \delta(E_\mu - E_F) \, \delta(E_\nu - E_F) \,, \qquad (4.29)$$

where E_F is the Fermi energy. The matrix element $M_{\mu\nu}$ is given by

$$M_{\mu\nu} = \frac{\hbar^2}{2m} \int d\boldsymbol{S} \, (\psi_\mu^* \nabla \psi_\nu - \psi_\nu \nabla \psi_\mu^*) \,. \qquad (4.30)$$

The integral is over any surface lying entirely within the vacuum barrier region separating the two systems. Tersoff and Hamann [120,121] have shown that the current can be expressed as

$$I \propto \sum_\nu |\psi(\boldsymbol{r}_t)|^2 \delta(E_\mu - E_\nu) = \rho(\boldsymbol{r}_t, E_F) \,, \qquad (4.31)$$

if the tip is modelled as a locally sperical potential centered at $\boldsymbol{r}_t$. In this so-called *Tersoff–Hamann picture*, the tunneling current is simply proportional to the local density of states *of the surface* $\rho(\boldsymbol{r}_t, E_F)$ at the position of the tip. The tunneling current is just given by a property of the surface alone, i.e., the surface-tip interaction does not influence the measured current in this model. This makes the simulation of STM images rather straightforward. All that is needed is the local density of state at a certain distance from the surface which is a standard information evaluated in electronic structure methods.

The tip-sample coupling in first-order perturbation theory can be extended beyond the Tersoff–Hamann theory in order to take into account higher angular momenta in the tip states or more realistic models for the tip structure [122]. Also nonperturbative approaches for the evaluation of STM tunneling probabilities have been developed [123]. But still the simple Tersoff–Hamann expression for the tunneling current is widely used because in most cases it is indeed sufficient to reproduce the most important features of STM images, as will be shown below.

The simulation of STM pictures is very important for the interpretation of imaged structures because there is not necessarily a simple relation between the STM picture and the real structure. A prominent example is given by STM images of graphite. In Fig. 4.21 a STM picture of highly oriented pyrolythic graphite (HOPG) is plotted. The bright spots correspond to atoms that are arranged in a hexagonal structure and not in the honeycomb structure that is characteristic for the (0001) plane of graphite. This means that in fact only every second carbon atom in the graphite basal plane is imaged.

A simulated STM image based on GGA-DFT calculations within the Tersoff–Hamann picture reproduces this peculiar feature of the graphite surface, as Fig. 4.21 demonstrates. This asymmetry in the imaging of the carbon atoms in the (0001) surface is explained by a purely electronic effect [124]. The graphite crystal is composed of two sublattices, sublattice Σ_1 with neighboring atoms directly above and below in adjacent layers and sublattice Σ_2

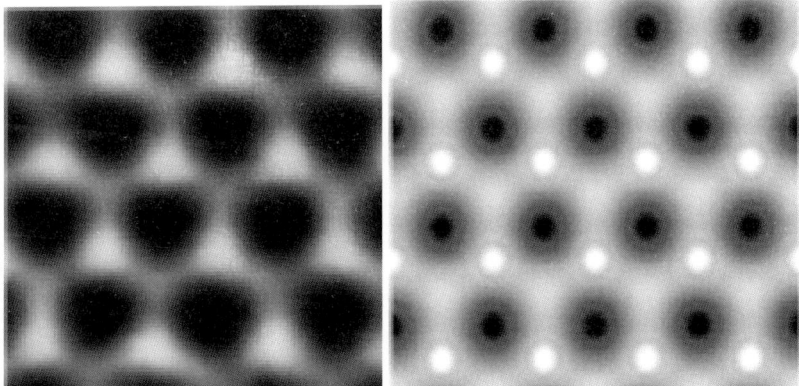

Fig. 4.21. Comparison of measured *(left panel)* and simulated *(right panel)* STM image of highly oriented pyrolythic graphite(0001) (courtesy of W. Heckl and T. Markert). Only every second atom of the graphite surface is imaged, leading to a seemingly hexagonal structure instead of the honeycomb structure. This feature is reproduced by the simulated image within the Tersoff–Hamann model using GGA-DFT

without such neighbors. The Fermi surface of bulk graphite lies close to the P line in the Brillouin zone which is defined by $\bm{k} = (\frac{1}{3}, \frac{1}{3}, x)$ in units of the reciprocal lattice vectors. Along this line, the electronic states on the atoms of the one sublattice are decoupled from the states on the other sublattice. Furthermore, atoms of sublattice Σ_2 do not interact with those of adjacent planes leading to localized dispersionless electronic states along the P line in the Brillouin zone. This causes a high density of states close to the Fermi energy in the surface Brillouin zone and therefore also a large tunneling current at small bias. The atoms of sublattice Σ_1, on the other hand, are interacting across the layers leading to dispersive states which have a lower density of states close to the Fermi energy. Therefore it is a density of states effect which causes a significantly smaller tunneling current at surface atoms that have nearest neighbors directly below them.

Of course, in the case of graphite we know the surface structure already without the simulation of STM images. However, if it comes to more complicated systems such as organic adsorbates on surfaces, the interpretation of STM images is no longer that straightforward. Then the simulation of STM images is rather helpful in order to identify the chemical nature of the imaged structures.

4.6 Surface Phonons

So far we have treated surfaces as static objects. However, surface atoms are not at rest at non-zero temperatures. Instead, they vibrate about their equilibrium positions. In periodic structures, these vibrations form waves, the phonons, that are characterized by their frequency, their momentum and their displacement pattern. Although phonons are a dynamical phenomenon, we focus on them in this chapter since first, they are directly related to the structure of surfaces, and secondly, their treatment does not necessarily require a dynamical treatment.

The vibrational frequencies of a system of L atoms can be determined in the harmonic approximation by solving the secular equation

$$\det \left| \frac{1}{\sqrt{M_I M_J}} \frac{\partial^2 E_{\text{el}}(\{\boldsymbol{R}\})}{\partial \boldsymbol{R}_I \partial \boldsymbol{R}_J} - \omega^2 \right| = 0 , \quad (4.32)$$

where $E_{\text{el}}(\{\boldsymbol{R}\})$ is the Born–Oppenheimer energy surface. This means that the frequencies are given by the Hessian of the Born–Oppenheimer energy scaled by the nuclear masses. Let us now consider a periodic crystal given by the Bravais lattice $\{\boldsymbol{R}_I\}$ with r atoms in the unit cell at positions $\boldsymbol{R}_{I\alpha} = \boldsymbol{R}_I + \boldsymbol{u}_{I\alpha}, \alpha = 1, \ldots, r$. The second derivatives

$$C_{ij}^{\alpha\beta}(\boldsymbol{R}_I, \boldsymbol{R}_J) = \frac{\partial^2 E_{\text{el}}(\{\boldsymbol{R}\})}{\partial \boldsymbol{R}_{I\alpha i} \partial \boldsymbol{R}_{J\beta j}} , \quad (4.33)$$

where i and j denote the cartesian coordinates, are also called *elastic force constants*. In a periodic crystal, the elastic force constants only depend on the distance $\boldsymbol{R}_{IJ} = \boldsymbol{R}_J - \boldsymbol{R}_I$. Because of this periodicity, it is sufficient to consider $3r \times 3r$ determinants in order to calculate the phonon frequencies $\omega(\boldsymbol{q})$ as a function of the wave vector $\boldsymbol{q}$ instead of taking a $3N \times 3N$ determinant into account. The phonon frequencies $\omega(\boldsymbol{q})$ are the solutions of the secular equation [4]

$$\det \left| D_{ij}^{\alpha\beta}(\boldsymbol{q}) - \omega^2(\boldsymbol{q}) \right| = 0 , \quad (4.34)$$

where $D_{ij}^{\alpha\beta}(\boldsymbol{q})$ is the dynamical matrix defined by

$$D_{ij}^{\alpha\beta}(\boldsymbol{q}) = \frac{1}{\sqrt{M_\alpha M_\beta}} \sum_{\boldsymbol{R}_{IJ}} C_{ij}^{\alpha\beta}(\boldsymbol{R}_{IJ}) \, e^{-i\boldsymbol{q}\cdot\boldsymbol{R}_{IJ}} . \quad (4.35)$$

Equivalent to the case of surface electronic states, surface phonons are vibrational modes that are localized at the outermost layers of the solid. Already at the end of the nineteenth century it was known that localized waves, the so-called *Rayleigh waves*, exist at the surface of isotropic continuous elastic media [125]. The Rayleigh waves can be calculated exactly using elasticity theory [12, 100]. They correspond to long wavelength acoustic phonons since their frequency depends linearly on the wave vector $\boldsymbol{q}_\|$ in the surface plane. Their displacements are confined to the *sagittal plane* which is the plane defined by the wave vector $\boldsymbol{q}_\|$ and the surface normal.

If the wavelength of the surface phonons becomes comparable to the interatomic distances, a continuous elastic description of the substrate is no longer justified. Instead, the discrete atomic nature of the solid and its surface has to be taken into account. There are basically three methods for the microscopic theoretical treatment of surface phonons using first-principles techniques: molecular dynamics simulations, frozen-phonon techniques, and the linear response formalism [126]. To extract phonon modes from classical trajectory calculations (see Sect. 6.1), one performs finite-temperature molecular dynamics runs of the substrate for a sufficiently long period of time. The frequency spectrum of the phonons is then given by the Fourier transform

$$g(\omega) = \frac{1}{T} \int dt\, W(t)\, \cos(\omega t) \quad (4.36)$$

of the velocity autocorrelation function

$$g(t) = \sum_{I=1}^{L} \frac{\langle \dot{\boldsymbol{u}}_I(t)\dot{\boldsymbol{u}}_I(0)\rangle}{\langle \dot{\boldsymbol{u}}_I(0)\dot{\boldsymbol{u}}_I(0)\rangle}, \quad (4.37)$$

where T is the total simulation time, $\dot{\boldsymbol{u}}_I(t)$ is the velocity of the I-th particle in the supercell at time t, $\langle \ldots \rangle$ denotes the ensemble average, and $W(t)$ is an window function. In order to determine phonon modes not only at the zone center, sufficiently large unit cells are required. Furthermore, rather large simulation times are needed to resolve low-frequency vibrational modes since the resolution of the Fourier transform is given by $\Delta\omega > 2\pi/T$. On the other hand, anharmonicities and temperature effects can be addressed with this method.

In the frozen-phonon approach, total energies and Hellmann–Feynman forces are evaluated as a function of atomic displacements from the equilibrium positions. The forces are fitted to a quadratic expansion in the distortions so that the harmonic contributions can be extracted and the dynamical matrix be determined. The unit cell that is used to calculate the forces has to match the wave vector of the considered phonon mode. These method has been used to address the phonon anomalies observed at hydrogen covered Mo(110) and W(110) surfaces [49].

The problem of matching supercell to determine phonons of a given wavevector can in fact be avoided. In density-functional perturbation schemes [127], dynamical matrices can be calculated for arbitrary wave vectors by using the same unit cell as for the ground state calculations. Density-functional perturbation theory (DFPT) is based on the Born–Oppenheimer approximation. To determine the dynamical matrix, the second derivatives of the energy with respect to the nuclear coordinates are needed. By differentiating the Hellmann–Feynman forces, a general expression for these second derivatives can be found [127],

$$\frac{\partial^2 E_{\rm el}(\{\boldsymbol{R}\})}{\partial \boldsymbol{R}_I \partial \boldsymbol{R}_J} = -\frac{\partial \boldsymbol{F}_I}{\partial \boldsymbol{R}_J} = \int d^3 R\, \frac{\partial n(\boldsymbol{r})}{\partial \boldsymbol{R}_J}\, \frac{\partial V_{\rm nucl-el}(\boldsymbol{r})}{\partial \boldsymbol{R}_I}$$
$$+ \int d^3 r\, n(\boldsymbol{r})\, \frac{\partial^2 V_{\rm nucl-el}(\boldsymbol{r})}{\partial \boldsymbol{R}_I \partial \boldsymbol{R}_J} + \frac{\partial^2 V_{\rm nucl-nucl}}{\partial \boldsymbol{R}_I \partial \boldsymbol{R}_J}. \quad (4.38)$$

This means that for the determination of the second derivatives in addition to the calculation of the ground-state eletron density $n(\mathbf{r})$, its linear response to a distortion of the nuclear geometry, $\partial n(\mathbf{r})/\partial \mathbf{R}_I$, is required.

In order to obtain the electron-density response $\partial n(\mathbf{r})/\partial \mathbf{R}_I$ within density-functional theory, we first linearize the terms appearing in the Kohn–Sham equations (3.49) with respect to wave function, density, and potential variation. For the electron density this gives

$$\Delta n(\mathbf{r}) = 2 \sum_{n=1}^{N} \psi_n^*(\mathbf{r}) \Delta \psi_n(\mathbf{r}) . \tag{4.39}$$

The variation of the Kohn–Sham orbitals $\Delta \psi_i(\mathbf{r})$ follows from standard first-order perturbation theory

$$(H_0 - \varepsilon_n)\,|\Delta\psi_n\rangle = -(\Delta v_{\text{eff}} - \Delta\varepsilon_n)|\psi_n\rangle . \tag{4.40}$$

In atomic physics, this equation is known as the Sternheimer equation. The variation of the effective potential can be written as

$$\Delta v_{\text{eff}}(\mathbf{r}) = \Delta v_{\text{ext}}(\mathbf{r}) + \int d^3 r' \Delta n(\mathbf{r}') \frac{e^2}{|\mathbf{r}-\mathbf{r}'|}$$
$$+ \left.\frac{dv_{\text{xc}}(n)}{dn}\right|_{n=n(\mathbf{r})} \Delta n(\mathbf{r}') \tag{4.41}$$

and the first-order variation of the Kohn–Sham eigenenergies is given by

$$\Delta\varepsilon_n = \langle\psi_n|\Delta v_{\text{eff}}|\psi_n\rangle . \tag{4.42}$$

The equations (4.39)-(4.42) form a set of coupled self-consistent equations for the perturbed system. The Kohn–Sham eigenvalue equation (3.49) has been replaced by the solution of a linear system (4.40). The crucial advantage of density-functional perturbation theory is that the response to perturbations of different wavelengths is decoupled [126, 127]. Only periodic displacements characterized by the wavevector $\mathbf{q}$ have to be considered. The derivatives entering the determination of the dynamical matrices can be obtained from the self-consistent solution of (4.39)–(4.42) with the first-order variations replaced by the derivatives [126]

$$\frac{\partial}{\partial \mathbf{R}_{I\alpha}(\mathbf{q})} = \sum_I \frac{\partial}{\partial \mathbf{R}_{I\alpha}} e^{-i\mathbf{q}\cdot\mathbf{R}_I} \tag{4.43}$$

The phonon spectrum of the buckled Si(100)-(2×1) surface determined by density-functional perturbation theory [128] is shown in Fig. 4.22. The surface phonons in this system have been addressed before by tight-binding calculations [130] with qualitatively very similar results. There is a large number of phonon modes localized at the surfaces which are indicated by the solid lines. This is due to the large structural changes of the reconstructed Si(100) surface. There are some special modes labeled in Fig. 4.22, such as the rocking

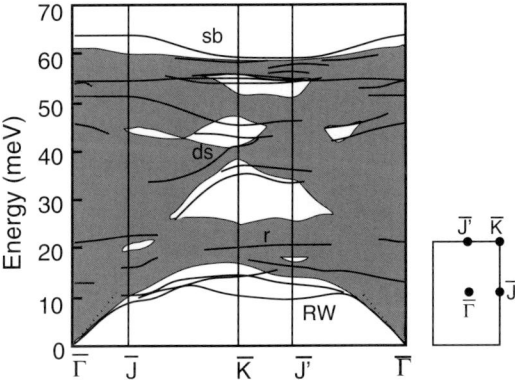

Fig. 4.22. Phonon dispersion curve of the buckled Si(100)–(2 × 1) surface. Vibrational modes located at the surfaces are plotted as solid lines while the shaded area corresponds to the bulk projected band structure (after [128]). The experimentally measured velocity of the Rayleigh wave (RW) [129] is indicated by the broken lines. The surface Brillouin zone and the special points are indicated in the right panel

mode (r), the dimer stretching (ds) and the dimer back bond (sb) mode. Of particular interest with respect to the buckled dimer structure of the Si(100) is the rocking mode which has an energy of about 20 meV. The displacement pattern of this mode corresponds to the rocking motion of the surface dimer. The eigenvectors and the frequencies of this phonon mode remain almost unchanged if they are calculated in a larger $c(4 \times 2)$ surface unit cell [128]. This mode is excited at room temperature ($k_B T \approx 25$ meV). Its existence supports the suggestion that the seemingly symmetric dimers observed in room-temperature STM experiments are in fact an averaged image over back and forth flipping dimers (see p. 76).

The energy of the dimer stretching mode (ds), about 43 meV, is directly related to the surface dimer bond strength. Actually, for Ge(100) the same mode has been found with a slightly lower energy [126] reflecting the fact that the dimer bond in Ge is weaker than in Si. The dimer back bond mode above the bulk continuum corresponds to an opposing motion of the dimer atoms with respect to the second layer atoms. Therefore it has the character of an optical phonon.

The Rayleigh wave (RW) is the lowest mode in the surface phonon spectrum. The broken lines in Fig. 4.22 correspond to $\hbar\omega = \hbar c_R q$ where c_R is the experimentally measured velocity of the Rayleigh wave [129]. The comparison with the calculated dispersion curve in the long-wavelength limit indicates a very good agreement between experiment and theory.

Exercises

4.1 Electronic Surface States in the One-dimensional Nearly-Free Electron Model

We describe an infinite solid in the one-dimensional nearly-free electron model as a chain of atoms with an effective potential for the electrons given by

$$V(z) = V_0 + V_G \cos(Gz),$$

where $G = 2\pi/a$ is the shortest reciprocal lattice vector of the chain.

a) Determine the dispersion $\varepsilon(k)$ and the eigenfunctions for wave-vectors k near the zone boundary, i.e. for $|G/2 - k| \ll G$.
Hint: Use degenerate perturbation theory

b) Now we consider a semi-infinite crystal. The potential barrier at the surface at $z = 0$ is modeled by a potential step of height W_0 so that the whole potential is given by

$$V(z) = V_0 + W_0 \Theta(-z) + V_G \cos(Gz) \Theta(z),$$

where $\Theta(z)$ is the step function.

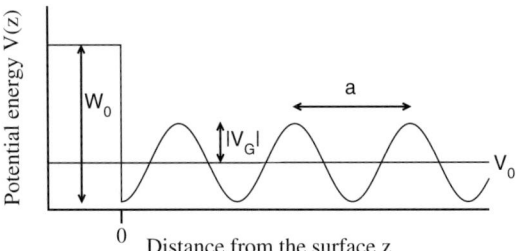

Show that now localized electronic states can exist at the surface with an energy that lies in the band gap. Which sign of V_G is necessary for the existence of a localized surface state?

4.2 Density of States

a) Determine the density of states of the free electron gas in one, two and three dimensions. Use periodic boundary conditions for N electrons along a line of length L, in a square of area $A = L^2$, and in a cube of volume $V = L^3$, respectively.
Hint: Use $n(\varepsilon)d\varepsilon = n_d(\boldsymbol{k})d^d\boldsymbol{k}$, where $n_d(\boldsymbol{k})$ is the density of the states in the d-dimensional $\boldsymbol{k}$-space.

b) Show that in the general case the density of states in the j-th band, ignoring the spin, is given by

$$n_j(\varepsilon) = V \int_{S_j(\varepsilon)} \frac{dS}{(2\pi)^3} \frac{1}{|\nabla_{\boldsymbol{k}} \varepsilon(\boldsymbol{k})|}, \qquad (4.44)$$

where $S_j(\varepsilon)$ is a surface of constant energy in the first Brillouin zone and $\nabla_{\boldsymbol{k}} \varepsilon(\boldsymbol{k})$ is the gradient of $\varepsilon(\boldsymbol{k})$ in $\boldsymbol{k}$-space.

c) Verify, that for free electrons in three dimensions the general expression (4.44) gives the same density of states as the one derived in **a)**.

4.3 Equilibrium Shape of Crystals

In equilibrium, a crystal assumes the shape with the lowest total surface energy under the constraint of constant volume.

a) Assume that the surface energy is independent of the orientation, as it is appropriate for liquids. Show that the equilbrium shape of a liquid droplet is the sphere.

b) Consider a three-dimensional crystal that is a rectangular prism having sides l_x, l_y and l_z and surface energies γ_x, γ_y and γ_z.

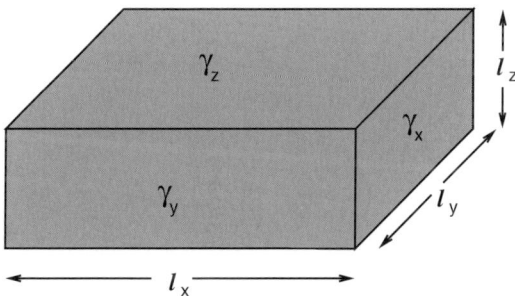

Show that the equilbrium condition for the crystal shape leads to [131]

$$\frac{\gamma_x}{l_x} = \frac{\gamma_y}{l_y} = \frac{\gamma_z}{l_z} = \text{const.} \qquad (4.45)$$

This means that in equilbrium the distance of each crystal face from the center of the crystal is proportional to its surface energy which is the basis of the Wulff construction (see Fig. 4.9).

4.4 Cohesive Energy in Tight-Binding

Use the tight-binding expression for the band width to show that the band energy contribution to the cohesive energy of a transition metal is proportional to the square root of the coordination number.

Hint: Replace the exact density of states by a rectangular density of states with band width W.

4.5 Localized Vibrations on a Semi-infinite Linear Chain

Consider a semi-infinite linear chain consisting of atoms with mass M and a spacing a. The first atom of the chain has a mass $M_s \neq M$. The atoms are coupled by springs with force constant k:

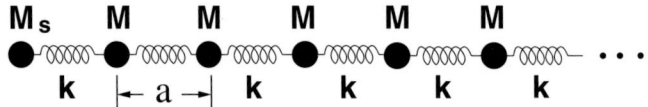

a) Determine the equations of motion for the displacements u_n for the n-th atom, $n = 1, \ldots$

b) Show that a localized vibrational mode exists for $M_s < M/2$ [12] and determine its frequency. Compare this frequency to the maximum "bulk" frequency for an infinite linear chain.

5. Adsorption on Surfaces

The study of adsorption is of central importance in the field of surface science. Adsorption processes are involved in almost all technological processes in which surfaces play a crucial role. Often they are an important step in the preparation of a device as, e.g., in the growth of a semiconductor device. But adsorption can also be of significant importance in industrially relevant processes. The most prominent example is heterogeneous catalysis since usually the reactants have to adsorb on the catalyst before they can react. But of course, also from a fundamental point of view the physical and chemical factors determining adsorption processes are most interesting. In this chapter I will first introduce the basic quantities necessary to describe adsorption. After classifying the different types of adsorption systems the necessary theoretical tools to treat these systems will be addressed. Furthermore, reactivity concepts will be discussed and their usefulness will be demonstrated in some case studies.

5.1 Potential Energy Surfaces

The central quantity in any theoretical description of adsorption is the potential energy surface (PES) of the system. It corresponds to the energy hyperplane over the configuration space of the atomic coordinates of the involved atoms. The PES directly gives information about adsorption sites and energies, vibrational frequencies of adsorbate and the existence of barriers for adsorption.

There is a long tradition in surface science of using one-dimensional potential curves to describe adsorption. The most prominent one goes back to Lennard-Jones [132] and is shown in Fig. 5.1. Two curves are plotted: the curve denoted by AB+S represents the potential energy of the molecule AB approaching the surface S. There is a shallow minimum $E_{\text{ad}}^{\text{AB}}$ before the curve rises steeply. The other curve A+B+S corresponds to the interaction of the two widely separated atoms A and B with the surface. Far away from the surface the energetic difference D between the two potential curves is equal to the dissociation energy of the free molecule AB. Close to the surface it is energetically more favorable to have two separate atoms interacting with the

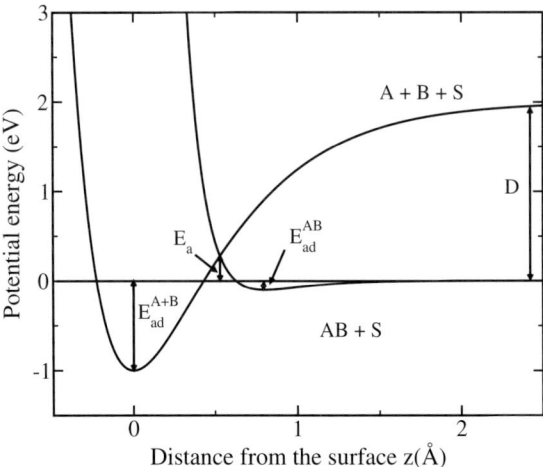

Fig. 5.1. Potential energy curves for molecular and dissociative adsorption according to Lennard-Jones [132]

surface than the intact molecule. This corresponds to a dissociative adsorption scenario. The energy gain upon adsorption of the two isolated atoms A and B is given by E_{ad}^{A+B}.

The exact location of the crossing point between the curves AB+S and A+B+S determines whether there is a barrier for dissociative adsorption or not. The scenario depicted in Fig. 5.1 illustrates the case of activated dissociative adsorption with the diabatic dissociation barrier given by E_a. The adiabatic barrier will be somewhat lower due to the avoided crossing between the adiabatic potential curves. If the crossing of the two curves is closer to the surface and thus at a potential energy < 0 eV, the molecule can dissociate spontaneously at the surface and we have non-activated dissociative adsorption.

Potential curves like the ones presented in Fig. 5.1 illustrate the energetics of the adsorption process. However, without any additional information we do not learn anything about the physical and chemical nature of the interaction between the surface and the adsorbates. The shallow molecular adsorption well in Fig. 5.1 usually corresponds to a physisorption well caused by van der Waals attraction while the steep rise of the potential energy is due to the Pauli repulsion between the molecular and substrate wave functions. The energy gain upon dissociative adsorption is typical for the so-called chemisorption which corresponds to the creation of a true chemical bond between adsorbate and substrate. This interaction can be further classified into ionic, metallic or covalent bonding.

In the following sections we will learn how to theoretically describe the interaction of atoms and molecules with surfaces. First we will address physisorption which still represents a challenge in density functional theory. In the

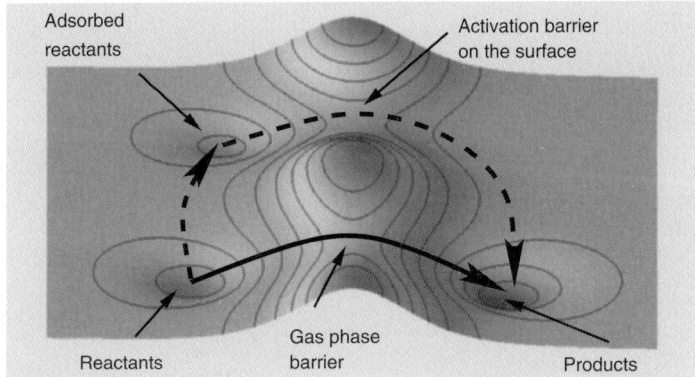

Fig. 5.2. Schematic illustration of the role of a catalyst using a two-dimensional representation of the potential energy surface. A catalyst provides a detour in the multi-dimensional PES with lower activation barriers

discussion of chemisorption we will focus on the analysis of the electronic structure which is crucial for the understanding of the nature of the chemical bond.

One note of caution should be added. One-dimensional representations of potential energy surfaces can be quite misleading. For example, it is often argued that the presence of a catalyst lowers activation barriers significantly. However, usually intermediate products are involved in heterogeneous catalysis which can only be illustrated in a multi-dimensional representation of a PES. This is demonstrated in Fig. 5.2. A catalytic reaction corresponds in principle to a detour in the multi-dimensional PES on the path from the reactants to the products. Along this detour, however, the activation barrier is much smaller than for example in the gas phase. Thus the reaction rate is enormously enhanced in the presence of a catalyst since the rate depends exponentially on the barrier height (see Sect. 7.1).

5.2 Physisorption

In the weakest form of adsorption no true chemical bond between surface and adsorbate is established. The bonding is rather due to the induced dipole moment of a nonpolar adsorbate interacting with its own image charges in the polarizable solid, which means that the attraction is caused by van der Waals forces. Although this bonding is usually rather weak ($\sim 0.1\,\text{eV}$), it is in fact crucial for the bonding in a wide range of matter. For example, the exceptional ability of geckos to climb up smooth vertical surfaces is caused by the van der Waals attraction between foot-hairs of the gecko and the surface [133]. Measurements indicate that a single foot of a gecko could produce 100 N of adhesive force which means that the feet of a gecko could lift a load of 40 kg.

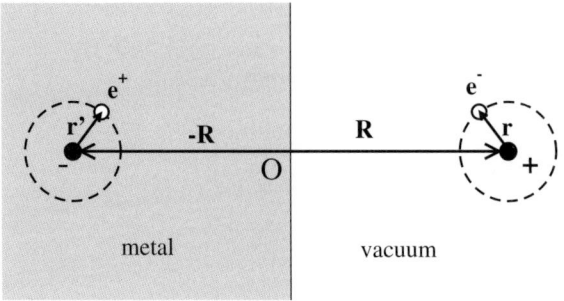

Fig. 5.3. Schematic illustration of a hydrogen atom in front of a perfect conductor interacting with its image charges

Here we are more concerned with van der Waals interaction of rare gases and molecules with filled electron shells with surfaces since it is the main source of the attraction between these species and surfaces.

I will first give a very elementary introduction into to the essential physics of the van der Waals interaction between an atom and a solid surface [134]. Let us first consider a hydrogen atom in front of a perfect conductor (Fig. 5.3). The positively charged nucleus is located at $\boldsymbol{R} = (0, 0, Z)$, and the electronic coordinates $\boldsymbol{r} = (x, y, z)$ are given with respect to the nucleus. This hydrogen atom is interacting with its image charges of both the nucleus and the electron in the conductor. The total electrostatic energy is then a sum of two repulsive and two attractive terms,

$$V_{\mathrm{im}} = -\frac{e^2}{2} \left[\frac{1}{|2\boldsymbol{R}|} + \frac{1}{|2\boldsymbol{R} + \boldsymbol{r} + \boldsymbol{r'}|} - \frac{1}{|2\boldsymbol{R} + \boldsymbol{r}|} - \frac{1}{|2\boldsymbol{R} + \boldsymbol{r'}|} \right]$$

$$= -\frac{e^2}{2} \left[\frac{1}{2Z} + \frac{1}{2(Z+z)} - \frac{2}{|2\boldsymbol{R} + \boldsymbol{r}|} \right]. \quad (5.1)$$

We assume that the atom is not too close to the surface. A Taylor expansion of (5.1) in powers of $|\boldsymbol{r}|/|\boldsymbol{R}|$ yields

$$V_{\mathrm{im}} = -\frac{e^2}{8Z^3} \left[\frac{x^2 + y^2}{2} + z^2 \right] + \frac{3e^2}{16Z^4} \left[\frac{z}{2}(x^2 + y^2) + z^3 \right] + O(Z^{-5}). \quad (5.2)$$

Let us first consider the leading term of (5.2). The nominator is proportional to the square of the electronic displacement from the nucleus. For the sake of simplicity we model the electronic motion in the atom by a three-dimensional oscillator:

$$V_{\mathrm{atom}} = \frac{m_e \omega_{\mathrm{vib}}^2}{2} \left(x^2 + y^2 + z^2 \right). \quad (5.3)$$

The frequency of the unperturbed oscillator is given by ω_{vib}. The atomic potential (5.3) is modified by the presence of the surface. The image charges lead to additional potential terms that are quadratic in the displacements.

This causes a change in the vibrational frequencies. The modified atomic potential is given by

$$V_{\text{atom}} = \frac{m_e \omega_{\text{vib}}^2}{2}\left(x^2 + y^2 + z^2\right) - \frac{e^2}{8Z^3}\left[\frac{x^2 + y^2}{2} + z^2\right] + \ldots$$

$$\approx \frac{m_e \omega_\parallel^2}{2}\left(x^2 + y^2\right) + \frac{m_e \omega_\perp^2}{2} z^2, \tag{5.4}$$

where the modified vibrational frequencies are

$$\omega_\parallel = \omega_{\text{vib}} - \frac{e^2}{16 m_e \omega_{\text{vib}} Z^3} \quad \text{and} \quad \omega_\perp = \omega_{\text{vib}} - \frac{e^2}{8 m_e \omega_{\text{vib}} Z^3}. \tag{5.5}$$

Here we have used $m_e \omega_{\text{vib}}^2 \gg e^2/(4Z)$. If we assume that the atomic oscillator remains in its quantum mechanical ground state, then the van der Waals binding energy in this simple picture is exactly given by the change in the zero-point energy of the atomic oscillator

$$V_{\text{vdW}}(Z) = \frac{\hbar}{2}(\omega_\perp(Z) + 2\omega_\parallel(Z) - 3\omega_{\text{vib}}) = \frac{-\hbar e^2}{8 m_e \omega_{\text{vib}} Z^3}. \tag{5.6}$$

This also demonstrates the long-range nature of the van der Waals interaction which is proportional to Z^{-3}.

The van der Waals potential (5.6) can be further simplified by introducing the atomic polarizability

$$\alpha = \frac{e^2}{m_e \omega_{\text{vib}}^2}. \tag{5.7}$$

Substituting (5.7) into (5.6) yields

$$V_{\text{vdW}}(Z) = -\frac{\hbar \omega_{\text{vib}} \alpha}{8 Z^3} = -\frac{C_v}{Z^3} \tag{5.8}$$

Here $C_v = \hbar \omega_{\text{vib}} \alpha / 8$ is the van der Waals constant that is directly related to the atomic polarizability.

By writing the fourth-order correction in the Taylor expansion (5.2) as $3 C_v Z_0 / Z^4$, the so-called *dynamical image plane* at Z_0 is defined

$$V_{\text{im}}(Z) = -\frac{C_v}{Z^3} - \frac{3 C_v Z_0}{Z^4} + O(Z^{-5}) = -\frac{C_v}{(Z - Z_0)^3} + O(Z^{-5}) \tag{5.9}$$

Note that the $1/Z^3$ long-range van der Waals attraction can also be rationalized from a $1/R^6$ atom-atom dispersion interaction summed over lattice atoms (see Exercise 5.1).

The derivation of the van der Waals force between an hydrogen atom and a perfect conductor given above basically corresponds to the interaction of two dipoles at distance $2Z$. However, a hydrogen atom in the ground state has no permanent dipole moment. Hence there is a rigorous quantum mechanical derivation necessary of the long-range interaction between a neutral atom and a solid surface. Such a derivation was given by Zaremba and Kohn [135].

They treated the interaction in perturbation theory under the assumption that there is no overlap between the wave functions of the atom and the solid. The full Hamiltonian is given by

$$H = H_a + H_s + V_{as}, \tag{5.10}$$

where the subscripts a and s denote the atom and the solid, respectively. The perturbation term V_{as} describes the electrostatic interaction between the atom and the solid

$$V_{as} = \int d^3r \, d^3r' \frac{\hat{\rho}_s(\mathbf{r})\hat{\rho}_a(\mathbf{r}')}{|\mathbf{r} - \mathbf{r}'|}, \tag{5.11}$$

where $\hat{\rho}$ corresponds to the total charge density of the positive ion core n^+ and the electron number operator $\hat{n}$,

$$\hat{\rho}_{s,a}(\mathbf{r}) = n^+_{s,a}(\mathbf{r}) - \hat{n}_{s,a}(\mathbf{r}). \tag{5.12}$$

It can be shown that the first-order contribution vanishes [135]. The second-order interaction energy $E^{(2)}$ is expressed in terms of the retarded response functions $\chi_{a,s}$ of the atom and the solid, respectively,

$$\begin{aligned}
E^{(2)} &= \sum_{\alpha \neq 0} \sum_{\beta \neq 0} \frac{|\langle \psi_0^a \psi_0^s | V'_{as} | \psi_\alpha^a \psi_\beta^s \rangle|}{(E_0^a - E_\alpha^a) + (E_0^s - E_\beta^s)} \\
&= -\int d^3r \int d^3r' \int d^3x \int d^3x' \frac{e}{|\mathbf{R} + \mathbf{x} - \mathbf{r}|} \frac{e}{|\mathbf{R} + \mathbf{x}' - \mathbf{r}'|} \\
&\quad \times \int_0^\infty \frac{d\omega}{2\pi} \chi_a(\mathbf{x}, \mathbf{x}', i\omega) \chi_s(\mathbf{r}, \mathbf{r}', i\omega).
\end{aligned} \tag{5.13}$$

Regrouping the terms and integrating over the atomic response function leads to a term proportional to the atomic polarizability α which already appeared in the simple qualitative derivation above. The remaining integrals can be expressed in terms of the dielectric function ϵ of the solid. Finally one arrives at the result that the interaction term $E^{(2)}$ indeed corresponds to the van der Waals atom-metal potential

$$E^{(2)}(Z) = V_{\text{vdW}}(Z) = -\frac{C_v}{(Z - Z_0)^3} + O(Z^{-5}), \tag{5.14}$$

where the van der Waals constant is given by

$$C_v = \frac{1}{4\pi} \int_0^\infty d\omega \, \alpha(i\omega) \frac{\epsilon(i\omega) - 1}{\epsilon(i\omega) + 1} \tag{5.15}$$

and the position Z_0 of the dynamical image plane by

$$Z_0 = \frac{1}{4\pi C_v} \int_0^\infty d\omega \, \alpha(i\omega) \frac{\epsilon(i\omega) - 1}{\epsilon(i\omega) + 1} \bar{z}(i\omega). \tag{5.16}$$

Table 5.1. Van der Waals coefficient C_V and dynamical image plane Z_0 for rare-gase atoms on various noble metals obtained by jellium calculations [135]. C_V is given in eV/Å^3 and Z_0 in Å

	He		Ne		Ar		Kr		Xe	
	C_V	Z_0	C_V	Z_0	C_V	Z_0	C_V	Z_0	C_V	Z_0
Cu	0.225	0.22	0.452	0.21	1.501	0.26	2.110	0.27	3.085	0.29
Ag	0.249	0.20	0.502	0.19	1.623	0.24	2.263	0.25	3.277	0.27
Au	0.274	0.16	0.554	0.15	1.768	0.19	2.455	0.20	3.533	0.22

Here $\bar{z}$ is the centroid of the induced charge density. This derivation confirms that the long-range interaction potential is expressible as a polarization energy. The polarization is due to the interaction of the instantaneous dipole on the atom caused by charge fluctuations with the induced image charge distribution in the solid.

The equations (5.15) and (5.16) provide a convenient scheme to evaluate the van der Waals interaction from first principles. For simple and noble metals the substrate can be reasonably well represented by the jellium model. Calculated values obtained in this way for rare-gas atoms on various noble metal surfaces are listed in Table 5.1 [135]. It is obvious that the van der Waals coefficient increases strongly from He to Ne for all considered metal surfaces. This increase is basically a direct consequence of the larger atomic polarizability of the heavier rare-gas atoms. As far as the dependence on the substrate is concerned, the van der Waals coefficients C_V reflect the increase in the dielectric function from Cu to Au. The positions of the image plane, on the other hand, depend only weakly on the atomic polarizabilities but decrease with increasing dielectric functions. The values for Z_0 are rather small in the order of 0.15–0.3 Å.

The van der Waals interaction (5.14) is purely attractive. However, closer to the surface the wave functions start to overlap with the substrate wave functions. There will be some electrostatic attraction towards the positive ion cores of the substrate. On the other hand, the orbitals of the approaching atom have to be orthogonal to the substrate wave functions which increases their kinetic energy. This Pauli repulsion is particularly strong for atoms with closed valence shells for which it dominates the interaction close to the surface. Thus there will be a balance between the short-range Pauli repulsion and the long-range van der Waals attraction leading to a physisorption minimum. In order to determine the physisorption equilibrium position for rare gases adsorbed on jellium, Zaremba and Kohn divided the total interaction into two parts: a short-range term described by Hartree–Fock theory and the longe-range van der Waals interaction. Thus the physisorption potential is given by

$$V(Z) = V_{\mathrm{HF}}(Z) + V_{\mathrm{vdW}}(Z). \tag{5.17}$$

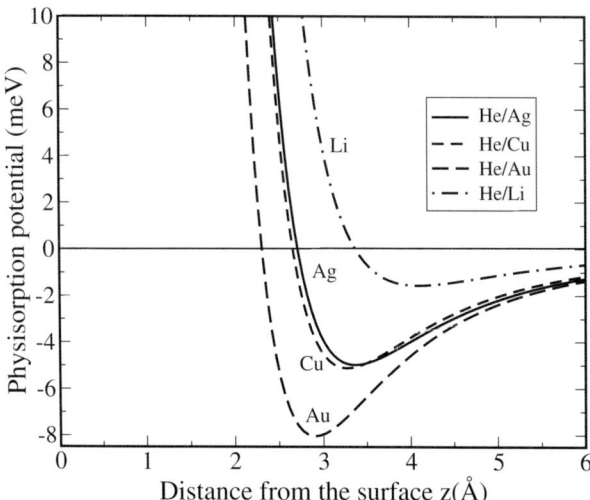

Fig. 5.4. Physisorption potential for He interaction with different jellium surfaces as a function of the distance from the jellium edge. The jellium electronic densities correspond to the noble metals Ag, Cu and Au and the simple metal Li, respectively. (After [136])

The physisorption potential for He interaction with jellium surfaces with densities corresponding to Ag, Cu and Au is shown in Fig. 5.4. It is obvious that the attraction due to the van der Waals interaction is rather weak leading to well depths below 10 meV for He. Furthermore Fig. 5.4 demonstrates that the divergence of the van der Waals attraction at $Z_0 \approx 0.2$ Å is irrelevant for physisorption systems since the Pauli repulsion sets in much further away from the surface.

There is, however, a certain inconsistency in the determination of the equilibrium physisorption energy and position using (5.17). In the derivation of the van der Waals attraction $V_{\rm vdW}(Z)$ (5.14) it was assumed that the wave function were not overlapping while the Pauli repulsion requires a wave function overlap. Therefore it would be desirable to have a consistent unified description of both van der Waals interaction and chemical interaction. Unfortunately, density functional theory using the LDA or GGA for the exchange-correlation functional does not properly describe the long-range van der Waals interaction. This is closely related to the fact that in the LDA and the GGA the exchange-correlation hole is still localized. Therefore the effective electron potential outside of a metal falls off exponentially and not proportional to $1/z$. Hence neither image forces nor the van der Waals interaction is appropriately reproduced. Still there have been calculations of the interaction of rare-gas atoms with surfaces using DFT within the GGA [137] and LDA [138]. These calculations yield reasonable potential well depths for rare-gas adsorption on metal surfaces. It has been argued that this is due to the fact that physisorption can induce a static dipole moment at the adsorbate [138] which is correctly described within LDA or GGA and which contributes significantly to the bond strength [139].

Still there have been several attempts to properly include the van der Waals interaction in density functional theory. Two recent approaches [140, 141] utilize the *adiabatic connection formula*

$$E_{xc}[n] = \frac{1}{2}\int d^3r d^3r' \frac{e^2}{|\mathbf{r}-\mathbf{r}'|}\int_0^1 d\lambda[\langle \tilde{n}(\mathbf{r})\tilde{n}(\mathbf{r}')\rangle_{n,\lambda} - \delta(\mathbf{r}-\mathbf{r}')\langle n(\mathbf{r})\rangle]. \tag{5.18}$$

For $\lambda = 0$ the Hamiltonian $H(\lambda)$ does not contain any longe-range interaction. This interaction is adiabatically switched on as a function of the coupling parameter λ so that for $\lambda = 1$ the Hamiltonian $H(\lambda)$ correponds to the true physical Hamiltonian. In (5.18), $\langle\ldots\rangle_{n,\lambda}$ means the expectation value in the ground state $H(\lambda)$ with a potential V_λ which keeps the ground-state density $n_\lambda(\mathbf{r})$ equal to the exact physical density $n_{\lambda=1}(\mathbf{r})$ for all λ. The advantage of the exact formula (5.18) is that approximate expressions for the interacting system can be used which can still be solved. The adiabatic connection formula then corresponds to an extrapolation to the exact expression.

Hult et al. use second-order perturbation theory equivalent to (5.13) and then introduce a local dielectric function

$$\epsilon(\omega; n(\mathbf{r})) = 1 - \kappa(n(\mathbf{r}))\frac{\omega_p^2(n(\mathbf{r}))}{\omega^2} \tag{5.19}$$

with the plasma frequency

$$\omega_p^2(n(\mathbf{r})) = \frac{4\pi e^2 n(\mathbf{r})}{m_e}, \tag{5.20}$$

thus defining a density functional. A cutoff function $\kappa(n(\mathbf{r}))$ has to be introduced because the local approximation (5.19) tends to overestimate the response in the low-density tails of the wave functions [141]. Kohn et al. avoid the introduction of a cutoff function by transforming the adiabatic connection formula into the time domain [140].

Both approaches give very satisfactory results for van der Waals constants. However, it would be much more desirable to find a more appropriate nonlocal form for the exchange-correlation functional that reproduces the correct long-range form of the effective one-particle potential. This would avoid the introduction of an explicit van der Waals density functional which requires some extra computational effort to include the long-range van der Waals attraction. Because of this extra effort so far the van der Waals functionals are usually not included in standard DFT implementations.

There is a relativistic modification of the van der Waals potential at distances larger than $Z \sim \lambda/2\pi$, where λ is the effective atomic transition wavelength that contributes to the polarizability. This distance of the order of 0.1 μm. At such a large distance the finite velocity of the photons cannot be neglected which causes retardation effects in the electrostatic interaction. Hence the van der Waals interaction falls off more rapidly with distance and becomes proportional to $-1/Z^4$ [142]. The retarded van der Waals or *Casimir-van der Waals* potential is in fact a manifestitation of the *Casimir effect* [143] which is a consequence of the zero-point energy of a quantized field. Although this effect leads to small changes in the attractive potential on an absolute scale, it has still been observed in the scattering of an ultracold beam of metastable neon atoms from silicon and glass surfaces [144].

5.3 Newns–Anderson Model

In contrast to physisorption, in chemisorption true chemical bonds between adsorbate and substrate are created. This means that there is a significant hybridization between the adsorbate and substrate electronic states which causes a modification of the electronic structure. Let us recall that within the supercell approach the one-electron eigenfunctions of the Kohn–Sham equations are delocalized Bloch functions. In the bulk, their eigenenergies as a function of the crystal momentum $\varepsilon(\mathbf{k})$ directly give the electronic band structure which is crucial for the electronic, structural and optical properties of the solid. However, adsorption at surfaces corresponds to the making of localized bonds between the substrate and the adsorbate. The band structure is not a convenient tool for a direct analysis and discussion of the nature of the chemical bonds. Often, an analysis of the local density of states $n(\mathbf{r}, \varepsilon)$ (LDOS), in particular the projected density of states, is better suited to analyze the nature and symmetry of chemical bonds between substrate and adsorbate [66].

In order to obtain the change in the density of states upon chemisorption, a full self-consistent electronic structure calculation of the interacting system has to be performed. However, to establish qualitative trends and basic mechanisms, it is often very useful to describe a complex system by a simplified Hamiltonian with a limited number of parameters. The dependence of the properties of the system on these parameters can then be studied in a well-defined way. In the next sections I will discuss approximative theories of chemisorption such as the Newns–Anderson model or the effective medium theory. When these methods were first introduced, they were also meant to provide semiquantitative results for chemisorption properties. Nowadays, due to the relative ease with which self-consistent electronic structure calculations can be performed, these approximative methods are mainly used for explanatory purposes.

The modification of the electronic structure upon adsorption in terms of the adsorbate-substrate coupling can be particularly well derived using the so-called Newns–Anderson or Anderson–Grimley–Newns model which was developed independently by Grimley [145] and by Newns [146] based on a model proposed by Anderson [147] for bulk impurities.

5.3 Newns–Anderson Model

Consider a substrate characterized by a quasi-continuum of Bloch states ϕ_k with eigenenergies ε_k and an adatom interacting with the substrate. The adatom shall be described by a single valence state ϕ_a with energy ε_a. The interaction can be described in its simplest form by the following model Hamiltonian

$$H = \varepsilon_a \hat{n}_a + \sum_k \varepsilon_k \hat{n}_k + \sum_k (V_{ak} \hat{b}_a^+ \hat{b}_k + V_{ka} \hat{b}_k^+ \hat{b}_a), \qquad (5.21)$$

with

$$\hat{n}_i = \hat{b}_i^+ \hat{b}_i, \quad i = a, k, \qquad (5.22)$$

where $\hat{n}_i$ is the number operator, and $\hat{b}_i^+$ and $\hat{b}_i$ are the creation and annihilation operator of the orbital ϕ_i, respectively. The interaction of the substrate and adatom states is given by the matrix elements V_{ak}. In (5.21) we have ignored the spin degree of freedom. Often the Coulomb repulsion U in the doubly occupied valence state is taken into account by adding a term $\frac{1}{2} U \hat{n}_{a\uparrow} \hat{n}_{a\downarrow}$ in the Newns–Anderson Hamiltonian. Usually the two-body operator $\hat{n}_{a\uparrow} \hat{n}_{a\downarrow}$ is treated in the Hartree-Fock approximation [12] in which it is replaced by an effective one-body operator. This leads to an effective decoupling of the spins and only causes a shift of the eigenenergies ε_a. Since the consideration of the Coulomb repulsion does not change the general conclusions, we will neglect it. Note that (5.21) in this simple form is a single-particle Hamiltonian, i.e., the Hamiltonian could in principle be written as a sum over independent one-particle states $H = \sum_i \varepsilon_i \hat{n}_i$. Still, a direct solution of the Schrödinger equation

$$H \mathbf{c}_i = \varepsilon_i \mathbf{c}_i \qquad (5.23)$$

by diagonalization is intractable due to the infinite number of substrate states. Nonetheless, the Newns–Anderson Hamiltonian (5.21) can be used to derive some fundamental aspects of the behaviour of the adatom valence state ϕ_a upon adsorption. We rewrite the projected density of states (4.6) as

$$n_a(\varepsilon) = \sum_i |\langle \phi_i | \phi_a \rangle|^2 \, \delta(\varepsilon - \varepsilon_i)$$

$$= -\frac{1}{\pi} \operatorname{Im} \sum_i \frac{\langle \phi_a | \phi_i \rangle \langle \phi_i | \phi_a \rangle}{\varepsilon - \varepsilon_i + i\delta} = -\frac{1}{\pi} \operatorname{Im} G_{aa}(\varepsilon), \qquad (5.24)$$

where G is the single particle Green function

$$G(\varepsilon) = \sum_i \frac{|\phi_i\rangle \langle \phi_i|}{\varepsilon - \varepsilon_i + i\delta}, \qquad (5.25)$$

where, as usual, δ is assumed to be a small positive number, $\delta = 0^+$, and G is formally defined by

$$(\varepsilon - H + i\delta) \, G(\varepsilon) = 1. \qquad (5.26)$$

By writing (5.26) in matrix form and eliminating $G_{ka} = \langle\phi_k|G|\phi_a\rangle$ it can be easily shown (see Exercise 5.2 [134]) that $G_{aa}(\varepsilon)$ can be written as

$$G_{aa}(\varepsilon) = \frac{1}{\varepsilon - \varepsilon_a - \Sigma(\varepsilon)}, \tag{5.27}$$

where the self-energy $\Sigma(\varepsilon) = \Lambda(\varepsilon) - i\Delta(\varepsilon)$ is given by

$$\Delta(\varepsilon) = \pi \sum_k |V_{ak}|^2 \, \delta(\varepsilon - \varepsilon_k) \tag{5.28}$$

and

$$\Lambda(\varepsilon) = \frac{1}{\pi} P \int \frac{\Delta(\varepsilon')}{\varepsilon - \varepsilon'} \, d\varepsilon'. \tag{5.29}$$

Here P denotes the principal part integral. Inserting (5.27) into (5.24) yields the projected density of states in terms of $\Lambda(\varepsilon)$ and $\Delta(\varepsilon)$,

$$n_a(\varepsilon) = \frac{1}{\pi} \frac{\Delta(\varepsilon)}{(\varepsilon - \varepsilon_a - \Lambda(\varepsilon))^2 + \Delta^2(\varepsilon)}. \tag{5.30}$$

Two limiting cases can now be conveniently discussed. Let us denote the substrate band width by W. If $V_{ak} \ll W$, then we may just take the average value $V_{\mathrm{av}} = \langle V_{ak}\rangle$ and insert it in (5.28)

$$\Delta(\varepsilon) \approx \pi \sum_k |V_{\mathrm{av}}|^2 \, \delta(\varepsilon - \varepsilon_k) = \pi \, V_{\mathrm{av}}^2 \, n_k(\varepsilon), \tag{5.31}$$

where $n_k(\varepsilon)$ is the density of states of the unperturbed substrate. In the wide-band limit, we may assume that $n_k(\varepsilon)$ and consequently also Δ is independent of the energy. Such a situation is typical for the sp-band of a simple metal. In this case Λ is zero, and the projected density of state simply corresponds to a Lorentzian of width Δ centered around ε_a. Physically this means that the adatom valence level is broadened into a resonance with a finite lifetime $\tau = \Delta^{-1}$. This scenario is called the *weak chemisorption case*.

We will deal with this scenario in much more detail in the next section. Here we only note that even if $\Delta(\varepsilon)$ varies slowly with energy on the scale of V_{ak}, the adatom projected density of state (5.24) will still essentially be of Lorentzian shape. The lifetime broadening of the resonance then reflects the local substrate density of state. As far as the dependence of Δ with respect to the adatom distance from the surface z is concerned, it is usually assumed that $\Delta \propto e^{-\alpha z}$ since the matrix elements V_{ak} fall off exponentially with the distance from the surface.

The other limiting case is given by $V_{ak} \geq W$. It is characterized by a weak continuous spectrum extending over the substrate energy band. In addition, increasing the coupling V_{ak} leads to two localized states outside the continuous spectrum given by the roots of $\varepsilon - \varepsilon_a - \Lambda(\varepsilon) = 0$. This situation is equivalent to the case of two interacting level resulting in bonding and antibonding states above and below the two original states. This *strong chemisorption case* is often observed in the adsorption of atoms and molecules on metal surfaces with d-bands which are usually narrow compared to the interaction strength. Examples for strong chemisorption will be given at the end of this chapter.

The equations (5.28) and (5.29) show that the knowledge of $\Delta(\varepsilon)$ which is sometimes called the *chemisorption function* is crucial for the determination of the adsorbate resonances. However, only with simplifying assumptions with respect to the substrate density of states, it may be directly evaluated [146]. Often its dependence on the energy and on the distance of the adatom from the surface is used as a variable parameter in simulations.

The same is true for the energetic location of the adatom resonance $\varepsilon_a^*(z) = \varepsilon_a(z) + \Lambda(z)$. In particular in weak chemisorption scenarios, the shift $\Lambda(z)$ is usually not explicitly considered, but rather the function $\varepsilon_a^*(z)$ is parametrized and the superscript suppressed [148]. Far away from the surface where there is negligible overlap with the substrate wave function, the dependence of both occupied and unoccupied adatom levels on the distance from the surface can be derived by simple electrostatic arguments. Before we do so, we first have to define the ionization energy I and the electron affinity A. The ionization energy I is defined as the energy to remove an electron from a neutral atom and bring it to infinity, i.e., to the vacuum level. It is always positive, otherwise neutral atoms would not be stable. The electron affinity is given by the energy that is gained when an electron is taken from infinity to the valence level of an atom, i.e. it corresponds to the energy difference between the neutral atom plus an electron at the vacuum level and the negatively ionized atom. The electron affinity can be both positive and negative depending on whether the negative ion is stable or not.

Ionization energy and electron affinity are not the same because of the additional Coulomb repulsion U between the electrons if another electron is added to the atom. Therefore the electron affinity is always smaller than the ionization energy, and the difference is given by U:

$$U = I - A. \tag{5.32}$$

Only for extended states the Coulomb repulsion might be negligible. In fact, for metals the ionization energy and the electron affinity are the same and equal to the metal work function.

The presence of a surface will modify both the ionization energy and the electron affinity. To see this, let us again consider a hydrogen atom in front of a perfect conductor, as in Fig. 5.3. For the sake of convenience, let us assume that the electron and the nucleus are collinear with respect to the surface normal and that their coordinates are both given with respect to the origin at the surface plane. The image potential (5.1) is then given by

$$V_{\text{im}} = -\frac{e^2}{2} \left[\frac{1}{2Z} + \frac{1}{2z} - \frac{2}{(Z+z)} \right]. \tag{5.33}$$

If we remove the electron from the adatom to infinity, we have to do work against the force $\partial V_{\text{im}}/\partial z$ due to the image charges of both the electron itself

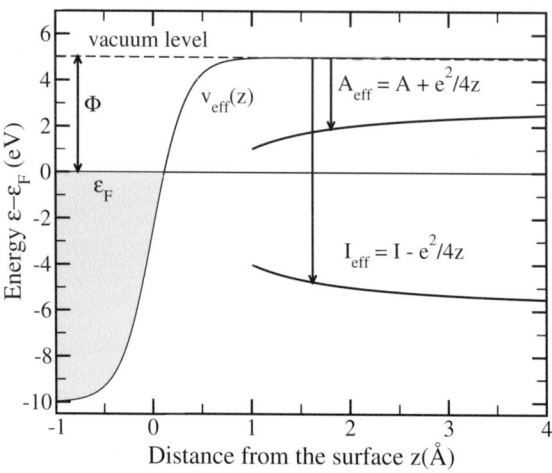

Fig. 5.5. Schematic sketch of the shift of the ionization energy I and the electron affinity A in front of a perfect conductor caused by the image potential. The metal work function is denoted by Φ

and the positive ion. Since the attraction of the electron to its own image is overcompensated by the repulsion with respect to the negatively charged image of the ion which stays at $-Z$, the ionization energy of the adatom in front of the surface is decreased by

$$\int_{z=Z}^{\infty} \frac{\partial V_{\text{im}}}{\partial z'} \, dz' = -\frac{e^2}{4z} . \tag{5.34}$$

Hence the effective ionization energy in front of a perfect conductor is given by

$$I_{\text{eff}}(z) = I - \frac{e^2}{4z} . \tag{5.35}$$

If, on the other hand, we want to add an electron to a neutral atom in front of a surface, we gain the additional energy due to the interaction of the electron with its own image charge. Therefore, the electron affinity is increased to

$$A_{\text{eff}}(z) = A + \frac{e^2}{4z} . \tag{5.36}$$

The influence of the image potential on the ionization and affinity levels is sketched in Fig. 5.5. Depending on whether the affinity or the ionization level crosses the Fermi energy of the metal when the atom approaches the surface, the adatom will become negatively or positively charged, respectively. However, the adatom may well be neutral, if the Fermi energy remains between the ionization and the affinity levels.

The considerations with respect to the ionization and affinity levels of a hydrogen atom can be extended to occupied and unoccupied atomic levels

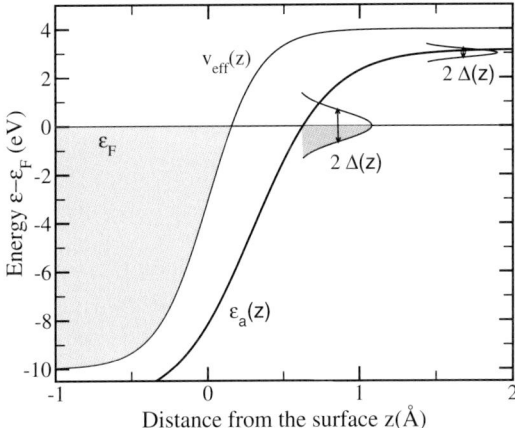

Fig. 5.6. Schematic sketch of the shift and the broadening of an adatom valence level $\varepsilon_a(z)$ upon approaching a surface. $v_{\text{eff}}(z)$ corresponds to the effective one-electron potential of the bare substrate. The shaded areas illustrate the filled levels

in general. Thus the energy of unoccupied levels tends to shift down in front of a conductor while occupied levels are shifted up. This is only true as long as there is negligible overlap with the substrate wave functions. Close to the surface, there is the additional modification of the adatom levels due to the interaction with the substrate states. A typical example of the shift and broadening of an affinity level is plotted in Fig. 5.6. Far away from the surface the affinity level is subject to the image interaction. Note that the increase of the electron affinity $A_{\text{eff}}(z) = A + e^2/4z$ translates into a downshift of the affinity level $\varepsilon_a(z) = \varepsilon_a^{\text{at}} - e^2/4z$. When $\varepsilon_a(z)$ crosses the Fermi level, the level will be filled and consequently the adsorbate becomes negatively charged. Close to the surface the pure $1/z$ dependence of the affinity level becomes modified due to the hybridization with the substrate states. In fact, the level approximately follows the effective one-electron potential $v_{\text{eff}}(z)$ of the bare substrate. In the next section we will see that this can be made plausible within first-order perturbation theory.

The Newns–Anderson model is rather useful for explanatory purposes. However, due to its approximative nature it is not suited for predictive purposes. For example, since the interaction between the substrate and adatom electrons is described only by hopping terms, electron correlation effects are not included [149]. Despite some efforts to consider some correlation effects within a self-energy matrix formalism [150] it has hardly been used for the determination of chemisorption properties. Recently it has been employed predominantly in the modeling of the dynamics of charge transfer processes in molecule-surface scattering [148]. For a reliable description of chemisorption and the determination of chemical trends it is necessary to perform

5.4 Atomic Chemisorption

Chemisorption corresponds to the creation of a true chemical bond between adsorbate and substrate, which means that the electronic structure of both the substrate and adsorbate are strongly perturbed by the interaction [151]. In order to discuss the energetic contributions to chemisorption within density functional theory, it can be useful to regroup the different energetic terms in the total energy expression (3.52):

$$\begin{aligned} E_{\text{tot}} &= \sum_{i=1}^{N} \varepsilon_i + E_{\text{xc}}[n] - \int v_{\text{xc}}(\boldsymbol{r})n(\boldsymbol{r})d^3\boldsymbol{r} - V_{\text{H}} + V_{\text{nucl-nucl}} \\ &= \sum_{i=1}^{N} \varepsilon_i + E_{\text{xc}}[n] - \int v_{\text{eff}}(\boldsymbol{r})n(\boldsymbol{r})d^3\boldsymbol{r} + V_{\text{H}} + V_{\text{el-nucl}} + V_{\text{nucl-nucl}} \\ &= \sum_{i=1}^{N} \varepsilon_i + E_{\text{xc}}[n] - \int v_{\text{eff}}(\boldsymbol{r})n(\boldsymbol{r})d^3\boldsymbol{r} + E_{\text{es}} \, . \end{aligned} \quad (5.37)$$

Here E_{es} corresponds to the total electrostatic energy of the system. The atomic adsorption energy is given by the energy difference between the energies of the separate constituents and the interacting system

$$E_{\text{ads}} = (E_{\text{tot}}(\text{substrate}) + E_{\text{tot}}(\text{atom})) - E_{\text{tot}}(\text{adatom/substrate}) \, . \quad (5.38)$$

Note that according to (5.38) adsorption energies are positive if the adsorption is stable. However, there is no consistency in the literature as far as the sign of the adsorption energy is concerned. Better check the sign convention always carefully. Throughout this book we will use the above convention (5.38).

There are several terms contributing to the total energy and consequently to the energy difference. There can be a delicate cancellation of different opposing effects. Hence it is no trivial task to determine the most crucial one for a particular adsorption system. In the following we will try to establish chemical trends in the adsorption properties by *disentangling* the different energetic contributions.

One of the first applications of self-consistent electronic structure calculations to adsorption properties was performed by Lang and Williams [152]. In this seminal paper DFT-LDA calculations of atomic adsorption on jellium surfaces were presented modeling the interaction of adsorbates with sp-bonded metals such as Al or Na. These calculations do not yield quantitative results but they are ideally suited in order to illustrate qualitative trends. In this particular treatment, the Kohn–Sham equations were not directly solved. It was rather taken advantage of the fact that the solution

5.4 Atomic Chemisorption 111

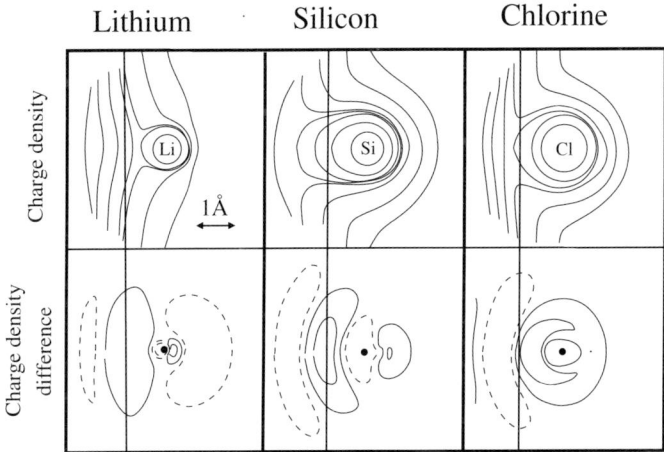

Fig. 5.7. Contours of constant charge density calculated using DFT for Li, Si and Cl adsorbed on a high-density jellium substrate. The solid vertical line indicates the jellium edge. Upper panel: Total charge density of states; lower panel: charge density difference, broken lines correspond to charge depletion. (After [152])

of Kohn–Sham equations can be regarded as being equivalent to solving a scattering Lippmann–Schwinger equation:

$$\psi_{ka}(\mathbf{r}) = \psi_k(\mathbf{r}) + \int d^3\mathbf{r}' G_k(\mathbf{r},\mathbf{r}')\delta v_{\text{eff}}(\mathbf{r}')\psi_{ka}(\mathbf{r}') \,. \tag{5.39}$$

Here the subscripts k and ka denote the unperturbed metal and the metal-adsorbate system, respectively, and $\delta v_{\text{eff}}(\mathbf{r})$ is the change of the effective potential due to the presence of the adsorbate. Equation (5.39) can be interpreted as describing the elastic scattering of metal states $\psi_k(\mathbf{r})$ by the adsorbate induced effective potential $\delta v_{\text{eff}}(\mathbf{r})$.

Figure 5.7 shows the calculated charge contours plots for a representative set of atoms adsorbed on a high-density jellium substrate simulating Al. The upper panel shows the total charge densities of the Li, Si and Cl on jellium. The charge distributions around Li and Si are still almost spherical while in the case of Si there is an elongated structure in the region between adatom and surface. Much more instructive is the lower panel of Fig. 5.7 where the charge density difference between the interacting system and the superposition of the bare atom and surface are plotted. These charge density difference plots illustrate the charge redistribution and the rehybridization due to the interaction of the reactants.

For Li, there is a charge transfer from the vacuum side of the adatom towards the metal. This can be simply understood by the larger electronegativity of Al compared to Li which is prototypical for positive ionic chemisorption. Chlorine, on the other hand, is more electronegative than Al, and this fact is reflected by the significant charge transfer from the substrate to the adsorbed

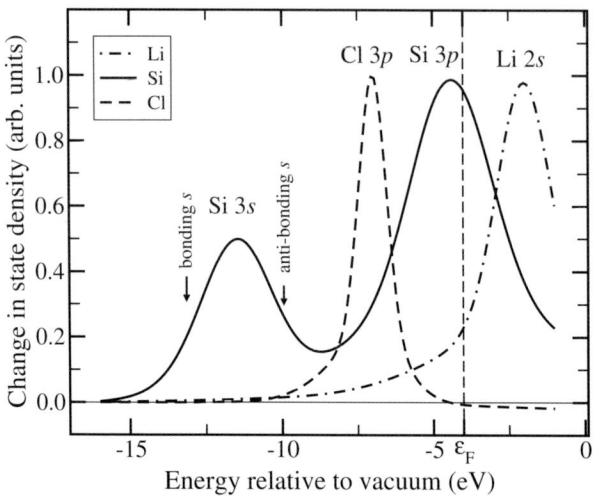

Fig. 5.8. Change of the density of states upon the adsorption of Li, Si and Cl on jellium with an electron density corresponding to Al. (After [152])

Cl atom. Thus chlorine adsorption provides a clear example of negative ionic chemisorption. Note that in the total charge density plots of lithium and chlorine adsorbed on jellium the contour lines in the metal are curved indicating the ionic attraction towards the positively charged Li and the repulsion away from the negatively charged Cl.

Adsorbed silicon, on the other hand, shows charge transfer from the region close to the nucleus to both the vacuum and the bond region. Such a charge accumulation in the bond region is typical for the formation of a covalent bond. However, the spatial information about the charge redistribution alone is not sufficient in order to gain insight into the delicate energetic balance between band-structure and electrostatic contributions. Additional information about density of states effects is provided by electronic structure calculations. The change of the density of states upon the adsorption of Li, Si and Cl is plotted in Fig. 5.8. Clearly visible are several peaks. These resonances which have been discussed in the previous section correspond to adatom levels that have been shifted and broadened due to the interaction with the jellium substrate.

The Li $2s$ derived state which is singly occupied in the free atom lies primarily above the Fermi energy ε_F. This confirms the charge transfer from the Li atom to the substrate and hence the positive ionic chemisorption. The Cl $3p$ derived state is basically fully occupied since it is almost entirely below ε_F indicating the negative ionic chemisorption. With respect to the illustration of the ionization and affinity levels in Fig. 5.5, Li provides an example where the ionization levels is shifted above the Fermi energy while for Cl the affinity level has crossed the Fermi energy.

The density of states of Si adsorbed on jellium shows two prominent peaks which can be associated with the Si 3s and 3p atomic levels. The Si 3p derived state is only half-filled. By analysing the local density of states attributed to the Si derived states Lang and Williams were able to show that the lower parts of the resonance add charge to the bond region while the upper parts substract charge from this region [152]. The lower parts can therefore be associated with a bonding contribution while the upper parts have an antibonding character. Hence a half-filled resonance level corresponds to a covalent interaction in weak chemisorption cases. For Cl adsorption, both the bonding and antibonding contributions are occupied. In this case it is the electrostatic attraction between the Cl core and the transferred electron that stabilizes the adsorption.

In fact there is an alternative derivation of the division of the resonance levels in bonding and antibonding contributions which is based on the scattering formulation (5.39) of atomic adsorption. If the metal states are elastically scattered at the adsorbate induced potential, there is a phase shift of the scattering states given by [139, 153]

$$\tan \delta_a(\varepsilon) = -\frac{\operatorname{Im} D_a(\varepsilon)}{\operatorname{Re} D_a(\varepsilon)}, \tag{5.40}$$

where $D_a(\varepsilon)$ is the determinant

$$D_a(\varepsilon) = \det[1 - G_M(\varepsilon)\delta v_{\text{eff}}]. \tag{5.41}$$

If one assumes that the system is enclosed in a sphere of Radius $R \to \infty$ so that the wave function vanishes for $r = R$, it is relatively easy to show (see Exercise 5.3 [12]) that the change in the density of states induced by the perturbing potential is related to the phase shift by

$$\delta n(\varepsilon) = \frac{g_a}{\pi} \frac{d\delta_a(\varepsilon)}{d\varepsilon}, \tag{5.42}$$

where g_a is the dimension of the representation of the symmetry group the adsorbate state ψ_a belongs to. Let now ε_a be the energy where $\operatorname{Re} D_a(\varepsilon)$ vanishes. If we expand $\operatorname{Re} D_a(\varepsilon)$ around ε_a, to first order we get

$$\operatorname{Re} D_a(\varepsilon) \approx \left.\frac{d\operatorname{Re} D_a(\varepsilon)}{d\varepsilon}\right|_{\varepsilon=\varepsilon_a} (\varepsilon - \varepsilon_a). \tag{5.43}$$

Inserting this Taylor expansion in the denominator of (5.40) allows us to simplify the expression of the phase shift to

$$\delta_a(\varepsilon) = -\arctan \frac{\Gamma}{\varepsilon - \varepsilon_a}, \tag{5.44}$$

with the constant Γ given by

$$\Gamma = \left[\frac{\operatorname{Im} D_a(\varepsilon)}{(d\operatorname{Re} D_a(\varepsilon)/d\varepsilon)}\right]_{\varepsilon=\varepsilon_a}. \tag{5.45}$$

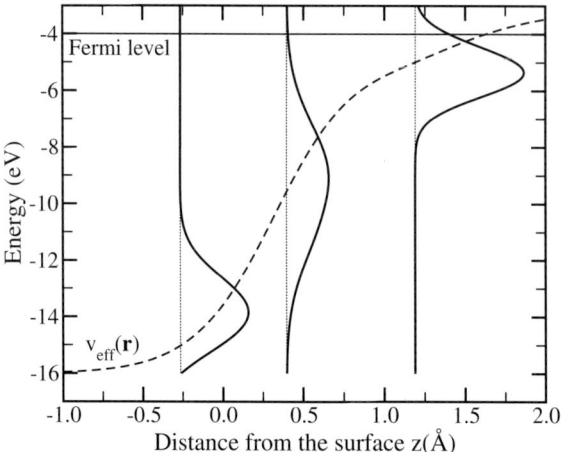

Fig. 5.9. Variation of the hydrogen-induced density of states as a function of the atomic distance from the surface (after [152,154]). The electron density corresponds to Al ($r_s = 2$). The effective one-electron potential $v_{\mathrm{eff}}(z)$ of the bare substrate is additionally plotted as the dashed line

Combining (5.42) and (5.44) gives a simple Lorentzian for the change of the density of state near a resonance

$$\delta n(\varepsilon) = \frac{g_a}{\pi} \frac{\Gamma}{(\varepsilon - \varepsilon_a)^2 + \Gamma^2} \qquad (5.46)$$

with the width of the resonance determined by Γ. However, the definition of Γ given by (5.45) does not provide a straightforward interpretation. Note the similarity with the expression (5.30) for the density of states projected onto an adsorbate level derived within the Newns–Anderson model.

This proves that the resonance occurs at energies ε_a where $\mathrm{Re}\, D_a(\varepsilon)$ vanishes. It follows that the phase shift

$$\tan \delta_a(\varepsilon) = -\frac{\Gamma}{\varepsilon - \varepsilon_a} \qquad (5.47)$$

increases through $\pi/2$ as the energy goes through ε_a from below which is a well-known phenomenon in resonance scattering. In elementary textbooks on quantum mechanics [7, 17] it is shown that a phase shift of $0 < \delta_a < \pi/2$ corresponds to a situation in which the wave function is pulled towards the scattering potential. This means that at the lower energy side of the resonance the phase shift of the wave function leads to an accumulation of charge density in the region of the adatom-substrate bond which is just a characterization of an interaction of bonding character. On the other hand, a phase shift of $\pi/2 < \delta_a < \pi$ pushes the wave function away from the scattering potential. As a consequence, at the higher energy side of the resonance the phase shift leads to a reduction of the electron density in the region of the adatom-substrate bond resulting in an antibonding contribution.

These considerations are not only relevant for resonance levels but indeed also for extended states which form energy bands. The lower half of an energy band can be associated with states of bonding character while the upper half corresponds to states of antibonding character. Thus for a transition metal with a half-filled d-band this band is of purely bonding character. This provides a natural explanation for the fact that transition metals with a half-filled d-band exhibit the largest cohesive and surface energies (see Fig. 4.10).

The jellium calculations do not only yield the resonance structure at the adsorption equilibrium position, they also allow to trace the shift and the broadening of valence levels as a function of the distance from the surface. This is illustrated in Fig. 5.9 for the interaction of a hydrogen atom with a high-density jellium surface. Close to the surface, the hydrogen induced resonance broadens considerably and shifts down in energy. In Fig. 5.9, the effective one-electron potential $v_{\text{eff}}(z)$ of the bare substrate is also plotted as the dashed line. The center of the resonance approximately follows this effective potential.

For the hydrogen atom penetrating into the jellium substrate the resonance narrows in fact again. According to (5.28), the width of the resonance is determined both by the strength of the interaction V_{ak} and by the density of substrate states that couple to the adatom. At the bottom of the metal band given by v_{eff} inside the jellium substrate the density of states decreases which overcompensates the increase in the interaction and thus causes the narrowing of the hydrogen resonance.

The fact that the resonance level tracks the effective potential can in fact be made plausible within first-order perturbation theory [152]. Let us consider an adatom valence state ϕ_a with eigenenergy ε_a^∞ in the gas phase. Upon chemisorption this state is shifted to $\varepsilon_a^\infty + \delta\varepsilon_a$. In first-order perturbation theory, the shift can be estimated by

$$\delta\varepsilon_a(z) = \langle \phi_a | \delta v_{\text{eff}}(\mathbf{r}) | \phi_a \rangle, \tag{5.48}$$

where the adsorbate induced change of the effective potential $\delta v_{\text{eff}}(\mathbf{r})$ can be expressed as

$$\delta v_{\text{eff}}(\mathbf{r}) = v_{\text{eff}}(\mathbf{r}; n_{ka}) - v_{\text{eff}}(\mathbf{r}; n_a). \tag{5.49}$$

$n_{ka}(\mathbf{r})$ and $n_a(\mathbf{r})$ are the electron density associated with the interacting system and the free atom, respectively. Ignoring any charge transfer, we can approximate

$$n_{ka}(\mathbf{r}) \approx n_k(\mathbf{r}) + n_a(\mathbf{r}). \tag{5.50}$$

For a given electron density distribution $n(\mathbf{r})$, the total electrostatic potential is given by

$$v_{\text{es}}(\mathbf{r}) = v_{\text{ext}}(\mathbf{r}) + v_{\text{H}}(\mathbf{r}) = e^2 \int d^3 r' \frac{n(\mathbf{r}') - n^+(\mathbf{r}')}{|\mathbf{r} - \mathbf{r}'|}, \tag{5.51}$$

where $n^+(\mathbf{r}')$ is the charge distribution of the positive ion cores.

Now we can write δv_{eff}, using (3.50) and (5.50), as

$$\delta v_{\text{eff}}(\mathbf{r}) = v_{\text{es}}(\mathbf{r}; n_k) + v_{\text{xc}}(\mathbf{r}; n_k + n_a) - v_{\text{xc}}(\mathbf{r}; n_a)$$
$$= v_{\text{es}}(\mathbf{r}; n_k) + \delta v_{\text{xc}}(\mathbf{r}). \qquad (5.52)$$

The exchange-correlation potential $v_{\text{xc}}(\mathbf{r}; n)$ is a function that increases monotonously with n in the local-density approximation. Doing Taylor expansions, we can thus write $\delta v_{\text{xc}}(\mathbf{r}) \approx v_{\text{xc}}(\mathbf{r}; n_k)$ for $n_k(\mathbf{r}) \gg n_a(\mathbf{r})$ and $\delta v_{\text{xc}}(\mathbf{r}) \approx 0$ for $n_k(\mathbf{r}) \ll n_a(\mathbf{r})$. Consequently, in these two limiting cases we have

$$\delta \varepsilon_a(z) \approx \begin{cases} v_{\text{eff}}(\mathbf{r}; n_k) & n_k(\mathbf{r}) \gg n_a(\mathbf{r}) \\ v_{\text{es}}(\mathbf{r}; n_k) & n_k(\mathbf{r}) \ll n_a(\mathbf{r}) \end{cases}. \qquad (5.53)$$

This suggests that the shift $\delta \varepsilon_a(z)$ indeed follows either the electrostatic or the effective surface potential in the two limiting cases and some mean value for intermediate situations. This is true for a hydrogen atom approaching a jellium surface, as Fig. 5.9 shows. However, the derivation (5.48)–(5.53) uses some rather crude assumptions. Hence it is not surprising that for more complex systems the simple relation (5.53) is not necessarily fulfilled.

5.5 Effective Medium Theory and Embedded Atom Method

So far we have analysed the atomic chemisorption with respect to the shift and broadening of electronic adatom resonances upon the interaction with the substrate which causes a change in the local electron density of states. There is an alternative theoretical model that is very helpful in deriving trends in atomic chemisorption properties, namely the *effective medium theory* [155, 156]. It is based on the main idea that an adsorbate can be considered as being embedded in an inhomogeneous electron gas set up by the substrate. The energy of an adsorbate at the position $\mathbf{r}$ where the substrate has the density $n(\mathbf{r})$ is then estimated as the embedding *cohesive* energy of the adsorbate in a homogeneous electron gas of the same density which is regarded as the *effective* medium:

$$E \approx E_c(n(\mathbf{r})) \qquad (5.54)$$

For each element, the embedding energy $E_c(n)$ is an universal function independent of the particular substrate that only has to be evaluated once and for all. This function determined by DFT-LDA calculations [157] for helium, hydrogen and oxygen is plotted in Fig. 5.10. Elements with a stable free negative ion, i.e. with a positive electron affinity like hydrogen and oxygen, show a minimum at negative energies which is due to the electrostatic attraction between the electrons and the positive ion core. Embedding a closed-shell atoms such as the rare-gas element helium in an electron gas is associated

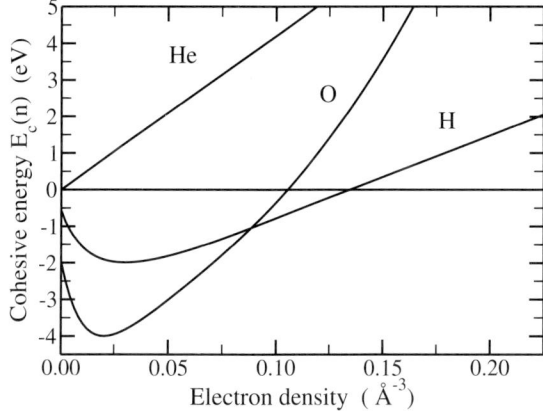

Fig. 5.10. Immersion energy for H, O and He embedded in infinite jellium obtained by LDA-DFT calculations as a function of the electron density (after [157]). The energy zero corresponds to the free atom

with an energy cost which is basically related linearly to the electron density. At sufficiently high densities, the embedding energy becomes positive for all elements. This is due to the rise in kinetic energy associated with the orthogonalisation of the electronic states to the occupied core states of the atoms.

In the limit of vanishing electron density, the embedding energy of atoms with stable negative ions does not approach the energy of the free atom (see Fig. 5.10). This is due to the fact that for arbitrarily small electron densities it is still energetically more favorable to transfer one electron from the infinite electron gas to the affinity level of the atom. Hence the limit of the embedding energy for $n \to 0$ is given by the electron affinity A.

Using the embedding energy $E_c(n)$ (5.54), it is very simple to construct the potential energy for, e.g., a hydrogen atom approaching and penetrating a solid by just reading off the energy corresponding to any density. This is demonstrated in Fig. 5.11. The hydrogen atom is attracted towards the surface until it reaches the position that corresponds to the optimum density. In Fig. 5.11 these positions are indicated by the dashed line. The chemisorption minimum is thus a direct reflection of the minimum of the $E_c(n)$ curve.

In the bulk, the electron density is in general higher than the optimum density for hydrogen embedding. Hence there is a one-to-one correspondence between energy and electron density. In the middle of a vacancy, on the other hand, the density can be smaller than the optimum density. In such a situation the hydrogen does not sit in the middle of the vacancy but rather off-center in the vacancy since the electron density is too low in the center of the vacancy.

There are further adsorption properties which can be easily qualitatively understood within the framework of the embedding function. For example,

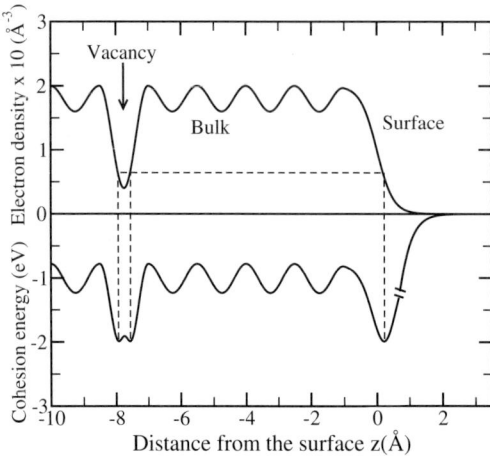

Fig. 5.11. Illustration of the potential energy of a hydrogen atom at a surface, in the bulk and in a vacancy as determined by the unperturbed substrate electron density through the hydrogen embedding energy. The dashed lines indicate the optimum density for hydrogen embedding. (After [156])

the adsorbate-substrate chemisorption bond length is the shorter, the lower the adsorbate coordination is because then there are less substrate atoms contributing to provide the optimum electron density.

The vibrational frequency of the adsorbate with respect to the substrate is determined by the curvature of the potential energy at the equilibrium position. Hence one can write, using that $dE_c(n)/dn$ vanishes at the chemisorption position,

$$\omega_{\text{vib}} \propto \sqrt{\frac{d^2 E_c(n(\mathbf{r}))}{dz^2}} = \sqrt{\frac{d^2 E_c(n)}{dn^2}} \frac{dn}{dz}. \tag{5.55}$$

For a given adsorbate, the factor $\sqrt{d^2 E_c(n)/dn^2}$ is the same for all substrates and adsorption sites. Consequently, the factor dn/dz is crucial for the particular frequency at a given site. If one assumes that the electron density associated with a single substrate atoms decays exponentially, $n_{\text{atom}} \propto \exp(-\beta r)$, the gradient is given by

$$\frac{dn}{dz} = \beta \, n_0 \sin \alpha \,, \tag{5.56}$$

where n_0 corresponds to the optimum density and α is the angle of the metal-adsorbate line with the surface plane. On low-index surfaces, this angle is usually the smaller, the larger the coordination of an adsorption site, so that we have

$$\omega_{\text{vib}}^{\text{top}} > \omega_{\text{vib}}^{\text{bridge}} > \omega_{\text{vib}}^{(111)\text{hollow}} > \omega_{\text{vib}}^{(100)\text{hollow}}. \tag{5.57}$$

Although the formulation of the potential energy with the simple embedding function (5.54), $E_{\text{tot}} = E_c(n)$, is rather convenient and instructive, there are severe shortcomings of this model. First of all, an adatom will find a position with the optimum density on *any* substrate. Consequently, the adsorption energy is independent of the substrate which means that no chemical trends are described within this simple form. Furthermore, an adatom will have no diffusion barrier on the surface in this simple model since it can always move on the contour line of the optimum density parallel to the surface which corresponds to a constant energy.

An improved description of the effective medium theory avoiding the problems mentioned above can be derived within perturbation theory [155, 158]. In the following, we will denote the deviations from homogeneity in the homogeneous electron gas, i.e., the effective medium, induced by the presence of the adatom by Δ while the differences between the real host and the effective medium are denoted by δ.

In zeroth order, the density $n(\boldsymbol{r})$ appearing in the embedding energy $E_c(n)$ is replaced by an average density $\bar{n}$ which is given by

$$\bar{n} = \frac{\int n(\boldsymbol{r}) \, \Delta\phi(\boldsymbol{r}) \, d^3\boldsymbol{r}}{\int \Delta\phi(\boldsymbol{r}) \, d^3\boldsymbol{r}}, \tag{5.58}$$

where the sampling function $\Delta\phi(\boldsymbol{r})$

$$\Delta\phi(\boldsymbol{r}) = \int \frac{\Delta n(\boldsymbol{r}') - Z\delta(\boldsymbol{r}')}{|\boldsymbol{r} - \boldsymbol{r}'|} \, d^3\boldsymbol{r}' \tag{5.59}$$

is the change in the electrostatic potential *in the jellium* induced by the atom. The atom-induced change of the charge density $\Delta\rho = \Delta n(\boldsymbol{r}) - Z\delta(\boldsymbol{r})$ also enters the first-order correction term

$$\Delta E^{(1)} = \int \phi_0(\boldsymbol{r}) \, \Delta\rho(\boldsymbol{r}) \, d^3\boldsymbol{r}, \tag{5.60}$$

where $\phi_0(\boldsymbol{r})$ is the unperturbed electrostatic potential of the substrate. Finally, it is only the second-order term where corrections due to the perturbation of the substrate enter. It describes the total change in the density of states both in the jellium and in the surface and thus corresponds to the covalent contribution to the bonding:

$$\Delta E^{(2)} = \delta \left(\int_{-\infty}^{\varepsilon_F} \Delta n(\varepsilon) \, \varepsilon \, d\varepsilon \right). \tag{5.61}$$

Summarizing (5.58)–(5.61), the total energy in the effective medium theory taking into account electrostatic and band-structure corrections is given by

$$E_{\text{tot}} = E_c(\bar{n}) + \int \phi_0(\boldsymbol{r}) \Delta\rho(\boldsymbol{r}) d^3(\boldsymbol{r}) + \delta \left(\int_{-\infty}^{\varepsilon_F} \Delta n(\varepsilon) \, \varepsilon \, d\varepsilon \right). \tag{5.62}$$

In fact, the total energy expression (5.62) can be further simplified by approximating the first and second order corrections by $\Delta E^{(1)} \approx \alpha_{\text{at}} \bar{n}$ and

$\Delta E^{(2)} \approx \alpha_v \bar{n} + \Delta E_{\text{hyb}}$ [158]. Adding the terms linear in the averaged density $\bar{n}$ to the embedding energy $E_c(\bar{n})$, the total energy can be written as

$$E_{\text{tot}} = E_c^{\text{eff}}(\bar{n}) + \Delta E_{\text{hyb}}.\tag{5.63}$$

The term ΔE_{hyb} describes the hybridization between the adatom level ε_a and the metal d-band. This interaction might be regarded as the hybridization of two levels. The adatom level which is fully occupied is shifted down (bonding) while the metal d levels are shifted up. To second order, these antibonding and bonding contributions can be written as

$$\Delta E_{\text{hyb}} = \sum_{d \text{ occ}} \frac{|V_{ad}|^2}{\varepsilon_d - \varepsilon_a} - \sum_{d} \frac{|V_{ad}|^2}{\varepsilon_d - \varepsilon_a}.\tag{5.64}$$

As shown in the last section, the position of the adatom level is given roughly by the effective potential at the adsorption site $\varepsilon_a \approx v_{\text{eff}}$. This energy corresponds to the bottom of the sp-band and is usually well separated from the d-band. Therefore, the differences in the denominator of (5.64) can be approximated by $\varepsilon_d - \varepsilon_a \approx c_d - v_{\text{eff}}$ where c_d is the center of the d-band. If we further assume that the matrix element V_{ad} is independent of ε_d, ΔE_{hyb} is given by the simple form

$$\Delta E_{\text{hyb}} = -2(1-f)\frac{|V_{ad}|^2}{c_d - v_{\text{eff}}},\tag{5.65}$$

where f is the filling factor of the d-band and the factor of two takes account of the assumed spin degeneracy. Roughly speaking, the term $E_c^{\text{eff}}(\bar{n})$ in (5.63) represents the interaction of the adatom level with the sp electrons of the substrate while ΔE_{hyb} describes the additional hybridization with the d-band [159].

Using a somewhat more elaborate formulation of the effective medium theory, the atomic adsorption energies of hydrogen and oxygen on the $3d$ transition metals have been calculated [156, 159]. A comparison of these results with experimental values is presented in Fig. 5.12. As far as the general chemical trend is concerned, it is obvious that the adsorption strength decreases by going from the left to right to the noble metals. The qualitative agreement between theory and experiment is quite satisfactory, for hydrogen the results even agree quantitatively. However, for oxygen there are differences of up to 4 eV.

Part of the discrepancy is due to the overbinding of LDA (see Table 3.1), since $E_c^{\text{eff}}(\bar{n})$ used for the results in Fig. 5.12 is based on LDA jellium calculation, but it cannot take account of the whole difference. A perturbative treatment of covalent bonding is not sufficient to obtain quantitative results in general. In fact, this is a general feature of the effective medium theory. While it gives almost quantitative results for system with non-directional bonds such as adsorption of atomic hydrogen on metal surfaces or the metal-metal bonding [160], all attempts have failed to extend the theory to give a reliable description of the directional covalent bonding of more complex

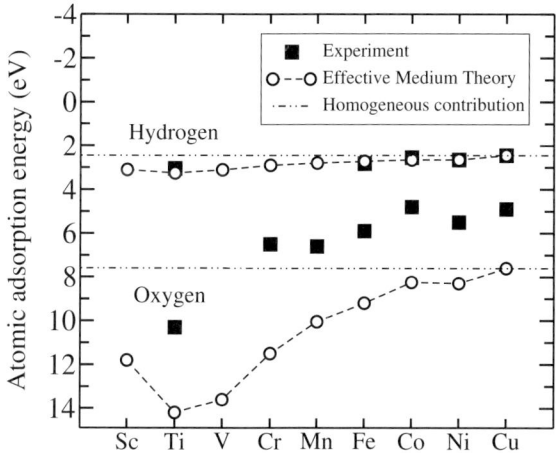

Fig. 5.12. Hydrogen and oxygen adsorption energy on 3d metals calculated within the effective medium theory (*open circles*) and compared to the experiment (*filled squares*). The dash-dotted lines correspond to the homogeneous contribution $E_c^{\text{eff}}(\bar{n}_0)$ in (5.63). (After [156])

atoms and molecules. Therefore its main purpose nowadays is to give qualitative insight into general trends observed in bonding.

There is another method based on the same ideas as the effective medium theoy, but with a more modest claim, the *Embedded Atom Method* (EAM) [161, 162]. The main idea underlying the EAM is to express the total energy as a sum of an embedding energy plus an electrostatic core-core repulsion

$$E_{\text{tot}} = \sum_i F_i(n_{h,i}) + \frac{1}{2} \sum_{i \neq j} \phi_{ij}(r_{ij}). \quad (5.66)$$

Here $n_{h,i}$ is the host density at atom i due to the remaining atoms of the system and $\phi_{ij}(r_{ij})$ is the core-core pair repulsion between atoms i and j separated by r_{ij}. The host density $n_{h,i}$ is estimated as a superposition of atomic densities, usually taken from independent quantum chemical calculations,

$$n_{h,i} = \sum_{j(\neq i)} n_j^a(r_{ij}), \quad (5.67)$$

while $\phi_{ij}(r)$ is represented by the interaction of two neutral, screened atoms

$$\phi_{ij}(r) = Z_i(r) Z_j(r)/r. \quad (5.68)$$

The embedding energy $F(n)$ and the effective charge $Z(r)$ in the embedded atom method are regarded as semiempirical parameters that are fitted to reproduce lattice and elastic constants, cohesive and vacancy formation energy, and energy difference between fcc and bcc phases. Figure 5.13 presents embedding energies and effective charges for Ni, Pd and H determined in this way [161]. Note that the parametrizations shown in Fig. 5.13 are by no means

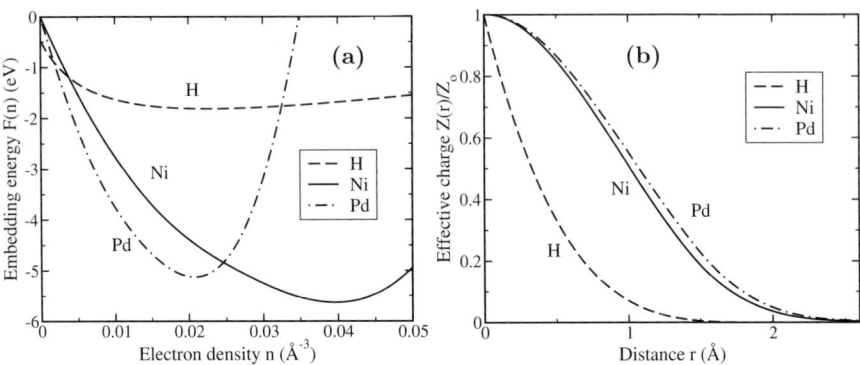

Fig. 5.13. Semiempirically determined functions for Pd, Ni, and H entering the embedded atom method. (a) Embedding energies $F(n)$ as a function of the background electron density; (b) Effective charges $Z(r)/Z_0$ as a function of the distance from the nucleus. (After [161])

unique. Depending on the parametric form chosen in a particular fit a different division of the total energy between the embedding and the repulsive part can result [162].

Using the functions plotted in Fig. 5.13, atomic hydrogen adsorption energies for different adsorption sites on Pd(100) and Pd(111) have been evaluated [161]. These EAM energies are compared to recent DFT results obtained with the PW91-GGA functional [34] in Table 5.2. Again, there is an almost quantitative agreement of the EAM results with the ab initio results. Only the splitting between the adsorption energies for the fcc and the hcp hollow adsorption sites which differ by the position of the second-layer metal atom is not reproduced since the effective charge is too short-ranged.

Table 5.2 demonstrates that the EAM gives a reasonable description of hydrogen chemisorption energies. However, it has been mainly used within the materials science community for bulk and surface properties of metals and alloys [163]. It is particularly suited for metals with empty or filled d bands where the bonding is not as directional as for transition metals with their partially filled d band.

Table 5.2. Hydrogen adsorption energies for different adsorption sites on Pd(100) and Pd(111) determined by the embedded atom method [161] and by DFT calculations with the PW91-GGA functional for a coverage of 1/4 (courtesy of A. Roudgar). h, b and t denote the adsorption sites at the hollow, bridge and top position, respectively

Method	Pd(100)			Pd(111)			
	h	b	t	fcc h	hcp h	b	t
EAM	0.53	0.45	0.10	0.53	0.53	–	0.03
GGA-DFT	0.468	0.426	−0.047	0.554	0.518	0.410	0.010

Like the effective medium theory, the embedded atom method does not satisfactorily describe covalent bonding. There have been extensions such as the *modified embedded atom method* (MEAM) [164, 165] and the *embedded-diatomics-in-molecules* (EDIM) formalism [166]. The MEAM includes angular forces so that the effects of covalent bonding can be included. The EDIM method combines the EAM with semiempirical valence bond theory. In the EDIM formalism, Coulomb and exchange integrals are expressed in a parametrized form. Usually the parameters entering both the EAM and the EDIM formalism are fitted to experimental results. Instead they might be used to reproduce first-principles calculations thus extending the range of ab initio derived applications.

5.6 Reactivity Concepts

It has been a long-term goal in the study of chemical reactions in the gas phase and at surfaces to gain an understanding of the reactivity of an interacting system from the properties of the isolated reactant systems alone. If such a reactivity concept were available, calculations of the interacting system would no longer be necessary. This would not only save a large amount of computer time but it would also help enormously, say, in the design of better catalysts. Unfortunately there is no single concept that has predictive power for all possible types of reactions. Often reactivity concepts are only reliably applicable as long as the perturbation of the isolated systems by the interaction is not too strong. Still, the reactivity concepts can be used to provide some guidelines for the understanding of the vast variety of possible reaction mechanisms. In this section some reactivity concepts will be introduced and their predictions tested against explicit calculations of adsorption energies and reaction barriers.

The starting point for any analysis of the reactivity of a particular system is the characterization of the electronic structure of the interacting fragments. Atoms and molecules in the gas phase have well-defined discrete energy levels. The classical theoretic measure of interaction is provided [66] by the second order perturbation theory expression

$$\Delta E = \frac{|V_{ij}|^2}{\varepsilon_i - \varepsilon_j}, \tag{5.69}$$

where ε_i and ε_j are the eigenenergies of particular levels of the isolated systems. For atoms and molecules, the levels are usually well-separated, and the interaction is governed by a particular subset of states. It turns out that for gas-phase reactions the interaction is often dominated by the *highest occupied molecular orbital* (HOMO) and the *lowest unoccupied molecular orbital* (LUMO) or a subset of states close to these orbitals. These states are called the *frontier orbitals* [66, 167].

At surfaces, there are no discrete energy levels corresponding to localized orbitals but rather energy bands. In particular for metals which do not have a band gap, the HOMO and the LUMO could be considered to be electronic states at the Fermi energy. And in fact, the local density of states at the Fermi energy has been successfully correlated with the reactivity of metal surfaces [168–170]. A more general theory has been derived that uses the concept of the local softness $s(\mathbf{r})$ to characterize the chemical reactivity. The softness can be expressed as [171]

$$s(\mathbf{r}) = \left.\frac{\partial n(\varepsilon, \mathbf{r})}{\partial \mu}\right|_{\varepsilon=\varepsilon_F} = \int d\mathbf{r}' K^{-1}(\mathbf{r}, \mathbf{r}') n(\varepsilon_F, \mathbf{r}). \tag{5.70}$$

This shows that the softness corresponds to an average of the local density of states $n(\varepsilon_F, \mathbf{r})$ at the Fermi level with the weighting function $K^{-1}(\mathbf{r}, \mathbf{r}')$. The kernel $K(\mathbf{r}, \mathbf{r}')$ is the transpose of a response function $\kappa(\mathbf{r}, \mathbf{r}')$ so that the softness can be regarded as a response to $n(\varepsilon_F, \mathbf{r})$. However, although the softness as a reactivity measure is more complex, it gives basically the same conclusions concerning the chemical reactivity as the local density of states at the Fermi level [171].

However, there are examples of metal surfaces, for example $Cu_3Pt(111)$, which have a very low density of states at the Fermi level but are rather reactive [172]. Still gas-phase reactivity concepts based on the frontier orbital concept can be applied to the strong chemisorption of atoms and molecules [66]. But instead of focusing on the Fermi level, it is more appropriate to regard the whole d-band as a single level located at the center of the d-band ε_d interacting with an adsorbate. This is justified since the d-band is usually relatively narrow due to the rather strong localization and small overlap between the wave functions of the d electrons.

The resulting interaction picture is illustrated in Fig. 5.14 for the case of the interaction of a hydrogen molecule with a d-band metal. The Fermi levels typical for a transition metal and for a noble metal, respectively, are indicated by the dash-dotted lines. In the gas phase, the bonding σ_g state of H_2 is fully occupied while the antibonding σ_u^* state is empty. Thus these states correspond to the HOMO and LUMO, respectively, of the adsorbate. Upon the interaction with the d-band of the metal, both states split into a bonding and antibonding state with respect to the surface-molecule bond. This splitting, however, is not symmetric. The up-shift of the antibonding state is larger than the down-shift of the bonding state. This is caused by the orthogonalization of the states which raises the kinetic energy and therefore leads to an energetic cost. Thus if both the bonding and the antibonding state are occupied, the total energy is raised leading to a repulsion. The energetic cost of the orthogonalization is the reason why the interaction of closed-shell atoms such as the rare gases with surfaces is usually repulsive except for the weak van der Waals attraction.

In the case of the interaction of H_2 with metal surfaces, we also have to consider the σ_u^* state which is split, too. The metal-σ_u^* interaction usually

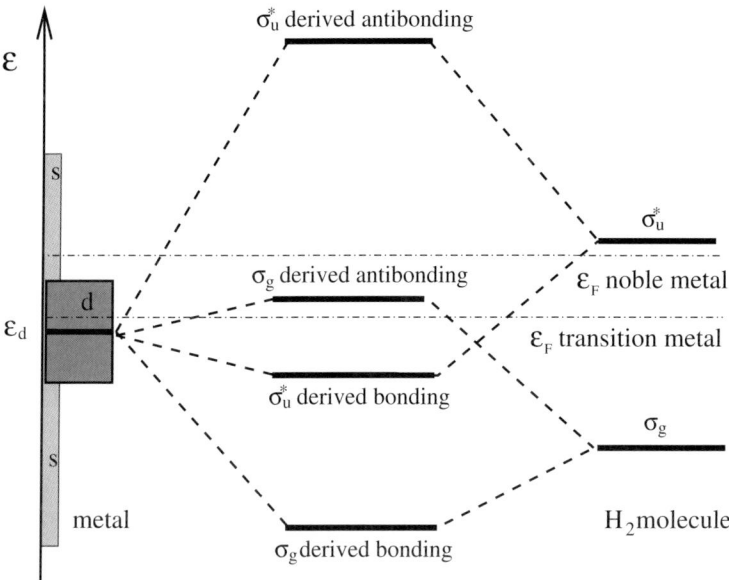

Fig. 5.14. Schematic drawing of the interaction of the H_2 σ_g and σ_u^* levels with a d-band metal surface

leads to an attraction because the σ_u^*-derived antibonding level remains unoccupied. Hence only the downshift of the bonding level contributes. Now it depends on the position of the Fermi level whether the overall interaction is purely attractive or not. For transition metals, the Fermi level lies within the d-band. For a situation like the one shown in Fig. 5.14, both the σ_g and the σ_u^*-derived bonding levels become occupied while the antibonding states stay empty. This causes an attractive interaction and is the reason for the high reactivity of transition metals. Due to the population of the σ_u^*-derived bonding level, in fact the H–H bond will be weakened and eventually broken. For a noble metal, on the other hand, the bonding *and* the anti-bonding state of the σ_g–d interaction are occupied making this interaction repulsive. This is the reason why noble metals are noble, i.e., less reactive than transition metals.

This scenario has been made quantitative in the d-band model by Hammer and Nørskov [172, 173]. Let us first consider the interaction of an atomic level with a transition metal surface which is illustrated in Fig. 5.15. This interaction is formally split into a contribution arising from the s and p states of the metal and a second contribution coming from the d-band. The s and p states lead to a broadening and a shift of the atomic level to lower energies. This broadening and shift is called *renormalization* of the energy level and can be modelled by the interaction with a jellium surface which has been discussed in detail in Sect. 5.4.

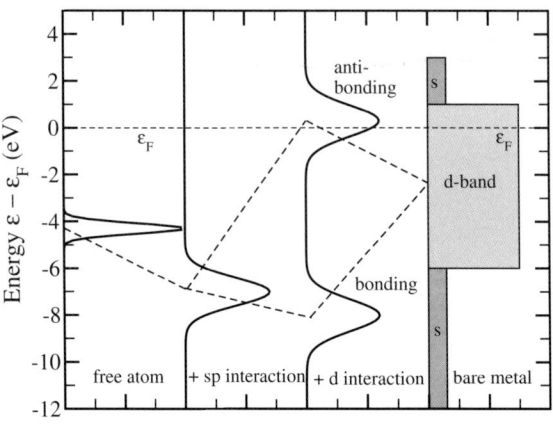

Fig. 5.15. Schematic drawing of the interaction of an atomic level with a transition metal surface

This renormalized level then splits due to the strong hybridization with the metal d-states in a bonding and an anti-bonding contribution. It depends on the position of the Fermi energy whether the anti-bonding state is fully or partially occupied or empty. Assuming that the anti-bonding state is below the upper edge of the d-band, the occupation of this state is related to the filling of the d-band. For a completely filled d-band, the overall effect of the adsorbate-d interaction is repulsive, since the up-shift of the anti-bonding state is larger than the down-shift of the bonding state.

The estimate for the additional chemisorption energy of a hydrogen atom due to the interaction with the d-band is given by [172]

$$\delta E_{\text{chem}} \approx -2(1-f)\frac{V^2}{\varepsilon_d - \varepsilon_{\text{H}}} + \alpha V^2 \,. \tag{5.71}$$

Here ε_d is the center of the d-band, ε_{H} is the *renormalized* H adsorbate resonance, and f is the filling factor of d-band. The coupling matrix element V is approximated as

$$V = \eta \frac{M_{\text{H}} M_d}{r^3} \,. \tag{5.72}$$

The potential parameters M_{H} and M_d are estimated from the isolated systems, and η is a metal-independent constant [172].

The first term in (5.71) describes the energy gain due to the interaction of the H resonance level with the d-band which depends on the filling of the d-band. The second term αV^2 gives the repulsion due to the energetic cost of the orthogonalization. For noble metals with $f = 1$ it should be the only active term. Still, irrespective of the filling factor a linear relationship between the d-band center upshift and the increase in the chemisorption strength of atomic hydrogen on metal surfaces has been found [174].

5.6 Reactivity Concepts

Table 5.3. Center of the d-band, energetic contributions and resulting reactivity measure (5.73) for the transition state configuration $(r_{H-H}, Z) = (1.2\,\text{Å}, 1.5\,\text{Å})$ in H_2 dissociative adsorption on various metals compared to the DFT total energies. All energies are given in eV. (From [172])

Metal	ε_d	V^2	$-2\frac{V^2}{\varepsilon_{\sigma_u^*}-\varepsilon_d}$	$2(1-f)\frac{V^2}{\varepsilon_d-\varepsilon_{\sigma_g}}$	αV^2	δE_{ts}	E_{ts}^{DFT}
Cu	−2.67	2.42	−1.32	0	1.02	−0.30	0.70
Cu:Cu$_3$Pt	−2.35	2.42	−1.44	0	1.02	−0.42	0.80
Pt:Cu$_3$Pt	−2.55	9.44	−5.32	−0.42	3.96	−1.78	−0.33
Pt	−2.75	9.44	−5.03	−0.44	3.96	−1.51	−0.28
Ni	−1.48	2.81	−2.27	−0.10	1.18	−1.19	−0.15
Ni:NiAl	−1.91	2.81	−1.93	−0.11	1.18	−0.86	0.48
Au	−3.91	8.10	−3.30	0	3.40	0.10	1.20

In the case of the interaction of hydrogen molecules with metal surfaces, both the renormalized H_2 bonding σ_g and the anti-bonding σ_u^* states have to be considered. The interaction with the σ_u^* state is always attractive since the σ_u^*-d antibonding level is too high in energy to become populated while for the σ_g state the filling of the d-band is again crucial. In total, the approximate reactivity measure due to the coupling of a hydrogen molecule to the metal d-bands takes the form

$$\delta E_d = -2\frac{V^2}{\varepsilon_{\sigma_u^*} - \varepsilon_d} - 2(1-f)\frac{V^2}{\varepsilon_d - \varepsilon_{\sigma_g}} + \alpha V^2. \tag{5.73}$$

This expression has been used to estimate the d-band contribution to the height of the barrier to dissociative adsorption on different metal surfaces [172]. The renormalized levels have been assumed to be at $\varepsilon_{\sigma_g} = -7\,\text{eV}$ and $\varepsilon_{\sigma_u^*} = 1\,\text{eV}$ relative to the Fermi level for all metals. The three terms of (5.73) are listed in Table 5.3. The parameter α has been used as the only fitting parameter independent of the metals. The resulting reactivity measure δE_{ts} is compared to the DFT activation barrier $\delta E_{ts}^{\mathrm{DFT}}$ for adsorption which has been calculated at an H-H bond length $r_{H-H} = 1.2\,\text{Å}$ and an H_2 center of mass distance from the surface of $Z = 1.5\,\text{Å}$.

In order to make the correlation clearer, E_{ts}^{DFT} has been plotted as a function of δE_{ts} in Fig. 5.16. There is indeed a very close correlation between δE_{ts} and E_{ts}^{DFT}. In fact, for the transition metal surfaces, the barrier is below zero indicating that hydrogen dissociates spontaneously on these surfaces. The noble metals, on the other hand, show the largest dissocation barriers.

We will now analyse the different contributions in more detail. It is obvious that the dominant attraction comes from the $\sigma_u^* - d$ interaction which is mainly due to the fact that the σ_u^* level is initially empty. The comparison between the different metal surfaces, in particular between Cu and Pt, shows that the position of the d-band center alone is not sufficient to explain the reactivity. The d-band center of Cu is in fact closer to the Fermi energy than

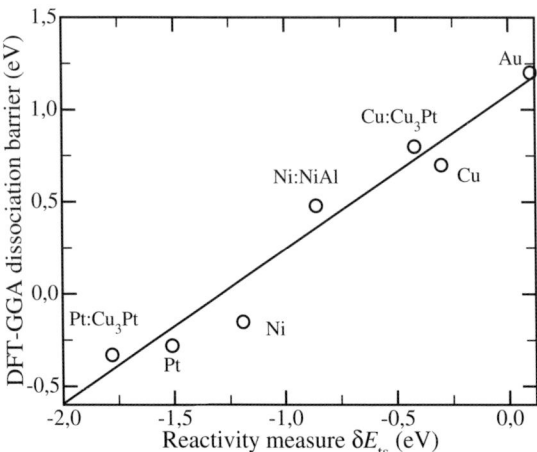

Fig. 5.16. DFT-GGA activation barrier as a function of the reactivity measure δE_{ts}. The barrier has been calculated at an H-H bond length and an H_2 center of mass distance from the surface of $(r_{\text{H-H}}, Z) = (1.2 \text{ Å}, 1.5 \text{ Å})$. (After [172])

the d-band center of Pt. Still Pt is more reactive because of the much larger matrix element V^2. Gold, on the other hand, has a similarly large matrix element, but the energy difference $\varepsilon_{\sigma_u^*} - \varepsilon_d$ appearing in the denominator of the attractive term is the highest of all the metal surfaces considered. Therefore the relatively large attractive term is overcompensated by the repulsive term causing a high dissociation barrier.

5.7 Adsorption on Low-Index Surfaces

After the presentation of the concepts in the bond-making at surfaces we will now discuss specific adsorption systems. Still the chosen systems will be used to introduce typical properties and general mechanisms occuring in adsorption at surfaces. Instead of addressing a broad variety of systems, we will first concentrate on atomic and molecular adsorption on well-defined low-index single-crystal surfaces that serve as model systems. With low-index surfaces, for example the (111), (100) and (110) surfaces of fcc crystals are meant.

One particular well-studied system, experimentally as well as theoretically, is the hydrogen adsorption on Pd surfaces. The interest was, among other reasons, motivated by the fact that bulk palladium can absorb huge amounts of hydrogen so that it was considered as a possible candidate for a hydrogen storage device needed for fuel cell technology. Although the absorption of hydrogen in the bulk is exothermic related to free hydrogen molecules, hydrogen adsorption on the surface is energetically even more favourable [175]. In Fig. 5.17a we have plotted the adsorption energy of atomic

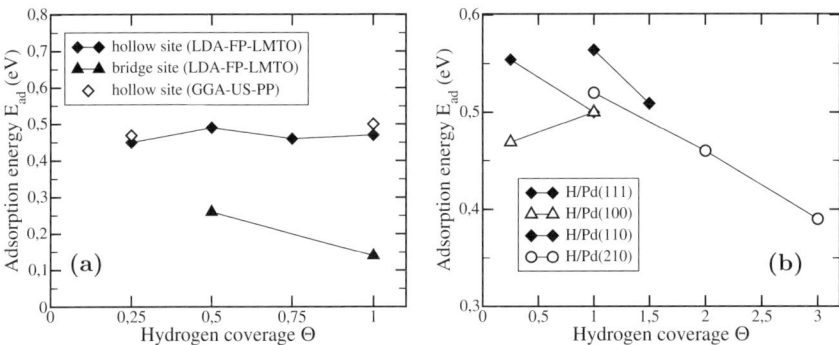

Fig. 5.17. Coverage dependence of the adsorption energy of atomic H on various Pd surfaces determined by DFT calculations. (**a**) H adsorption energy on Pd(100) on the hollow and bridge site determined by LDA-FP-LMTO calculations [175] and by GGA-US-PP calculations [176]. (**b**) H adsorption energies on the most favorable adsorption sites on Pd(100) and Pd(111) [176], on Pd(110) [177], and on Pd(210) [67] (note the different energy scale)

H on Pd(100) in the fourfold hollow position as a function of the coverage determined by LDA-FP-LMTO calculations [175] and by GGA calculations using ultrasoft pseudopotentials (US-PP) [176]. It is obvious that the adsorption energy depends only weakly on the coverage. In fact, the adsorption energy is largest for $\theta = 0.5$ corresponding to a $c(2 \times 2)$ structure, and also the (1×1) hydrogen overlayer has a higher adsorption energy per hydrogen atom than the (2×2) structure.

Figure 5.17a) also includes the hydrogen adsorption energies on the bridge site as a function of coverage. First of all it is evident that hydrogen prefers highly coordinated sites since the adsorption energy on the two-fold coordinated bridge site is significantly lower than on the fourfold-hollow site. This trend is also true for the hydrogen adsorption on Pd(111) and Pd(110) where also the highly coordinated adsorption sites are preferred [178]. However, in contrast to the fourfold hollow site on Pd(100), on the bridge site the adsorption energy decreases with increasing coverage (Fig. 5.17a) indicating repulsion between the H atoms. This repulsion can actually be traced back to the dipole-dipole interaction between adsorbed hydrogen atoms. On the

Table 5.4. Adsorption energies E_{ads}, adsorption heights h_0 and adsorbate-induced work function change $\Delta \Phi$ calculated for the adsorption of (1×1)H monolayer on Pd(100). (From [175])

Site	E_{ads} (eV)	h_0 (Å)	$\Delta \Phi$ (meV)
hollow	0.47	0.11	180
bridge	0.14	1.01	390

bridge site the hydrogen atoms are located 1 Å above the surface layer (see Table 5.4). Due to a partial charge transfer from the surface there is a dipole moment associated with the hydrogen atoms. This is confirmed by the calculated hydrogen-induced work function change of 390 meV which is also given in Table 5.4. The dipole-dipole interaction is repulsive leading to the decrease in the adsorption energies for increasing coverage.

At the fourfold hollow site, on the other hand, the hydrogen atom is located at almost the same height as the surrounding Pd atoms. Therefore there is almost no dipole moment associated with the hydrogen atoms which is reflected by the much lower work function change. The hydrogen atoms are effectively screened by the surrounding Pd atoms leading to a much smaller coverage dependence of the adsorption energy.

The fourfold hollow adsorption site on Pd(100) shows in fact an exceptional coverage dependence. This is demonstrated in Fig. 5.17b where the atomic hydrogen adsorption energies for the most favorable adsorption sites on Pd(100) and Pd(111) [176], Pd(110) [177] and Pd(210) [67] determined by GGA-DFT calculations have been plotted as a function of the coverage. Note that since the coverage θ is related to the surface unit cells whose area is not the same for the different surfaces, the same coverage does not necessarily correspond to the same density. Still the general trend, with the exception of Pd(100), is obvious: hydrogen atoms on the surface repel each other so that the hydrogen adsorption energy decreases for increasing coverage.

This trend is also true for Pd(110). As far as the hydrogen adsorption on Pd(110) is concerned, there is another very interesting phenomenon occuring, namely adsorbate-induced reconstructions. Strongly interacting adsorbates like sulphur, oxygen, carbon and nitrogen can induce a restructuring of the surface [179]. As far as (110) metal surfaces are concerned, $5d$ fcc metal surfaces such as Au(110) or Pt(110) undergo a spontaneous reconstruction, whereas clean $3d$ and $4d$ metal surfaces such as Ni(110) and Pd(110) are stable in the unreconstructed (1×1) structure. However, already a relatively weakly chemisorbed species such as hydrogen induces a surface reconstruction of the Pd(110) surface. The hydrogen induced reconstruction is very sensitive to the hydrogen coverage. At coverages up to $\theta = 1$, unreconstructed Pd(110) surfaces have been found [180]. Experimentally, however, it is very hard to determine the positions of hydrogen atoms on the reconstructed surfaces because they scatter electrons only very weakly. Therefore hydrogen is almost invisible for experimental methods using electron diffraction. This calls for theoretical support, and indeed, the hydrogen induced polymorphism has been studied in detail by DFT calculations [177, 181].

For one monolayer H on unreconstructed Pd(110), DFT calculations find that the (1×1) structure shown in Fig. 5.18a is not stable [177]. Hydrogen atoms adsorb rather in a (2×1) structure illustrated in Fig. 5.18b which is more stable by 29 meV/atom. The driving force is again the dipole-dipole repulsion between the adsorbed hydrogen atoms. In the (2×1) structure

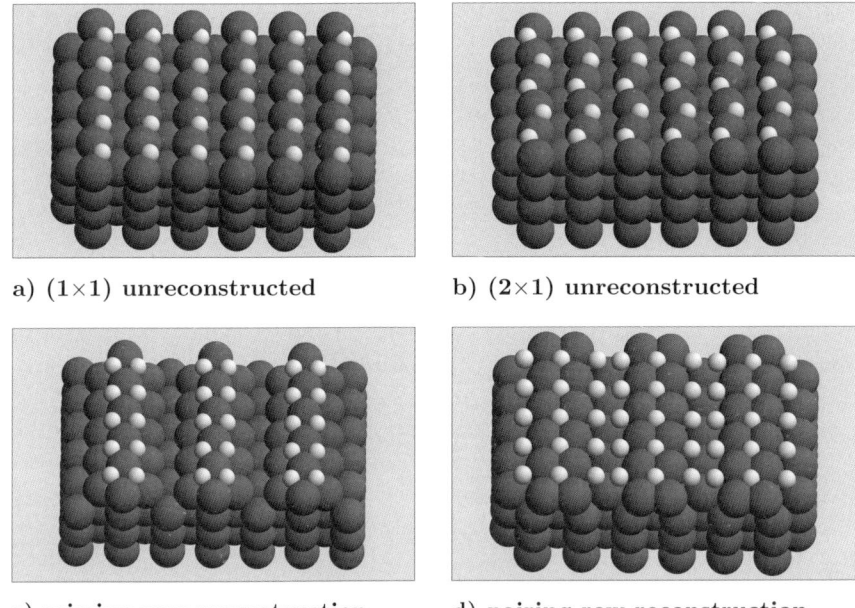

a) (1×1) unreconstructed   b) (2×1) unreconstructed

c) missing-row reconstruction   d) pairing-row reconstruction

Fig. 5.18. Hydrogen adsorption structures on Pd(110). (**a**) (1 × 1) unreconstructed structure for a hydrogen coverage of $\theta = 1$; (**b**) (2×1) unreconstructed surface for $\theta = 1$; (**c**) hydrogen-induced missing-row reconstruction for $\theta = 1$; (**d**) hydrogen-induced pairing-row reconstruction for $\theta = 1.5$. (After [177])

consisting of zigzag chains of H atoms the distance between the hydrogen atoms on the surface is maximized. In addition, the hydrogen atoms are effectively screened by the Pd atoms in the top layer in this configuration which further reduces the repulsion. This (2×1) structure is indeed verified by the experiment [180].

Later experiments found that already at a hydrogen coverages at and below $\theta = 1$ adsorbate-induced reconstructions can occur [179]. Several different hydrogen adsorbate structures on reconstructed surfaces have been considered in the calculations. The most stable one for $\theta = 1$ is shown in Fig. 5.18c which is energetically more favorable by 62 meV/atom than the (2×1) superstructure on the unreconstructed surface. The structure corresponds to a missing-row reconstruction where every second row of Pd atoms on the (110) surface is missing. Upon this reconstruction, close-packed (111) facets are formed at the slopes of the V-shaped troughs. The particular hydrogen configuration maximizes the distance between the hydrogen atoms in the same trough thus minimizing their electrostatic repulsion.

Although this hydrogen-induced missing-row reconstruction is most stable, it is kinetically hindered which means that it is separated from the (2×1)

superstructure by a sufficiently high energetic barrier so that the unreconstructed surface is metastable at low temperatures and only reconstructs when the temperature is increased [177].

Of course the question arises about the driving force that makes the missing-row reconstruction stable. The electronic structure calculations show that there is a general downshift of the energy levels upon hydrogen adsorption on the reconstructed surface compared to the unreconstructed surface. Thus it is the better adsorbate-substrate interaction that stabilizes the missing-row reconstruction [177].

Experimentally, a (1×2) pairing-row reconstruction has also been observed [180]. Its structure is illustrated in Fig. 5.18d. It is obtained from the unreconstructed surface by pushing two adjacent Pd rows together. Furthermore, Fig. 5.18d illustrates the hydrogen adsorption pattern at a coverage of $\theta = 1.5$. There is also a stable pairing-row reconstruction for $\theta = 1$ which is equivalent to the structure for $\theta = 1.5$ with just the hydrogen atoms in the fourfold hollow positions between the paired rows missing. However, at both coverages the pairing-row reconstructions are only meta-stable with respect to the missing-row reconstructions which are energetically still more favorable.

As far as the electronic structure of the hydrogen-covered Pd(110) surface in the pairing-row reconstruction is concerned, there is no indication of a significant downshift of the energy levels compared to the unreconstructed case. Hence electronic structure effects cannot be responsible for this reconstruction. An inspection of Fig. 5.18d shows that by pushing the Pd rows together, the hydrogen atoms in the threefold coordinated sites can increase their distance with respect to each other and thereby reduce their mutual repulsion. Hence the driving force for this reconstruction is mainly due to the electrostatic adsorbate-adsorbate repulsion.

Usually hydrogen molecules adsorb dissociatively on metal surfaces [182]. In order to determine whether the dissociation of molecules on surfaces is activated or non-activated, the potential energy surface (PES) of the molecule approaching the surface has to be determined. In Fig. 5.19 contour plots along two-dimensional cuts through the six-dimensional coordinate space of H_2 interacting with (100) metal surfaces, so-called elbow plots, determined by GGA calculations [76, 183] are shown. The coordinates in the figure are the H_2 center-of-mass distance from the surface Z and the H–H interatomic distance d. The lateral H_2 center-of-mass coordinates in the surface unit cell and the orientation of the molecular axis, i.e., the coordinates X, Y, θ, and ϕ are kept fixed for each 2D cut and depicted in the insets. Figure 5.19a and c represent so-called h–b–h dissociation paths which means that the H_2 center of mass is located above the bridge site and the two hydrogen atoms are oriented towards the hollow sites. The dissociation paths shown in Fig. 5.19b and d correspond to the h–t–h configuration, i.e., here the H_2 center of mass is located above the top site.

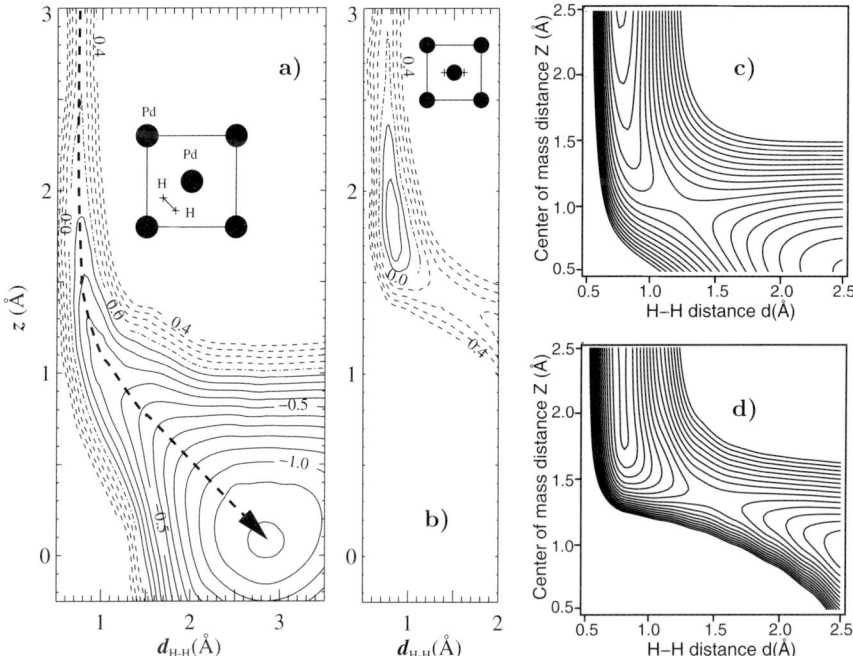

Fig. 5.19. Contour plots of the potential energy surface along two-dimensional cuts through the six-dimensional coordinate space of H$_2$ in front of (100) metal surfaces determined by DFT-GGA calculations. The contour spacing is 0.1 eV per H$_2$ molecule. (**a,b**) Elbow plots for the h–t–h and h–b–h geometry of H$_2$/Pd(100), respectively. The molecular configuration is shown in the insets (after [183]); (**c,d**) the corresponding elbow plots for the same geometries in the system H$_2$/Cu(100) [76]

Contour plots of the potential energy surface of two prototype systems for the interaction of molecules with surfaces, H$_2$/Pd(100) and H$_2$/Cu(100), are presented in Fig. 5.19. When the H$_2$ molecule approaches the Pd(100) surface with its center of mass above the bridge site and the H atoms pointing towards the four-fold hollow sites, the molecule can dissociate spontaneously without any hindering barrier. However, dissociative adsorption corresponds to a bond making–bond breaking process that depends sensitively on the local chemical environment. Indeed, if the molecule comes down over the on-top site, the shape of the PES looks entirely different. First the molecule is attracted towards the surface, in fact even more than above the bridge site, but then it encounters a barrier of above 0.15 eV towards dissociation. The minimum in Fig. 5.19b is closely related to the molecular dihydride PdH$_2$ configuration which is stable in the gas phase [184]. It does not, however, correspond to a local minimum but rather to a saddle point in the multi-dimensional PES. The H$_2$ molecule can still move further downhill in the energy landscape if the center of mass of the H_2 molecule moves laterally away from the on-top position [185].

While the dissociation of H_2 on Pd(100) can happen spontaneously, on Cu(100) it is hindered by significant energetic barriers. The minimum dissociation barrier is about 0.5 eV along a h–b–h path which is shown in Fig. 5.19c. Above the top site (Fig. 5.19d), the barrier is 0.7 eV which is only slightly higher compared to the bridge site. The difference in the barrier height for H_2 dissociation between the two systems Pd and Cu can be well understood within the d-band model (5.73). Cu with the completely filled d-band is much less reactive than the transition metal Pd. Interestingly enough, apart from the difference in the barrier heights the shape of the minimum energy paths shows similar features. For Pd(100) as well as for Cu(100), the dissociation path above the top site is energetically less favorable, the minimum path is further away from the surface compared to the h–b–h path, and the curvature is stronger for the h–t–h path.

It is important to note that the PES does not only depend on the lateral position of the H_2 molecule, i.e., the PES is not only corrugated. It is also highly anisotropic which means that the interaction of the H_2 with the metal surfaces strongly depends on the orientation of the molecular axis. Only molecules with their axis parallel to the surface can dissociate, for molecules approaching the Pd surface in an upright orientation the PES is purely repulsive [183]. For H_2 adsorption on Pd(111) and Pd(110) the potential energy surfaces look similar [186, 187].

Besides the interaction of H_2 with metal surfaces, the chemisorption of CO on transition metals belongs to the most extensively studied systems in surface science. In particular, the system CO/Pt(111) has attracted a lot of attention, certainly also motivated due to its relevance for the processes occuring in the car exhaust catalyst.

In the spirit of the frontier orbital concept (see p. 123), Blyholder has proposed [188] that only the 5σ and the 2π orbitals play a role in the bonding of CO to the surfaces. These states correspond to the HOMO and LUMO in this system, respectively. The CO 5σ orbital which is completely filled in the gas phase becomes partially empty upon the interaction with the metal surface ("donation") whereas the originally empty 2π orbitals become partially filled ("backdonation"). The lower lying 4σ and 1π levels of CO remain filled upon chemisorption and thus do not contribute to the bonding.

The binding of CO to the on-top site of Pt(111) has been addressed by LDA-DFT calculations [189]. CO adsorbs perpendicularly on Pt(111) with the C-end down. A gross population analysis similar to the Mulliken population analysis well-known in quantum chemistry showed that the 5σ and the 2π states are occupied by 1.47 and 0.52 electrons thus confirming the Blyholder model.

The bonding and antibonding character of the levels has been analysed in more detail by determining the *crystal orbital overlap population* (COOP)

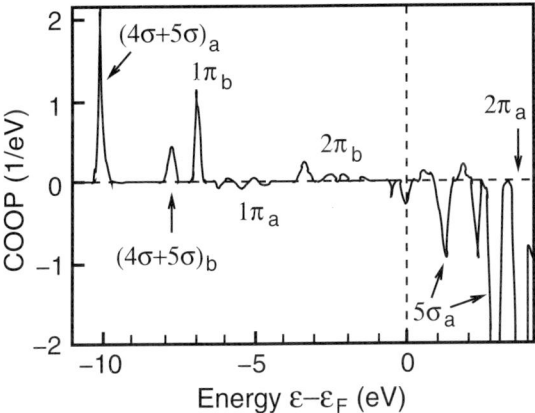

Fig. 5.20. Crystal orbital overlap population (COOP) curve with respect to the bond between CO and Pt(111) determined by LDA-DFT calculations [189]. Positive and negative values correspond to bonding and antibonding contributions which have been labeled by b and a, respectively

[66]. Positive values of the COOP indicate bonding contributions to the bond whereas negative values correspond to antibonding contributions. The COOP determined by the LDA-DFT calculations [189] is plotted in Fig. 5.20. Upon the interaction of CO with Pt(111), the 4σ and 5σ states hybridize with each other. There are two states derived from this hybridzation which are denoted by $4\sigma + 5\sigma$. Still, a detailed analysis of the electronic structure revealed that the bonding contribution from the $(4\sigma + 5\sigma)$ states is mainly due to the interaction of the 5σ orbital with the Pt substrate bands. Figure 5.20 shows that there are in fact 1π derived states that lie higher in energy than the 5σ states. This means that the ordering of these levels is reversed compared to the gas phase which has also been observed for CO adsorption on other metal surfaces [139]. The bonding and antibonding contributions of the 1π derived peaks effectively cancel each other leading to a non-bonding character, as it is expected from completely filled states. However, there is a net bonding contribution from the 2π states which are partially occupied. This partial occupation also leads to a slight elongation of the CO bond length from its calculated gas-phase value of 1.131 Å to 1.143 Å, since the 2π level is of antibonding character with respect to the C–O bond. The COOP analysis again confirms that mainly the 5σ and the 2π orbitals are involved in the CO bonding to Pt(111).

However, there are disturbing results as far as the energetics of the CO bonding to Pt(111) is concerned [190]. Experimentally it is well established that the CO molecule adsorbs at the on-top site on Pt(111) [191]. For this site, GGA-DFT yields a CO binding energy of 1.5 eV [192]. However, no mattter whether DFT calculations are based on LDA or GGA functionals, whether pseudopotentials are employed or all electrons are taken into account, all these calculations yield the fcc hollow site as the most favorable adsorption site [190]. On the average, the fcc hollow site is prefered by about 0.2 eV compared to the atop-site. No "stone was left unturned" [190] in order to determine the reason for the discrepancy between theory and experiment, but neither defect structures nor contaminations nor relativistic or spin effects nor zero-point energies can account for the difference. This is an obvious warning that the results of present-day DFT implementations still have to be accepted with caution.

So far, I have discussed the atomic and molecular adsorption on low-index crystal surface. For years, surface science studies have concentrated on such adsorption systems, and many important concepts have been gained from these investigations, as demonstrated in this section. However, surfaces under realistic conditions are neither microscopically flat nor are they clean and free of adsorbates. This difference between the ideal surfaces subject to most of the academic research and realistic surfaces has been termed the *structure gap*. There has been a long, ongoing debate whether the insight gained from studies of clean, low-index crystal surfaces can be directly applied to surfaces present, e.g., in heterogeneous catalysis. However, the surface science approach can still be used in order to close the structure gap, namely by studying adsorbate-covered and structured, defect-rich surfaces under well-defined conditions so that the influence of the particular co-adsorbate or the specific surface defect on adsorption properties can be derived. There is an increasing number of studies devoted to complex structures which will be addressed in the next two sections.

5.8 Adsorption on Precovered Surfaces

The modification of the reactivity of a surface structure by the presence of adsorbates is of great technological relevance, especially in heterogeneous catalysis. These processes do usually not occur under vacuum conditions so that the presence of co-adsorbates cannot be avoided. On the other hand, further reactants might be deliberately added to a reaction chamber since adsorbates can either promote or poison a particular reaction on the surfaces. The most prominent example for the reduction of the activity is the poisoning of the platinum-based car-exhaust catalyst by lead present in the gasoline. But not only lead, also sulfur causes a reduction of the efficiency of the car-exhaust catalyst.

In fact, the poisoning effect of sulfur is not restricted to oxidation reactions on platinum surfaces. On Pd(100), sulfur adsorption leads to a significant reduction of the hydrogen dissociation probability [193, 194]. In order to examine the microscopic effects of sulfur preadsorption, the potential energy surface of the H_2 dissociation on the $p(2 \times 2)$ and $c(2 \times 2)$ sulfur-covered Pd(100) surface has been determined in great detail by DFT studies [195,196]. On $p(2 \times 2)$S/Pd(100), H_2 dissociation is no longer non-activated as on the

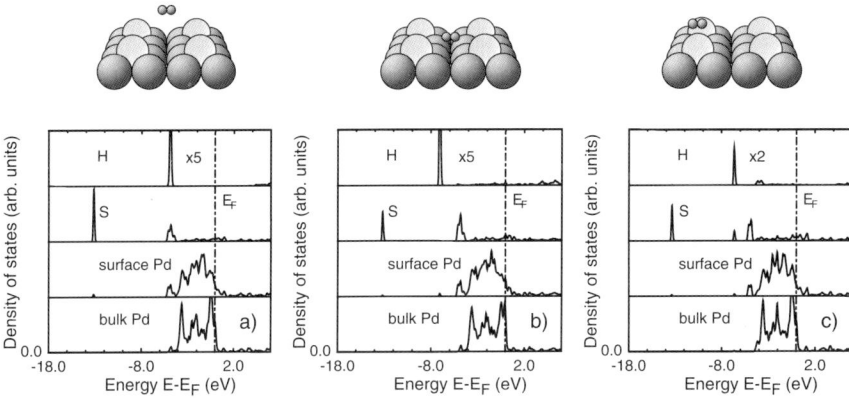

Fig. 5.21. Local density of states for three different configurations in the hydrogen dissociation process on (2×2) sulfur-covered Pd(100). The upper panels indicate the corresponding position of the hydrogen molecule. (After [196])

clean Pd(100) surface but hindered by a barrier of 0.1 eV. The closer the hydrogen molecules comes to the adsorbed sulfur atoms, the higher the barriers are for hydrogen dissociation indicating a strong repulsion between the adsorbed sulfur and H_2. On the $c(2 \times 2)$ sulfur-covered Pd(100) surface, the density of sulfur atoms is so high that the H_2 dissociation is effectively blocked by dissociation barriers larger than 2 eV [196].

The electronic factors influencing the reactivity of the sulfur-covered Pd(100) surface have been determined in detail by analysing the local density of state of the interacting system. In Fig. 5.21, the density of states projected onto the hydrogen, sulfur and palladium atoms is plotted for three different configurations. Figure 5.21a corresponds to the non-interacting system with the hydrogen molecule still far away from the surface. The prominent peak in the hydrogen 1s density of states is given by the bonding σ_g state. It seems to be in resonance with a sulfur-related state at the same energy, however, this is just coincidental. This is confirmed by Fig. 5.21b which shows the projected density of states for the hydrogen molecule located at the minimum barrier for dissociation. The hydrogen and sulfur-related states are no longer in resonance indicating that there is no direct interaction between the hydrogen molecule and the sulfur atoms for this configuration. In fact, it turns out that the building up of the minimum barrier is an indirect effect of the sulfur adsorption [196]: sulfur adsorption leads to a downshift of the Pd d-states in the surface which makes the surface more repulsive with respect to hydrogen dissociation according to the d-band model (5.73) [172, 173]. On the other hand, when the H_2 molecule directly approaches a sulfur atom (Fig. 5.21c), there is a strong hybridization and splitting in bonding and antibonding states between the H_2 and the sulfur orbitals. Since both the bonding

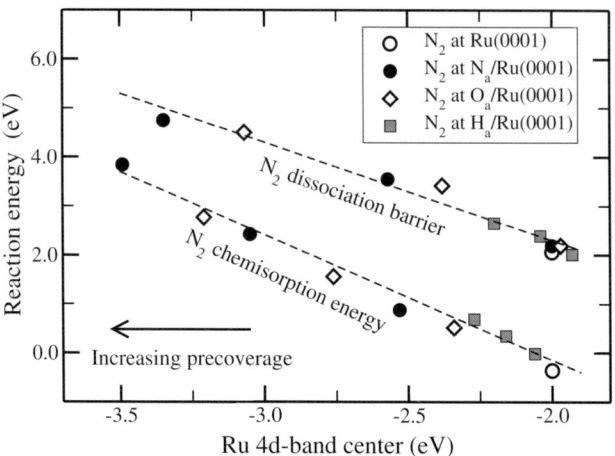

Fig. 5.22. Dependence of the dissociation barriers and the chemisorption energies on the Ru d-band center for N_2 adsorption on clean and precovered Ru(0001) surfaces. The considered precoverages correspond to 1/4, 1/2 and 3/4 monolayers. (After [197])

and anti-bonding contribution are fully occupied, the interaction is strongly repulsive [172].

The d-band model has also been used to analyse the influence of coadsorbates on the dissociative adsorption of N_2 on Ru(0001) [197], a system which is of relevance for the ammonia synthesis. Precovering the Ru(0001) surface with atomic nitrogen, oxygen or hydrogen shifts the center of the Ru d-band to lower energies, i.e., leads to a poisoning of the surface. This is illustrated in Fig. 5.22 where the dissociation barriers and the atomic chemisorption energies are plotted as a function of the Ru d-band center. The considered precoverages correspond to 1/4, 1/2 and 3/4 monolayers. It is apparent that nitrogen preadsorption has the strongest influence on the electronic structure of the Ru surface atoms while the effect of hydrogen adsorption is relatively moderate. Both the dissociation barriers and the atomic chemisorption energies show a linear relationship with respect to the Ru d-band center, regardless of the chemical nature of the coadsorbate. This confirms the universal role of the d-band center for the comparison of the reactivity of related adsorption systems. Consequently, there is also a linear relationship between the N_2 dissociation barrier and the atomic N chemisorption energies. This so-called *Brønsted–Evans–Polanyi relation* has in fact been found for a number of transition metal surfaces [198].

5.9 Adsorption on Structured Surfaces

The activity of realistic catalysts is often assumed to be dominated by so-called *active sites*, i.e., sites with a specific geometric configuration that modifies their electronic and chemical properties. In order to identify the nature of these active sites, surfaces with well-defined defect structures have been investigated. Vicinal surfaces are particularly well-suited since they can be relatively easily prepared in the experiment, they are accessible by electronic structure calculations, and they allow to determine the role of steps in the interaction of atoms and molecules with surfaces.

In fact, many adsorbates bind much stronger to step sites than to sites on a flat terrace. Again, the CO/Pt system will serve as a model system. Variations of 1 eV in the adsorption energies of CO on on-top sites of several flat, stepped, kinked and reconstructed Pt surface have been found by DFT-GGA calculations [192] revealing a strong structure sensitivity of the binding strength. As far as stepped surfaces are concerned, the Pt(211) and Pt(11,7,5) surfaces have been considered. Both surfaces have (111) terraces of similar width, but while the (211) surface is close-packed along the steps, the (11,7,5) surface has a open kinked structure along the steps (see Fig. 5.23). And indeed, the lowest-coordinated Pt atoms which are the kink atoms of the (11,7,5) surface show the strongest binding to CO with bonding energies that are about 0.7 eV stronger than on the flat Pt(111) terrace. These findings have again been rationalized using the d-band model [192]. The lower the coordination, the larger the d-band shift and consequently the higher the adsorption energy.

On the other hand, the Pt(100) surface exhibits in equilibrium a Pt(100)-hex reconstruction which is an otherwise flat (100) surface covered by a hexagonally packed, buckled Pt overlayer. This overlayer is buckled because the Pt density in the overlayer is 4% higher than in the Pt(111) surface. This

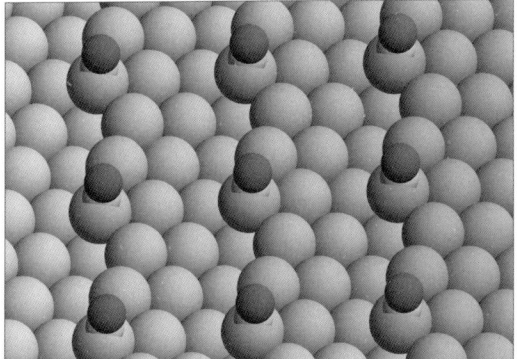

Fig. 5.23. CO adsorption on top of the kink sites at the steps of a Pt(11,7,5) surface. CO binds with the C end down

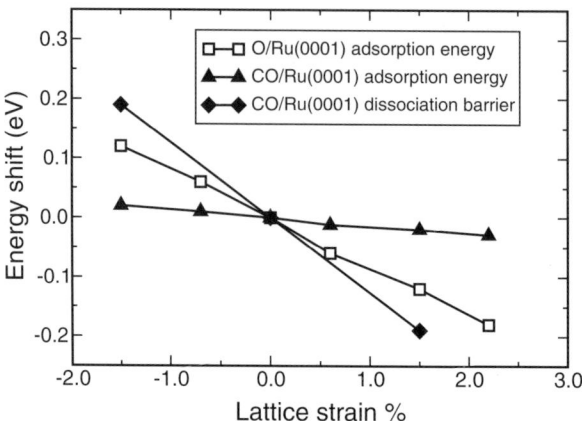

Fig. 5.24. Dependence of the O and CO adsorption energy and the CO dissociation barrier on Ru(0001) on the lattice strain. The results have been obtained by DFT-GGA calculations [199]

larger density has the same effect as a higher coordination. Because of the increased overlap the d-band broadens and shifts to lower energies making it less reactive. This is exactly what has been found for the binding of CO on the Pt(100)-hex(1×5) surface which is weaker by 0.1 eV compared to the Pt(111) surface.

Similar effects can be induced by applying either compressive or tensile stress to surfaces. By implementing subsurface argon bubbles at a Ru(0001) surface, laterally stretched and compressed surface regions have been created [200, 201]. This modified surface was then exposed to oxygen and CO. STM images confirmed that oxygen atoms and CO molecules adsorb preferentially in the regions of the expanded lattice.

These findings have been rationalized by DFT-GGA calculations. The adsorption energies of O and CO on Ru(0001) and the CO dissociation barrier have been calculated as a function of the lattice strain (Fig. 5.24) [199]. In general, the surface reactivity increases with lattice expansion. This can be explained by the accompanying upshift of the d-band center. Increasing the distance between the substrate atoms lowers the overlap between the metal orbitals which narrows the width of the d-band. If the band is more than half-filled, charge conservation than leads to the upshift of the d-band.

Figure 5.24 demonstrates that the qualitative trend due to the strain is the same for the O and CO adsorption energy and the CO dissociation barrier, but the quantitative effect can be quite different. Still the strain effect could be used as a means to tailor the catalytic activity of transition metals since, within certain limits, the lattice constant can be modified by the heteroepitxial growth of a metal overlayer on another metal with a lattice mismatch.

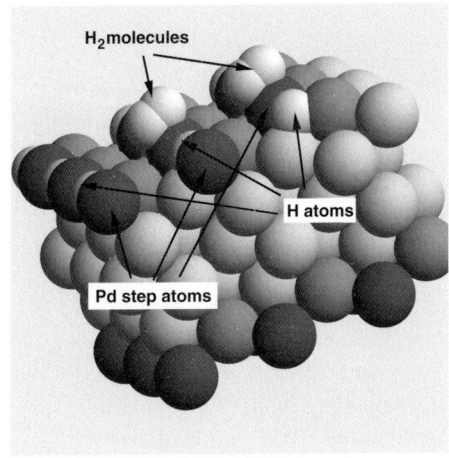

Fig. 5.25. Side view of the Pd(210) surface with a monolayer H precoverage and the molecular H_2 adsorption state [67]

Stepped surfaces do not only show higher adsorption energies, they can also induce unusual adsorption structures. While hydrogen usually adsorbs dissociatively at metal surfaces [182], as already mentioned, experiments have found the coexistence of chemisorbed hydrogen atoms and molecules on Pd(210) [202]. The microscopic nature of the adsorbate states has been identified by DFT-GGA calculations [67, 202].

The (210) surface is a relatively open surface that can be regarded as a stepped surface with a high density of steps. The geometry of this surface is shown in Fig. 5.25. Vicinal fcc(n10) surfaces have (100) terraces with steps running along the [001] direction. These steps are forming open (110)-like microfacets. The most favourable adsorption site for atomic hydrogen is in fact the long-bridge position between two Pd step atoms, as indicated in Fig. 5.25. Still two other atomic adsorption sites are available in the (210) surface unit cell.

On the hydrogen precovered Pd(210) surface, H_2 can chemisorb molecularly over the Pd step atoms with a binding energy of 0.27 eV [67, 202]. However, this molecular state which is also illustrated in Fig. 5.25 is only stabilized by the presence of atomic hydrogen on the surface. Without any precoverage, H_2 does spontaneously dissociate on Pd(210). The preadsorbed atomic hydrogen does not significantly disturb the interaction of the H_2 molecules with the step Pd atoms but hinders the H_2 dissociation on Pd(210). In fact, the molecular adsorption state corresponds locally to the apparent well on Pd(100) shown in Fig. 5.19b which is changed from a saddle point to a local minimum on Pd(210) due to the presence of the atomic hydrogen. Such unique features of structured surfaces might be useful for catalyzing certain reactions in which, e.g., relatively weakly bound hydrogen molecules are required.

5.10 Reactions on Surfaces

The dissociative adsorption of diatomic molecules corresponds to a simple reaction on a surface since it involves a bond-breaking and bond-making process. Technologically relevant reactions on surfaces, however, are usually much more complex (see Sect. 7.6). The phase space of reactions on surfaces involving more than two adsorbate atoms is already so high-dimensional that studying these reactions by ab initio eletronic structure calculations is computationally very demanding.

Nonetheless, in order to understand catalytic reactions, the study of dissociative adsorption on surfaces is not sufficient. In fact, DFT studies have already addressed more complex reactions on surfaces and thus contributed to the elucidation of catalytic reaction channels. I will focus on a relatively simple, but still technologically immensely important reaction, the CO oxidation on Pt(111). Transition metals such as Pt, Pd, or Rh are the active components in car exhaust catalytic converters which remove CO and other pollutants from exhaust emission. The CO oxidation on Pt(111) represents the model systems for the understanding of the post-combustion oxidation of CO. Note that the reaction $CO + \frac{1}{2}O_2 \rightarrow CO_2$ is strongly exothermic ($\Delta H \approx 3\,\mathrm{eV}$), but it is hindered by high energetic barriers in the gas phase. The role of the catalyst is to provide a route for the CO oxidation with a much smaller activation barrier (see Fig. 5.2).

Two different CO oxidation paths on Pt(111) have been considered by DFT calculations: CO oxidation by adsorbed *atomic* oxygen [203, 204] and by adsorbed molecular oxygen [204]. Figures 5.26a–c show the initial, transition and final state of the CO oxidation by adsorbed atomic oxygen on Pt(111) [203], $CO^{ads} + O^{ads} \rightarrow CO_2^{ads}$. The minimum energy path has been found using a constraint minimization scheme. In the initial state, atomic oxygen is located in a three-fold hollow position whereas CO is adsorbed perpendicularly at the on-top position. Along the minimum energy path, the CO moves via the bridge site to the on-top position adjacent to the O atom, thereby pushing the O atom towards the neighbouring bridge site (Fig. 5.26b). Finally the adsorbed O atom and the CO molecule move together and form a new bond whereby the bond between the O atom and the Pt substrate is broken.

The transition state geometry of CO oxidation on Pt(111) can be characterized by a bonding competition effect [203]. The oxygen atom prefers an adsorption site with high coordination. On the other hand, there is a strong indirect repulsion mediated by the surface d-band between the oxygen atom and the CO molecule if they share a surface atom that they are bind to [205]. Hence the transition state geometry corresponds to a compromise between allowing the reacting atoms to interact with as many distinct surface atoms as possible while still approaching each other. In fact, this bonding competition effect also determines the transition state geometries for CO dissociation on a number of other transition metal surfaces [205, 206]. In addition, it con-

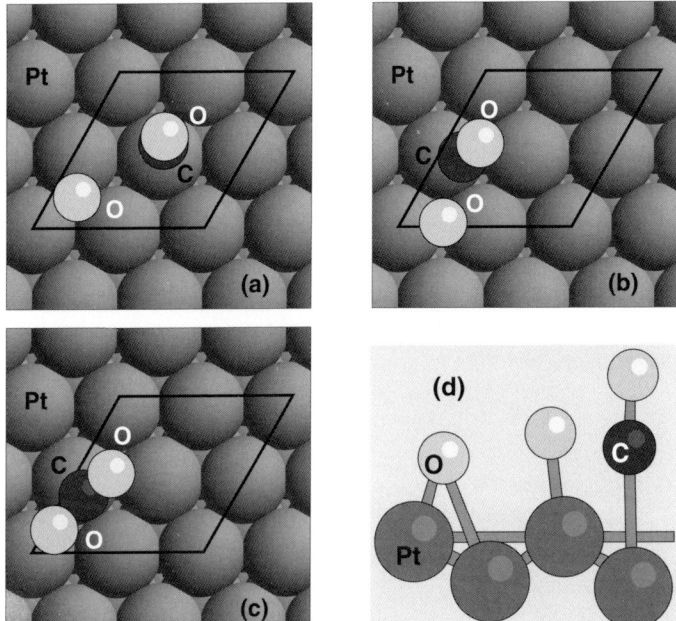

Fig. 5.26. CO oxidation on Pt(111) studied by DFT calculations. Panels (**a–c**) show top views of the initial, transition and final state, respectively, along the reaction path of the CO oxidation by adsorbed atomic oxygen [203], whereas panel (**d**) shows a side view of the transition state of the CO oxidation by adsorbed molecular oxygen [204]. In panels (**a–c**) the surface unit cell used in the calculations is indicated

tributes to the high reactivity of stepped surfaces since the reacting atoms can often maintain a high coordination along reaction paths close to the steps due to the open structure, as was confirmed in a DFT study of the NO+CO reaction catalyzed by flat and stepped palladium surfaces [207].

Returning to the CO oxidation on Pt(111), we now focus on the CO oxidation by an adsorbed O_2 molecule, $CO^{ads} + O_2^{ads} \rightarrow CO_2^{ads} + O^{ads}$. The transition state geometry is plotted in Fig. 5.26d. Prior to the CO oxidation, the adsorbed O_2 molecule dissociates. At the transition state, one O atom is located above a bridge site, the other one at an on-top position [204]. The CO molecule then reacts with this more weakly bound O atom to form CO_2. This process is hindered by a barrier of 0.46 eV [204] and can be associated with the α peak observed in temperature programmed desorption (TPD) experiments (see p. 203) at $T \approx 150$ K [208]. Above this temperature, O_2 starts to dissociate on Pt(111) so that the CO can only react with adsorbed oxygen atoms. Since these atoms are more strongly bound than the oxygen molecules, in fact the activation barrier for CO oxidation by adsorbed atoms is higher than for adsorbed molecules. This barrier has been calculated to be about 1.0 eV [203] or 0.75 eV [204] which corresponds to the β peaks above 200 K observed in the TPD experiments. The relatively large discrepancy between the two calculated barriers indicates that the determination of activation barriers for the catalytic oxidation by DFT methods is still far from being trivial.

Exercises

5.1 Atom-Surface van der Waals Attraction

Show that the atom-atom long-range attraction

$$V_{aI} = -\frac{C_6}{R_{aI}^6} \tag{5.74}$$

between a gas atom α and the substrate atoms I of a semi-infinite cubic lattice with lattice constant a leads to a long-range atom-surface interaction potential of

$$V_{\text{vdW}}(Z) = -\frac{\pi}{6} \frac{C_6}{Z^3} \frac{1}{a^3} \tag{5.75}$$

Hint: Express the sum over the lattice atoms as an integral and use

$$\int_0^\infty \frac{x^{\mu-1}}{(1+\beta x^p)^\nu} dx = \frac{1}{p} \beta^{-\mu/p} B(\mu/p, \nu - \mu/p) \tag{5.76}$$

with $B(x,y) = \Gamma(x)\Gamma(y)/\Gamma(x+y)$ [209].

5.2 Newns–Anderson Model

a) Prove explicitly that the projected density of states $n_a(\varepsilon)$ in a system described by the Newns–Anderson Hamiltonian

$$H = \varepsilon_a \, \hat{n}_a + \sum_k \varepsilon_k \, \hat{n}_k + \sum_k (V_{ak} \, \hat{b}_a^+ \hat{b}_k + V_{ka} \, \hat{b}_k^+ \hat{b}_a) , \tag{5.77}$$

is given by

$$n_a(\varepsilon) = \frac{1}{\pi} \frac{\Delta(\varepsilon)}{(\varepsilon - \varepsilon_a - \Lambda(\varepsilon))^2 + \Delta^2(\varepsilon)} . \tag{5.78}$$

with

$$\Delta(\varepsilon) = \pi \sum_k |V_{ak}|^2 \, \delta(\varepsilon - \varepsilon_k) \tag{5.79}$$

and

$$\Lambda(\varepsilon) = \frac{P}{\pi} \int \frac{\Delta(\varepsilon')}{\varepsilon - \varepsilon'} d\varepsilon' . \tag{5.80}$$

b) For a semi-elliptical band of width W centered at $\varepsilon = 0$, determine the density of states (normalized to unity). Determine $\Delta(\varepsilon)$ and $\Lambda(\varepsilon)$ under the assumption that $V_{ak} \equiv V$ is independent of k.

5.3 Adsorbate-Induced Change of the Density of States

The effect of an adsorbate on the electronic structure of a substrate can be modeled by free electrons scattered by a perturbing spherical potential. Show that in this model the adsorbate-induced change of the density of states can be estmated by

$$\delta n(\varepsilon) = \frac{1}{\pi} \frac{d\delta_l(\varepsilon)}{d\varepsilon}, \tag{5.81}$$

where l is the angular momentum and δ_l the associated phase shift.

Hint: The radial part of the wave function of a particle scattered at a spherical potential of finite range is asymptotically given by

$$u_l(r \to \infty) = \frac{D_l \sin(kr - l\pi/2 + \delta_l)}{kr}. \tag{5.82}$$

Assume that the system is enclosed in a sphere of Radius $R \to \infty$ so that the wave function vanishes for $r = R$.

5.4 Adsorbate Vibrations

In the harmonic approximation for a rigid substrate, the adsorption potential is in general given by

$$V_{\text{ads}}(\boldsymbol{R}_e, \boldsymbol{u}) = E_{\text{ads}}(\boldsymbol{R}_e) + Ax^2 + By^2 + Cz^2 + Dxy + Eyz + Fzx, \tag{5.83}$$

where $\boldsymbol{R}_e$ is the equilibrium position and x, y, and z are the components of the displacement vector $\boldsymbol{u}$.

a) Using symmetry, simplify (5.83) so that it is appropriate for high-symmetry sites, i.e., the on-top, bridge, three-fold and four-fold hollow site.

b) Assuming that the optimum density n_0 can be described by a superposition of exponentially decaying atomic densities of the nearest neighbors

$$n_0(\boldsymbol{R}_e) = e^{-\beta R_e} \sum_{\text{n.n.}} n_{\text{atom}}, \tag{5.84}$$

show that the perpendicular vibration frequency in the effective medium theory using the embedding energy E_c is given by

$$\omega_{\text{vib}} = \sqrt{\frac{1}{M_a} \frac{d^2 E_c(n)}{dn^2}\bigg|_{n=n_0}} \beta \, n_0 \sin \alpha \tag{5.85}$$

where α is the angle of the metal-adsorbate line with the surface plane.

c) Verify that ω_{vib} decreases when the coordination number increases.

6. Gas-Surface Dynamics

In the preceding chapter we have been concerned with the solution of the electronic Schrödinger equation for fixed nuclear coordinates. By repeating total-energy calculations for many different nuclear configurations, energy minima and whole potential energy surfaces for chemical reactions at surfaces can be determined. However, this static information is often not sufficient to really understand how a reaction proceeds. Furthermore, in the experiment the potential energy surface (PES) is never directly measured but just reaction rates and probabilities. For a real understanding of a reaction mechanism a dynamical simulation has to be performed. In addition, this allows a true comparison between theory and experiment and thus provides a reliable check of the accuracy of the calculated PES on which the dynamics simulation is based.

In this chapter methods to perform dynamical simulations will be introduced. In principle the atomic motion should be described by a quantum mechanical treatment, but often classical mechanics is sufficient. I will therefore first present classical methods and then review quantum mechanical methods.

6.1 Classical Dynamics

One can perform classical molecular dynamics studies by integrating the classical equations of motion, either Newton's equation of motion

$$M_i \frac{\partial^2}{\partial t^2} \boldsymbol{R}_i = -\frac{\partial}{\partial \boldsymbol{R}_i} V(\{\boldsymbol{R}_j\}), \qquad (6.1)$$

or Hamilton's equation of motion

$$\dot{q} = \frac{\partial H}{\partial p} \quad \dot{p} = -\frac{\partial H}{\partial q}. \qquad (6.2)$$

The solution of the equations of motion can be obtained by standard numerical integration schemes like Runge–Kutta, Bulirsch–Stoer or predictor-corrector methods (see, e.g., [16]). Very often the rather simple Verlet algorithm [210, 211] is used which is easily derived from a Taylor expansion of the trajectory.

6. Gas-Surface Dynamics

$$\boldsymbol{r}_i(t+h) = \boldsymbol{r}_i(t) + h\left.\frac{d\boldsymbol{r}_i}{dt}\right|_{h=0} + \frac{h^2}{2}\left.\frac{d^2\boldsymbol{r}_i}{dt^2}\right|_{h=0} + \frac{h^3}{6}\left.\frac{d^3\boldsymbol{r}_i}{dt^3}\right|_{h=0} + \ldots$$

$$= \boldsymbol{r}_i(t) + h\,\boldsymbol{v}_i(t) + \frac{h^2}{2}\frac{\boldsymbol{F}_i(t)}{m} + \frac{h^3}{6}\left.\frac{d^3\boldsymbol{r}_i}{dt^3}\right|_{h=0} + \ldots \quad (6.3)$$

Here we have introduced the velocity $\boldsymbol{v}_i$. Furthermore, we have used Newton's equation of motion to include the force $\boldsymbol{F}_i$ acting on the i-th particle. Analogously we can derive

$$\boldsymbol{r}_i(t-h) = \boldsymbol{r}_i(t) - h\,\boldsymbol{v}_i(t) + \frac{h^2}{2}\frac{\boldsymbol{F}_i(t)}{m} - \frac{h^3}{6}\left.\frac{d^3\boldsymbol{r}_i}{dt^3}\right|_{h=0} + \ldots \quad (6.4)$$

Adding (6.3) and (6.4) yields the Verlet algorithm [210]

$$\boldsymbol{r}_i(t+h) = 2\boldsymbol{r}_i(t) - \boldsymbol{r}_i(t-h) + \frac{h^2}{2}\frac{\boldsymbol{F}_i(t)}{m} + O(h^4). \quad (6.5)$$

The accuracy of the numerical integration of the equation of motion can be checked by testing the energy conservation. In order to evaluate the kinetic energies, the velocities at time t are needed. Note that they do not explicitly appear in (6.5). They can be estimated by

$$\boldsymbol{v}_i(t) = \frac{\boldsymbol{r}_i(t+h) - \boldsymbol{r}_i(t-h)}{2h}. \quad (6.6)$$

However, the kinetic energy evaluated with (6.6) belongs to the time step prior to the one used for the positions (6.5) which enter the evaluation of the potential energy. This problem can be avoided in the so-called *velocity* Verlet algorithm [211]

$$\boldsymbol{r}_i(t+h) = \boldsymbol{r}_i(t) + h\,\boldsymbol{v}_i(t) + \frac{h^2}{2}\frac{\boldsymbol{F}_i(t)}{m}$$

$$\boldsymbol{v}_i(t+h) = \boldsymbol{v}_i(t) + h\,\frac{\boldsymbol{F}_i(t+h) + \boldsymbol{F}_i(t)}{2m}, \quad (6.7)$$

which is mathematically equivalent to the Verlet algorithm (Problem 6.1).

In order to perform molecular dynamics simulations with the Verlet algorithm, a specific time step has to be chosen. Of course, the error associated with each time step is the smaller, the shorter the time step. On the other hand, a shorter time steps means more iterations for a given trajectory or simulation time which increases the computational cost. Furthermore, the error of each time step may accumulate. Hence the chosen time step will represent a compromise. The change in the total energy during one molecular run should be well below 1%. As a rule of the thumb, the time step should be ten times smaller than the shortest vibrational or rotational period of a given system. If, e.g., hydrogen belongs to the simulation ensemble, then usually the H-H intramolecular vibration corresponds to the fastest time scale with a vibrational period of $\tau_{\text{vib}} \approx 8$ fs, hence the time step should be shorter than 0.8 fs.

In a conservative system, the energy is conserved in a molecular dynamics run. In a thermodynamical sense this means that the phase space trajectory belongs to the microcanonical ensemble. Often it is desirable to include dissipation effects in the gas-surface dynamics simulations. The simplest way to achieve this is to add a friction term to the Hamiltonian. If, however, the substrate should not only act as an energy sink but rather as a heat bath in order to model thermalization and accommodation processes, both energy loss and energy gain processes have to be taken into account. This can be achieved by a number of techniques. The most prominent ones are the *generalized Langevin equation* approach [212] and the *Nosé thermostat* [213, 214]. In both approaches the molecular dynamics simulations sample the canonical ensemble at a specified temperature.

6.2 Quantum Dynamics

There are two ways to determine quantum mechanical reaction probabilities: by solving the time-dependent or the time-independent Schrödinger equation. Both approaches are equivalent [215] and should give the same results. The question which method is more appropriate depends on the particular problem. Time-independent implementations are usually more restrictive as far as the form of the potential is concerned, but often the choice of the method is a matter of training and personal taste.

In the most common time-independent formulation, the concept of defining one specific reaction path coordinate is crucial. Starting from the time-independent Schrödinger equation

$$(H - E)\, \Psi = 0, \tag{6.8}$$

one chooses one specific reaction path coordinate s and separates the kinetic energy operator in this coordinate

$$(\frac{-\hbar^2}{2\mu}\partial_s^2 + \tilde{H}E)\, \Psi = 0. \tag{6.9}$$

Here $\tilde{H}$ is the original Hamiltonian except for the kinetic energy operator in the reaction path coordinate. Usually the use of curvelinear reaction path coordinates results in a more complicated expression for the kinetic energy operator involving cross terms, but for the sake of clarity I have neglected this in (6.9). As the next step one expands the wave function in the coordinates perpendicular to the reaction path coordinate in some suitable set of basis functions,

$$\Psi = \Psi(s, \ldots) = \sum_n \psi_n(s)\, |n\rangle. \tag{6.10}$$

Here n is a multi-index, and the expansion coefficients $\psi_n(s)$ are assumed to be a function of the reaction path coordinate. Now we insert the expansion of

Ψ in (6.9) and multiply the Schrödinger equation by $\langle m|$, which corresponds to performing a multi-dimensional integral. Since the basis functions $|n\rangle$ are assumed to be independent of s, we end up with the so-called coupled-channel equations,

$$\sum_n \left\{ \left(\frac{-\hbar^2}{2\mu}\partial_s^2 - E\right) \delta_{m,n} + \langle m|\tilde{H}|n\rangle \right\} \psi_n(s) = 0 \,. \tag{6.11}$$

Instead of a partial differential equation – the original time-independent Schrödinger equation (6.8) – we now have a set of coupled ordinary differential equation. Still a straightforward numerical integration of the coupled-channel equations leads to instabilities, except for in simple cases, due to exponentially increasing so-called closed channels. These problems can be avoided in a very stable and efficient coupled-channel algorithm [216–218] that will be briefly sketched in the following.

For the solution Ψ defined in (6.10), which represents a vector in the space of the basis functions, the initial conditions are not specified. This function can also be considered as a matrix

$$\Psi = (\psi)_{nl} \,, \tag{6.12}$$

where the index l labels a solution of the Schrödinger equation with an incident plane wave of amplitude one in channel l and zero in all other channels. Formally one can then write the solution of the Schrödinger equation for a scattering problem in a matrix notation as

$$\begin{aligned}\Psi(s \to +\infty) &= e^{-iqs} - e^{iqs}\, r \,, \\ \Psi(s \to -\infty) &= e^{-iqs}\, t \,.\end{aligned} \tag{6.13}$$

Here $q = q_m \delta_{m,n}$ is a diagonal matrix, r and t are the reflection and transmission matrix, respectively. Now one makes the following ansatz for the wave function,

$$\Psi(s) = (1 - \rho(s))\, \frac{1}{\tau(s)}\, t \,. \tag{6.14}$$

Equation (6.14) defines the *local reflection matrix* $\rho(s)$ (*LORE*) and the *inverse local transmission matrix* $\tau(s)$ (*INTRA*). The boundary values for these matrices are (except for phase factors which, however, do not affect the transition probabilities):

$$(\rho(s); \tau(s)) = \begin{cases} (r; t) & s \to +\infty \\ (0; 1) & s \to -\infty \end{cases} . \tag{6.15}$$

From the Schrödinger equation first order differential equations for both matrices can be derived [216] which can be solved by starting from the known initial values at $s \to -\infty$; at $s \to +\infty$ one then obtains the physical reflection and transmission matrices. Thus the numerically unstable boundary value problem has been transformed into a stable initial value problem.

In the time-dependent or wave-packet formulation, the solution of the time-dependent Schrödinger equation

$$i\hbar \frac{\partial}{\partial t} \Psi(\boldsymbol{R}, t) = H\,\Psi(\boldsymbol{R}, t) \tag{6.16}$$

can formally be written as

$$\Psi(\boldsymbol{R}, t) = e^{-iHt/\hbar}\,\Psi(\boldsymbol{R}, t=0)\,, \tag{6.17}$$

if the potential is time-independent. The most common methods to represent the time-evolution operator $\exp(-iHt/\hbar)$ in the gas-surface dynamics community are the split-operator [219, 220] and the Chebychev [221] methods. In the split-operator method, the time-evolution operator for small time steps Δt is written as

$$e^{-iH\Delta t/\hbar} = e^{-iK\Delta t/2\hbar}\,e^{-iV\Delta t/\hbar}\,e^{-iK\Delta t/2\hbar} + O(\Delta t^3)\,, \tag{6.18}$$

where K is the kinetic energy operator and V the potential term. In the Chebyshev method, the time-evolution operator is expanded as

$$e^{-iH\Delta t/\hbar} = \sum_{j=1}^{j_{\max}} a_j(\Delta t)\,T_j(\bar{H})\,, \tag{6.19}$$

where the T_j are Chebyshev polynomials and $\bar{H}$ is the Hamiltonian rescaled to have eigenvalues in the range $(-1, 1)$. Both propagation schemes use the fact that the kinetic energy operator is diagonal in $\boldsymbol{k}$-space and the potential is diagonal in real-space. The wave function and the potential are represented on a numerical grid, and the switching between the $\boldsymbol{k}$-space and real-space representations is efficiently done by Fast Fourier Transformations (FFT) [16].

Quantum dynamical studies are still computationally very demanding. This prevents the explicit dynamical consideration of surface degrees of freedom. At most, one surface oscillator has been taken into account to model the influence of the substrate vibrations on scattering or adsorption probabilities [222–224]. However, usually one is not interested in the explicit dynamics of the substrate vibrations in the context of gas-surface dynamics as long as there is no strong surface rearrangement due to the interaction with atoms and molecules. In such a case, substrate phonons are then rather treated as a heat bath that lead to thermalization and dissipation effects. In order to describe these effects in terms of an open system dynamics, the system can be partioned into a subsystem (often refered to as just *the system*) and the surrounding bath. The *reduced density matrix*

$$\rho = \sum_{m,n} w_{mn}\,|m\rangle\langle n|\,, \tag{6.20}$$

is then defined in the Hilbert space of the system Hamiltonian H. If the functions $|n\rangle$ form a basis of this space, the diagonal matrix elements $w_{nn} = \langle n|\rho|n\rangle$ are interpreted as the population of the state $|n\rangle$, whereas the off-diagonal elements $w_{mn} = \langle m|\rho|n\rangle$ are related to the phase coherence between the states $|m\rangle$ and $|n\rangle$ that leads to interference effects [225]. Physical oservables are obtained as usual in a density matrix formulation by forming the trace

$$\langle A \rangle = \mathrm{tr}\,(A\rho)\,. \tag{6.21}$$

The time-evolution of the reduced density matrix is given by the *Liouville–von Neumann equation* [225, 226]

$$\frac{\partial \rho}{\partial t} = -\frac{i}{\hbar}\,[H,\rho] + \mathcal{L}_B \rho, \tag{6.22}$$

where dissipation effects are taken into account through the Liouville bath operator $\mathcal{L}_B$. Without the dissipation term, (6.22) is equivalent to the time-dependent Schrödinger equation. However, solving (6.22) for a closed system is not advisable, since a $N \times N$ matrix has to be determined if the wave functions are expanded in a set of N basis functions while in the ordinary time-dependent Schrödinger equation the wave function is just represented by a N-dimensional vector. Still, the computational effort associated with the density-matrix formalism is necessary in order to include dissipation effects in the quantum dynamics which is not only essential for scattering and reactive processes at surface, but also for reactions induced by electronic transitions (see Sect. 8.5). The dissipative term can describe vibrational or electronic relaxation effects as well as so-called dephasing processes and the corresponding time scales, T_1, T_2 and T_2^*, respectively.

There is no unique way to choose the bath operator $\mathcal{L}_B$. Usually one invokes the *Markov approximation*, which means that one assumes that the change of density matrix at time t is a function of the reduced density matrix at that time only, i.e. there are no memory effects of the past history. Even in this approximation, $\mathcal{L}_B$ is not fully specified. There is a phenomenological form proposed by Lindblad [227] which guarantees that at all times the diagonal elements of ρ properly correspond to state populations. The Lindblad semigroup functional is given by

$$\mathcal{L}_B \rho = \sum_k \left(C_k \rho C_k^+ - \{C_k^+ C_k, \rho\} \right), \tag{6.23}$$

where $\{A, B\} = AB + BA$ denotes the anticommutator. The C_k are the Lindblad operators. The dissipative channel is labeled by the subscript k. In the Lindblad approach, different diagonal elements are coupled by energy relaxation processes on a timescale T_1. The off-diagonal elements decay because of energy relaxation on the timescale $T_2 = 2T_1$ as well as because of pure dephasing processes due to elastic processes on a time scale T_2^*. A shortcoming of the Lindblad approach is that the operators are not connected in a physically transparent way to the interaction Hamiltonian between system and bath [225].

In the Redfield approach [228], the bath operator $\mathcal{L}_B$ is derived from the system-bath interaction using second-order perturbation theory. This leads to the Redfield equations

$$\mathcal{L}_B\rho = \sum_l \left([G_l^+ \rho, G_l] + [G_l, \rho G_l^-] \right), \tag{6.24}$$

where the bath modes are labelled by l and the system dependence of the system-bath operators is represented by the operators G_l. Still, the Redfield approach suffers from another deficiency, namely the possible violation of the positivity of the density matrix. There have been further proposals for the construction of $\mathcal{L}_B$ [225] which will not be discussed here.

Often negative imaginary potentials, so-called *optical potentials* have been used in dissipative dynamics. However, it is important to realize that there is a difference between the use of a friction term in classical dynamics and the use of an optical potential in quantum dynamics. Whereas a friction term leads to momentum relaxation processes, an optical potential reduces the norm of the wave function.

It is a wide-spread believe that classical dynamical methods are much less time-consuming than quantum ones. This is certainly true if one compares the computational cost of one trajectory to a quantum calculation. If integrated quantities such as the sticking probability are to be determined, then the statistical error of the result is only related to the number of computed trajectories and not to the dimensionality of the problem. For example, if the sticking probability S lies in the range $0.1 \leq S \leq 1$, then usually 10^3–10^4 trajectories are sufficient to obtain a sufficiently accurate result independent of the complexity of the system.

However, for the evaluation of detailed microscopic distribution function in scattering or desorption processes at surfaces the statistical requirements are much more demanding. Then the delocalized nature of the wave functions in the quantum dynamics can be advantageous. Instead of many trajectory calculations one quantum calculation might be sufficient. Thus quantum calculations correspond in a sense to the simultaneous determination of many trajectories. The crucial difference between quantum and classical dynamics is that in the quantum dynamics the averaging is done coherently while it is done incoherently in the classical dynamics. In addition, in wave-packet calculations dynamical simulations are performed for a whole range of energies in one run, and in a time-independent coupled-channel method the microscopic transitions probabilites of all open channels are determined simultaneously. Consequently, quantum dynamical simulations do not necessarily have to be more time-consuming that classical calculations.

6.3 Parametrization of ab initio Potentials

In order to perform dynamical simulations on a potential energy surface derived from first-principles calculations, one needs a continuous description of the potential. This is especially true for quantum dynamical simulations. Since the wave functions are delocalized, they always probe a certain area of the PES at any time. The total energy calculations, however, just provide total energies for discrete configurations of the nuclei. In classical molecular dynamics simulation, the gradients of the potential are only needed for one particular configuration at any time. This makes ab initio molecular dynamics simulations possible in which the forces necessary to integrate the classical equations of motions are determined by electronic structure calculations in each step [229–232]. In spite of the efficiency of modern electronic structure codes, the evaluation of the forces for every time step of a MD run is computationally still so demanding that these ab initio molecular dynamics runs have been limited to the simulation of well below 100 trajectories. This number is usually much too small to obtain sufficient statistics for the reliable determination of reaction probabilities or distributions.

On the other hand, molecular dynamics simulations on a suitable analytic representation of a potential energy surface can be extremely fast. Hence it is desirable to adjust the first-principles energies to an analytical or numerical continuous representation of the PES. This is a highly non-trivial task. On the one hand the representation should be flexible enough to accurately reproduce the ab initio input data, on the other hand it should have a limited number of parameters so that it is still controllable. Ideally a good parametrization should not only accurately interpolate between the actually calculated points, but it should also give a reliable extrapolation to regions of the potential energy surface that have actually not been determined by the ab initio calculations.

The explicit form of the chosen analytical or numerical representation of the ab initio potential varies from application to application. Often the choice is dictated by the dynamics algorithm in which the representation is used. Most applications have been devoted to the interaction of a diatomic molecule with the surface [233, 234]. The angular orientation of the molecule has usually been expanded in spherical harmonics and the center-of-mass coordinates parallel to the surface in a Fourier series [235–238]. For the PES in the plane of the molecular distance from the surface and the interatomic separation a representation in reaction path coordinate has been employed [236–239], but also two-body potentials have been used [235]. Before detailed ab initio potentials became available, the LEPS form was often used to construct a global PES [240]. This parametrization contains only a small number of adjustable parameters which made it so attractive for model calculations, but which makes it at the same time relatively unflexible. A modified LEPS potential has still been successfully used to fit an ab initio PES of the interaction of atomic hydrogen with the the hydrogenated Si(100) surface [241].

6.3 Parametrization of ab initio Potentials

Ab initio total energies are often mainly determined at high-symmetry points of the surface in order to reduce the computational cost. It is true that these high-symmetry points usually reflect the extrema in the PES. However, due to this limitation the fitted continuous PES can only contain terms that correspond to these high-symmetry situations. On the one hand this often saves computer time also in the quantum dynamics because certain additional selection rules are introduced which reduces the necessary basis set [235, 236]. On the other hand, of course this represents an approximation. The question, how serious the neglect of terms with lower symmetry is, remains open until these terms have been determined and included in actual dynamical calculations.

Most of the corrugation in molecule-surface potential energy surfaces can already be derived from the atom-surface interaction. This observation has been used in corrugation-reducing procedures [242, 243]. The first step is the ab initio determination of the interaction of both the atomic and the molecular species with the surface. From the atomic data, a three-dimensional reference function is constructed which is substracted from the molecular potential energy surface. The remaining function is much smoother than the original potential energy surface and therefore much easier to fit. This method has been successfully used for a continuous representation of the $H_2/Pd(111)$ [242] and the $H_2/Ni(111)$ interaction [243].

If more than just the molecular degrees of freedom should be considered in a parametrization of an ab initio PES, analytical forms become very complicated and cumbersome. As an alternative, the interpolation of ab initio points by a neural network has been proposed [244–246]. Neural networks can fit, in principle, any real-valued, continuous function to any desired accuracy. They require no assumptions of the functional form of the underlying problem. On the other hand, there is no physical insight that is used as an input in this parametrization. Hence the parameters of the neural network do not reflect any physical or chemical property.

This deficiency is avoided if the results of first-principles electronic structure calculations are used to adjust the parameters of a tight-binding formalism [63]. A tight-binding method is more time-consuming than an analytical representation or a neural network since it requires the diagonalization of a matrix. However, due to the fact that the quantum mechanical nature of bonding is taken into account [59] tight-binding schemes need a smaller number of ab initio input points to perform a good interpolation *and* extrapolation [247]. This is demonstrated in Fig. 6.1 that shows the PES of $H_2/Pd(100)$ obtained by a tight-binding fit to the ab initio data [183]. The plots should be compared with the ab initio PES in Fig. 5.19 The filled circles denote the points that have been used to obtain the fit. While for the h–b–h cut (see p. 132) a relatively large number of input points were necessary for the fit (Fig. 6.1a), for the h–t–h three points were sufficient for a satisfactory agreement with the ab initio data (Fig. 6.1a). This is caused by the fact that

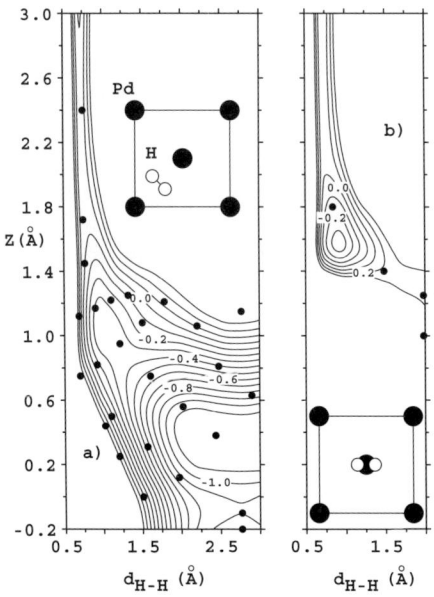

Fig. 6.1. Potential energy surface of H_2/Pd(100) evaluated by a tight-binding Hamiltonian that was adjusted to reproduce ab initio results of [183] (compare with Fig. 5.19). The filled circles indicate the points that have been used to obtain the fit. The insets demonstrate the lateral and angular configuration of the H_2 molecule. (After [247])

the parameters of the tight-binding scheme, the Slater-Koster integrals [58], have a well-defined physical meaning.

An important issue is to judge the quality of the fit to an ab initio PES. Usually the root mean squared (RMS) error between fit and input data is used as a measure of the quality of a fit. If this error is zero, then everything is fine. However, normally this error is larger than zero. The systematic error of the ab initio energies is usually estimated to be of the order of 0.1 eV. Often it is said that the RMS error of the fit should be of the same order. But the dynamics of molecular dissociation at surfaces can be dramatically different depending on whether there is a barrier for dissociation of height 0.1 eV or not [248]. Hence for certain regions of the PES the error has to be much less than 0.1 eV, while for other regions even an error of 0.5 eV might not influence the dynamics significantly. Another example occurs in a reaction path parametrization. If the curvature in the parametrization is off by a few percent, the energetic distribution of barrier heights is not changed and the dynamical properties are usually not altered significantly. However, the location of the barriers is changed and consequently the RMS error can become rather large. Hence one has to be cautious by just using the RMS error as a quality check of the fit. Unfortunately there is no other simple error function for the assessment of the quality of a fit. If it is possible, one should perform a dynamical check. Obviously, if the dynamical properties calculated on a fitted PES agree with the ones calculated on the original PES, the quality of the fit should be sufficient.

6.4 Scattering at Surfaces

If a beam of atoms or molecules is hitting a surface that has a small adsorption well for the particular particles, most of them will be scattered back into the gas phase. Especially for the case of light atoms and molecules when the de Broglie wave length of the particles is of the order of the lattice spacing, the quantum nature of the scattering event has to be taken into account which leads to elastic scattering and diffraction.

Let us first consider a beam of atoms with initial wave vector $\boldsymbol{K}_i$ that is scattered elastically at a periodic surface. The component of the wave vector parallel to the surface $\boldsymbol{K}_f^{\parallel}$ after the scattering is given by

$$\boldsymbol{K}_f^{\parallel} = \boldsymbol{K}_i^{\parallel} + \boldsymbol{G}_{mn}, \tag{6.25}$$

where $\boldsymbol{G}_{mn}$ is a vector of the two-dimensional reciprocal lattice of the periodic surface. Since there is no energy transfer to the surface in elastic scattering, the total kinetic energy of the atoms is conserved:

$$\frac{\hbar^2 \boldsymbol{K}_f^2}{2M} = \frac{\hbar^2 \boldsymbol{K}_i^2}{2M}. \tag{6.26}$$

By (6.25) and (6.26) all possible final scattering angles are specified. This leads to a discrete, finite set of scattering channels. Note that for a given incident energy, angle and mass of the atoms the scattering angles are entirely determined by the geometry of the surface. The interaction potentials only influences the intensity of the scattering peaks, but not their position. That is why diffraction at surfaces can be used to determine the structure of surfaces.

Now we allow for energy transfer processes to the surface, i.e., we consider the inelastic scattering of atoms. The main source for inelastic effects is the excitation and deexcitation of substrate phonons. These phonons also carry momentum so that the conservation of parallel momentum leads to

$$\boldsymbol{K}_f^{\parallel} = \boldsymbol{K}_i^{\parallel} + \boldsymbol{G}_{mn} + \sum_{\text{exch.phon.}} \pm \boldsymbol{Q}, \tag{6.27}$$

where $\boldsymbol{Q}$ is a two-dimensional phonon-momentum vector parallel to the surface. The plus-signs in the sum correspond to the excitation or emission of a phonon while the minus-signs represent the deexcitation or absorption of a phonon. The excitation and deexcitation of phonons with momentum $\boldsymbol{Q}$ and mode index j also modifies the energy conservation relation:

$$\frac{\hbar^2 \boldsymbol{K}_f^2}{2M} = \frac{\hbar^2 \boldsymbol{K}_i^2}{2M} + \sum_{\text{exch.phon.}} \pm \hbar \omega_{\boldsymbol{Q},j}. \tag{6.28}$$

If only one phonon is emitted or absorbed in the collision process, the sums in (6.27) and (6.28) reduce to one term. If one-phonon processes are dominant in scattering, the phonon spectrum of a surface can be measured. This has been extensively used in helium atom scattering.

158   6. Gas-Surface Dynamics

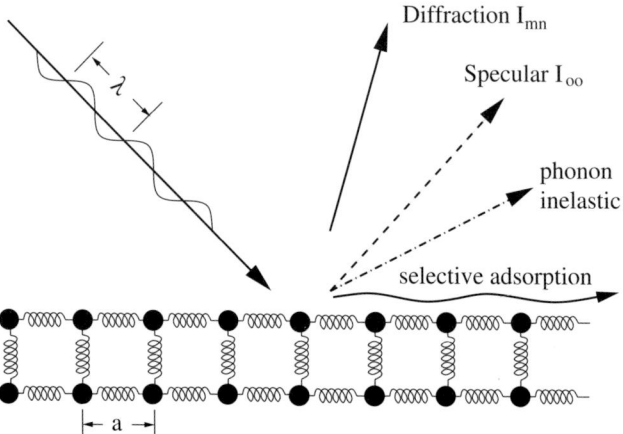

Fig. 6.2. Summary of the different collision processes in nonreactive scattering (after [262])

A schematic summary of possible collision processes in nonreactive scattering is presented in Fig. 6.2. With I_{mn} the intensity of the elastic diffraction peak mn according to (6.25) is denoted. The scattering peak I_{00} with $\boldsymbol{K}_f^{\parallel} = \boldsymbol{K}_i^{\parallel}$ is called the *specular peak*. The excitation of phonons usually leads to a reduced normal component of the kinetic energy of the back-scattered atoms or molecules. Thus the reflected beam is shifted in general to larger angles with respect to the surface normal compared to the angle of incidence. The resulting supraspecular scattering is indicated in Fig. 6.2 as the phonon-inelastic reflection event.

The coherent scattering of atoms or molecules from surfaces has been known as a tool for probing surface structures since 1930 [249]. The diffraction pattern yields direct information about the periodicity and lattice constants of the surface. Furthermore, if one measures the intensity of the specular peak as a function of the angle of incidence, then at specific angles resonances appear [250]. They are due to so-called *selective adsorption resonances* which are also indicated in Fig. 6.2. These resonances occur when the scattered particle can make a transition into one of the discrete bound state of the adsorption potential. This can only happen if temporarily the motion of the particle is entirely parallel to the surface. The interference of different possible paths along the surface causes the resonance effects. Energy and momentum conservation yields the selective adsorption condition

$$\frac{\hbar^2 \boldsymbol{K}_i^2}{2M} = \frac{\hbar^2 (\boldsymbol{K}_i^{\parallel} + \boldsymbol{G}_{mn})^2}{2M} - |E_l|, \qquad (6.29)$$

where E_l is a bound level of the adsorption potential. Usually these selective adsorption resonances only occur for relative weak adsorption potentials that are not strongly corrugated, i.e., mainly for physisorption potentials.

The bound state energies can be obtained without detailed knowledge of the scattering process. Typically one assumes a Morse potential

$$V(z) = D_0 \left(e^{-2\alpha(z-z_o)} - 2e^{-\alpha(z-z_o)} \right), \tag{6.30}$$

or some other parametrization of the interaction potential and can then derive the well depth, position and range of the adsorption potential from an adjustment of the parameters of the potential to reproduce the experimentally observed binding energies [100]. In particular helium atom scattering (HAS) has been used intensively to study surface crystallography and the shape of physisorption potentials (see, e.g., [251] and references therein). Helium atom scattering has furthermore been used extensively in order to determine the surface phonon spectrum in one-phonon collisions via (6.27) and (6.28) [251].

Hydrogen molecules have been utilized less frequently in order to study interaction potentials [252]. The coherent elastic scattering of molecules is more complex than atom scattering because in addition to parallel momentum transfer the internal degrees of freedom of the molecule, rotations and vibrations, can be excited during the collision process. Then the total energy balance in the scattering is

$$\frac{\hbar^2 \mathbf{K}_f^2}{2M} = \frac{\hbar^2 \mathbf{K}_i^2}{2M} + \Delta E_{\text{rot}} + \Delta E_{\text{vib}} + \sum_{\text{exch.phon.}} \pm \hbar \omega_{\mathbf{Q},j}. \tag{6.31}$$

Usually the excitation of molecular vibrations in molecule-surface scattering is negligible, in contrast to the phonon excitation. This is due to the fact that the time-scale of the molecular vibrations is usually much shorter than the scattering time or the rotational period. Therefore the molecular vibrations follow the scattering process almost adiabatically. Molecular rotations, on the other hand, can be excited rather efficiently in the scattering at highly anisotropic surfaces. This rotational excitation leads to additional peaks in the diffraction spectrum, the *rotationally inelastic diffraction* peaks.

Experimentally, rotationally inelastic diffraction of hydrogen molecules has been first observed in the scattering at inert ionic solids such as MgO [253] or NaF [254]. At metal surfaces with a high barrier for dissociative adsorption rotationally inelastic diffraction peaks are usually hard to resolve except in the case of HD scattering, where the displacement of the center of mass from the center of the charge distribution leads to a strong rotational anisotropy [255].

In the case of hydrogen scattering from reactive surfaces, where non-activated dissociative adsorption is possible, the repulsive interaction is not mediated by the tail of the metal electron density, but occurs rather close to the surface. Due to the chemical nature of this interaction the potential is strongly corrugated and anisotropic with regard to the molecular orientation. Thus there should be large intensities in the off-specular and rotationally inelastic diffraction peaks. This is indeed the case, as six-dimensional quantum coupled-channel calculations for the scattering of $H_2/Pd(100)$ have

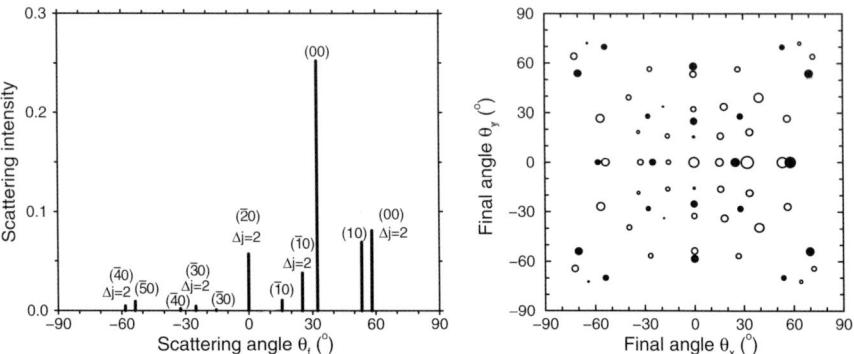

Fig. 6.3. Six-dimensional quantum results of the rotationally inelastic scattering of H_2 on Pd(100) for a kinetic energy of 76 meV at an incidence angle of 32° along the [10] direction of the square surface lattice. The left panel shows the inplane diffraction spectrum where all peaks have been labelled according to the transition. The right spectrum shows all diffraction peaks where the open and filled circles correspond to the rotationally elastic and rotationally inelastic scattering, respectively. The radius of the circles is proportional to the logarithm of the scattering intensity. (After [256])

shown [256]. In this calculations a parametrization of the ab initio PES plotted in Fig. 5.19 [183] has been used.

One typical calculated angular distribution of H_2 molecules scattered at Pd(100) is shown in Fig. 6.3. The total initial kinetic energy is $E_i = 76$ meV. The incident parallel momentum corresponds to $2\hbar G$ along the $\langle 0\bar{1}1 \rangle$ direction. This leads to an incident angle of $\theta_i = 32°$. The molecules are initially in the rotational ground state $j_i = 0$. (m, n) denotes the parallel momentum transfer $\Delta \mathbf{G}_{\parallel} = (mG, nG)$. The specular peak is the most pronounced one, but the first order diffraction peak (10) is only a factor of four smaller (note, that in a typical HAS experiment the off-specular peaks are about two orders of magnitude smaller than the specular peak). The results for the rotationally inelastic diffraction peaks $j = 0 \rightarrow 2$ have been summed over all final azimuthal quantum numbers m_j. The excitation probability of the so-called cartwheel rotation with $m = 0$ is for all peaks approximately one order of magnitude larger than for the so-called helicopter rotation $m = j$, since the polar anisotropy of the PES is stronger than the azimuthal one.

This steric effect in scattering could already be expected from detailed balance arguments. According to the principle of detailed balance [257–259], in a equilibrium situation the flux impinging on a surface from the gas-phase, which is rotationally isotropically distributed, should equal the desorption plus the scattering flux. Since in desorption the helicopter rotations are preferentially occupied (see p. 184), in scattering the cartwheel rotations have to be preferentially excited.

The intensity of the rotationally inelastic diffraction peaks in Fig. 6.3 is comparable to the rotationally elastic ones. Except for the specular peak they are even larger than the corresponding rotationally elastic diffraction peak with the same momentum transfer (m, n). Note that due to the initial conditions the rotationally elastic and inelastic $(\bar{2}0)$ diffraction peaks fall upon each other.

The out-of-plane scattering intensities are not negligible, which is demonstrated in the right panel of Fig. 6.3. The open circles correspond to rotationally elastic, the filled circles to rotationally inelastic diffraction. The radii of the circles are proportional to the logarithm of the scattering intensity. The sum of all out-of-plane scattering intensities is approximately equal to the sum of all in-plane scattering intensities. Interestingly, some diffraction peaks with a large parallel momentum transfer still show substantial intensities. This phenomenon is well known from helium atom scattering and has been discussed within the concept of so-called rainbow scattering.

The experimental observation of diffraction in reactive systems is not trivial. Because of the reactivity, during the experiment an adsorbate layer builds up very rapidly. These destroy the perfect periodicity of the surface and thus suppress diffraction effects. In order to keep the surface relatively clean, one has to use rather high surface temperatures so that adsorbates quickly desorb again. High surface temperatures, on the other hand, also smear out the diffraction pattern. Still experimentalists managed to clearly resolve rotationally inelastic peaks in the diffraction pattern of D_2/Ni(110) [260] and D_2/Rh(110) [261] in addition to rotationally elastic peaks.

However, for a complete description of the diffraction of light atoms and molecules from surfaces the excitation of substrate phonons has to be taken into account. This is not possible in a full-dimensional quantum dynamical framework. Instead, standard approximative methods known from scattering theory [215] have to be used. Much effort has been devoted to the theoretical description of helium atom scattering [262] since this is a well-established technique to measure surface and adsorbate structures and surface phonon dispersion curves [251].

Adsorbed monolayers of heavy rare gas atoms have served as a benchmark system for the study of single- and multiphonon excitation in atom-surface scattering [262]. These monolayers have non-dispersive Einstein surface vibrational modes which can be easily identified in energy transfer scattering spectra. Figure 6.4 shows a comparison of measured and calculated multiphonon spectra in the scattering of He at Xe/Cu(111) and Xe/Cu(100) [263]. Whereas Xe forms a commensurate adlayer on Cu(111) above 50 K, on Cu(100) there is an incommensurate hexagonal Xe adlayer. The calculations were performed in the so-called *exponentiated distorted-wave Born approximation* [262], the interaction potential was determined empirically from the simulation of single-phonon scattering in the same systems. For Xe/Cu(111), the spectrum shows well-defined, uniformly spaced peaks with a distance of 2.62 meV. The peaks

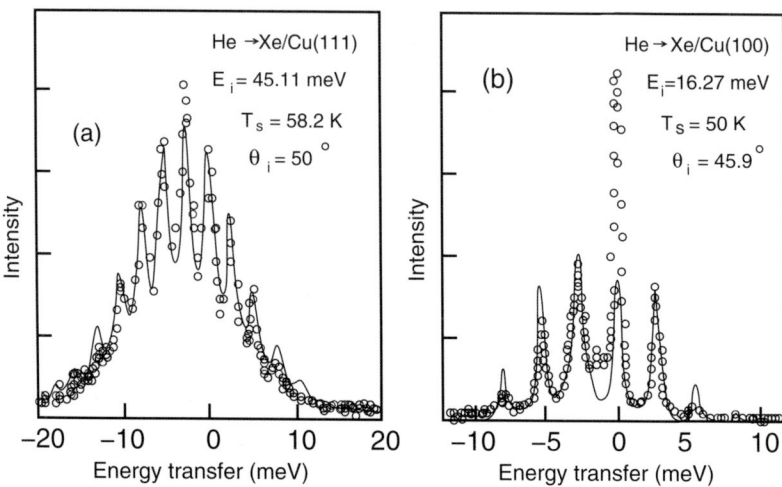

Fig. 6.4. Comparison of measured (circles) and calculated (solid line) multiphonon spectra in the scattering of He at Xe covered Cu surfaces. (a) He→Xe/Cu(111) scattering, (b) He→Xe/Cu(100) scattering [263]

correspond to the uncorrelated multiple absorption and emission of a dispersionless collective Xe vibrational mode perpendicular to the surface, called the S-mode. The broad background is caused by other adlayer modes that show a stronger dispersion.

In the case of the incommensurate Xe adlayer on Cu(100), it is not possible to set up a finite surface unit cell and the corresponding dynamical matrix. Instead, the scattering is modelled by a floating adlayer on a rigid substrate. There is also a dispersionless S-mode evident with energy 2.71 meV (Fig. 6.4b). The hump seen in the experiment near the elastic line is due to the excitation of the Cu(001) Rayleigh wave which is of course not reproduced in the model with a fixed Cu substrate. Otherwise the agreement between theory and experiment for both systems is rather satisfactory, except for the elastic line. The intensity of the elastic line can also be reproduced if diffuse scattering contributions are taken into account [262].

6.5 Atomic and Molecular Adsorption on Surfaces

The sticking or adsorption probability is defined as the fraction of atoms or molecules impinging on a surface that are not scattered back, i.e. that remain on the surface. In principle, at surfaces with non-zero temperatures every adsorbed particle will sooner or later desorb again because of thermal fluctuations. Hence there is no unambiguous definition of the sticking probability since it depends on the time-scale of the required residence time on the surface. However, often the residence times of adsorbed atoms or molecules

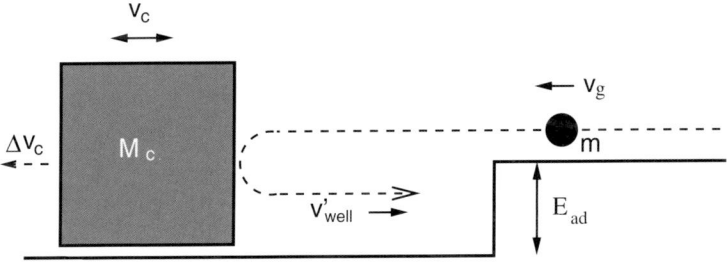

Fig. 6.5. Schematic illustration of the hard-cube model. An atom or molecule with mass m is impinging in an attractive potential with well depth E_{ad} on a surface modeled by a cube of effective mass M_c. The surface cube is moving with a velocity v_c given by a Maxwellian distribution

are rather long compared to microscopic time scales so that this unambiguity does not really pose a problem for the definition of the sticking probability.

First we consider atomic adsorption and molecular adsorption which means that the molecule stays intact on the surface. In the following we will refer to both processes by the term atomic adsorption. In order to stick on a surface in an attractive potential well, an atom has to transfer its kinetic energy to the substrate. We define $P_E(\epsilon)$ as the probability that an incoming particle with kinetic energy E will transfer the energy ϵ to the surface. If an atom impinging on a surface looses more energy than its kinetic energy in the gas phase, then it cannot escape the adsorption well and will remain trapped at the surface. Hence the atomic or molecular sticking probability can be expressed as

$$S(E) = \int_E^\infty P_E(\epsilon) \, d\epsilon \,. \tag{6.32}$$

The atomic excess energy has to be transferred to substrate excitation, i.e., either to phonons or electron-hole pairs. Hence any theoretical description of atomic or molecular adsorption has to consider dissipation to the continuous excitation spectrum of the substrate.

In the simplest approximation, the scattering of an atom at a surface can be treated as a binary elastic collision between a gas phase atom (mass m) and a stationary substrate atom (mass M). Using energy and momentum conservation, the energy transfer $\Delta = E_i - E_f$ to the substrate is given by the *Baule formula* [264] (see Problem 6.3)

$$\Delta = \frac{4\mu}{(1+\mu)^2} E_i \,, \tag{6.33}$$

where μ is the mass ratio $\mu = m/M$.

This simple Baule model is also the basis of the somewhat more sophisticated *hard-cube model* (HCM) [265, 266] for the estimation of trapping probabilities in atomic adsorption. In this model that is illustrated in Fig. 6.5, the surface is described by a cube of effective mass M_c which is moving freely with a velocity distribution $P_c(v_c)$. An atom impinging on the surface with an attractive well of depth E_ad will hit the hard cube with a velocity

$$v_\text{well} = -\sqrt{v_g^2 + \frac{2E_\text{ad}}{m}}. \tag{6.34}$$

Taking energy and momentum conservation into account, the velocity of the atom in the potential well after the collision is given by

$$v'_\text{well} = \frac{\mu - 1}{\mu + 1} v_\text{well} + \frac{2}{\mu + 1} v_c, \tag{6.35}$$

where now μ is the mass ratio $\mu = m/M_c$. Particles with a velocity $v'_\text{well} < \sqrt{2E_\text{ad}/m}$ cannot escape the potential well and will remain at the surface. For a given velocity v_g, v'_well depends on v_c. Using (6.35), an atom that hits the surface cube with a velocity

$$v_c < \frac{\mu + 1}{2} \sqrt{\frac{2E_\text{ad}}{m}} - \frac{\mu - 1}{2} v_\text{well} = v_\text{lim} \tag{6.36}$$

will get trapped. This means that the trapping probability is determined by

$$P_\text{trap}(v_g) = \int_{-\infty}^{v_\text{lim}} P_c(v_c) dv_c. \tag{6.37}$$

Now we use a weighted Maxwellian velocity distribution

$$P_c(v_c) = \frac{\alpha}{\sqrt{\pi} v_g} (v_g - v_c) \exp\{-\alpha^2 v_c^2\}, \tag{6.38}$$

with $\alpha = \sqrt{M_c/2k_B T_s}$. This distribution takes the fact into account that collisions are more probable when the surface cube is moving toward the incoming atom than when it is moving away from it [265]. The trapping probability is then given by [266] (Problem 6.4)

$$P_\text{trap}(v_g) = \frac{1}{2} + \frac{1}{2} \text{erf}(\alpha v_\text{lim}) + \frac{\exp\{-\alpha^2 v_\text{lim}^2\}}{2\sqrt{\pi} \alpha v_\text{well}}. \tag{6.39}$$

In Fig. 6.6 trapping probabilities evaluated according to (6.39) are plotted as a function of the kinetic energies for different parameters of the model. All curves show a typical behavior, namely the decrease of the sticking probability with increasing kinetic energy. This is due to the fact that the energy transfer to the surface becomes less efficient at higher kinetic energies. Of course, the higher the kinetic energy is, the more energy is transferred to the surface. But the fraction of particles that loose more energy than their initial kinetic energy becomes smaller at higher kinetic energy.

The examples plotted in Fig. 6.6 illustrate general trends in atomic adsorption. Heavier atoms will transfer more energy to the surface than lighter atoms and therefore stick more easily at the surface. Larger adsorption energies will also enhance the energy transfer and consequently the trapping

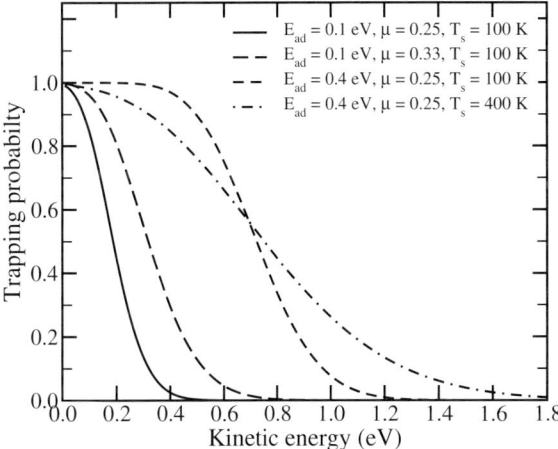

Fig. 6.6. Trapping probability as a function of the kinetic energy evaluated according to the hard cube model (6.39) for different adsorption energies E_{ad}, mass ratios $\mu = m/M_c$ and surface temperatures T_s

probability. Increasing the surface temperature leads to a broader velocity distribution. Thus a temperature rise corresponds to an averaging over a wider range of kinetic energies which leads to a decrease for negative curvature of the sticking curve, i.e. at high sticking probabilities, and an increase for positive curvature, i.e. at low sticking probabilities.

The hard-cube model has often been used in order to derive adsorption well depths from measured trapping probabilities, also for molecular adsorption (see, e.g., [266, 267]). In the hard-cube model the surface is assumed to be flat and structureless which means that in any scattering and adsorption process the incident parallel momentum would be conserved. For the sticking probability this leads to the *normal energy scaling*, i.e., the sticking probability is a function of the normal component $E_i \cos^2 \theta_i$ of the incident energy alone, where θ_i is the angle of incidence.

However, real surfaces are not structureless as far as the interaction of atoms and molecules is concerned. Adsorption corresponds to the making of a chemical bond which strongly depends on the local environment. This leads to corrugation in the potential energy surface, i.e., the potential depends on the lateral position of the interacting particle on the surface. Trapping probalities often scale as $E_i \cos^n \theta_i$ with $n < 2$. An exponent of $n = 0$ corresponds to *total energy scaling* which is usually associated with a highly corrugated potential energy surface.

In the case of molecular adsorption, the dynamics of adsorption become even more complex due to the presence of the internal degrees of freedom of the molecule, rotations and vibrations. During the trapping process, energy can be very efficiently stored in these internal degrees of freedom thus enhanc-

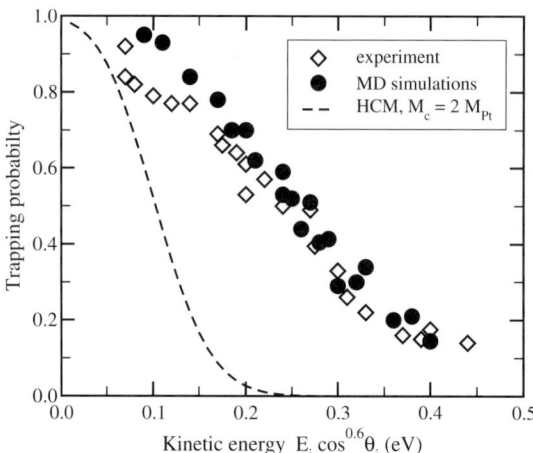

Fig. 6.7. Trapping probability of ethane (C_2H_6) on Pt(111) at a surface temperature of $T_s = 95$ K as a function of $E_i \cos^{0.6} \theta_i$ where E_i is the initial kinetic energy and θ_i the angle of incidence. Diamonds: experiment (from [268], not all measured data points are plotted), circles: molecular dynamics simulations (from [269]), dashed line: hard-cube model (HCM) at normal incidence for a well depth of 0.28 eV and cube mass of $2 \cdot M_{Pt}$ to account for finite size effects

ing the trapping probability. This has been carefully analysed in a molecular dynamics study of the trapping of ethane, C_2H_6, on Pt(111) [269]. Experimentally it was found that the adsorption probability scales as $E_i \cos^{0.6} \theta_i$ [268]. This is demonstrated in Fig. 6.7 where the adsorption probabilities plotted as a function of $E_i \cos^{0.6} \theta_i$ roughly fall on one line.

Additionally we have plotted the results of the prediction of the hard-cube model for the trapping probability using a realistic well-depth of 0.28 eV [269] and a cube mass of $2 \cdot M_{Pt}$. This higher mass takes into account that the impinging molecule interacts with more than a single substrate atom due to its finite size. First of all the hard-cube model would not be able to reproduce the observed energy scaling. But it also predicts sticking probabilities that are much smaller than the ones measured in the experiment. In order to reproduce the experimental trapping probability, an unrealistically deep adsorption well of more than 0.6 eV has to be assumed. However, it is not possible to get a good fit of the measured data over the whole energy range with the hard-cube model. This shows how dangerous it is to derive potential parameters from low-dimensional model calculations.

In order to understand the role of the corrugation and the internal degrees of freedom of the molecule, classical molecular dynamics simulations have been performed [269] within periodic boundary conditions using a three-layer Pt slab with 36 atoms per layer in the supercell and friction and random forces applied to the bottom layer. The ethane molecule was described as a

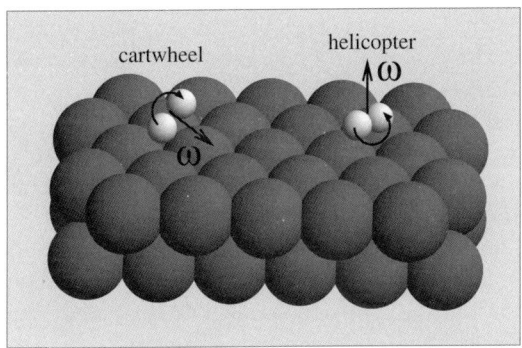

Fig. 6.8. Illustration of the difference between cartwheel motion (angular momentum vector ω parallel to the surface) and helicopter motion (ω parallel to the surface normal) for a molecule rotating at a surface

pseudo-diatomic molecule consisting of two methyl groups. The parameters of the interaction potential were determined empirically.

The classical equations of motion were integrated numerically with a time step of 2 fs. Each trapping probability was obtained by averaging over $N = 2000$ trajectories yielding a statistical uncertainty in the calculated trapping probabilities of $1/\sqrt{2000} \approx 0.02$. The results of the molecular dynamics simulations are also plotted in Fig. 6.7. They reproduce the measured data quite well. The scatter in the theoretical data is not due to the statistical uncertainty but rather to the scaling of the data for different angles of incidence.

To analyze the role of the different degrees of freedom in the trapping process, the partitioning of the energy was examined in detail for the first bounce on the surface. The rotational motion was further devided into cartwheel rotations with the angular momentum vector ω perpendicular to the surface normal and into helicopter rotation with ω parallel to the surface normal. These two types of rotational motion are illustrated in Fig. 6.8.

The average energy storage in the different degrees of freedom after the first bounce at the surface for those molecule that remain trapped on the surface are listed in Table 6.1. Two initial energies, 0.2 eV and 0.4 eV, and two different angles of incidence, 0° and 45° have been chosen. It is obvious that the excitation of surface phonons at the first bounce plays an important role as a dissipation channel. However, due to the corrugation and anisotropy of the potential the impinging molecule can transfer initial kinetic energy into rotational motion and motion parallel to the surface. This energy is then not available for the backscattering into the gas phase. In fact, it was found that the trapping probability is determined to within 10% by the first surface collision. That is how the transient energy transfer into rotational and parallel motion leads to an enhancement of the sticking probability.

Indeed, the excitation of cartwheel rotations and phonons are the primary dissipation channels determining the trapping. Conversion of the incident energy into motion parallel to the surface and into helicopter rotations is less important. The relative influence of the cartwheel excitation even increases with increasing kinetic energy so that at high energy it is mainly the excitation of cartwheel rotation that determines the trapping probability. This is confirmed by the findings that the measured slope of the trapping probability in Fig. 6.7 cannot be reproduced by the hard-cube model alone since the phonon-mediated trapping mechanism becomes too quickly ineffective at higher energies. Thus only within multi-dimensional theoretical treatent the rather large sticking probabilities at higher energies can be understood.

So far we have treated the atomic and molecular adsorption with purely classical dynamical methods. Let us focus on the low-energy regime in the following. At low kinetic energies all the trapping probabilities that have been presented so far became close to one. In fact, no matter how small the adsorption well, no matter how small the mass ratio between the impinging atom and the substrate oscillator, for $E \to 0$ and $T_s \to 0$ the sticking probability will always reach unity if there is no barrier before the finite adsorption well. This is due to the fact that every impinging particle will transfer energy to the substrate at zero temperature. In the limit of zero initial kinetic energy any energy transfer will be sufficient to keep the particle in the adsorption well. However, this behavior is intimately linked to classical physics. Quantum mechanically, however, there is a non-zero probability for elastic scattering at the surface, i.e. without any energy transfer. Hence the sticking probabilities should become less than unity in the zero-energy limit, in particular for light atoms impinging on a surface. In fact this has been observed in the sticking of rare gas atoms at cold Ru(0001) surfaces [270, 271].

Any theoretical description trying to reproduce elastic scattering has to take into account the quantum nature of the phonon system. In the following we model the surface as an ensemble of independent quantum surface

Table 6.1. Average of the calculated energy storage in different modes after the first bounce from ethane incident on Pt(111) for four different initial conditions for those molecules that remained trapped on the surface. All energies in meV. (From [269])

Incident energy	200	200	400	400
Angle of incidence	0°	45°	0°	45°
Trapping probability	0.63	0.78	0.13	0.35
Cartwheel	72	70	215	134
Helicopter	17	13	19	13
Parallel	37	95	50	182
Perpendicular	−68	−85	−66	−71
Phonons	142	107	182	142

oscillators. Since the oscillators are assumed to be independent, we can capture the essential physics by just considering the two-dimensional problem of an atomic projectile interacting via linear coupling with a single surface oscillator. The total Hamiltonian has the following form:

$$H_0 = -\frac{\hbar^2}{2M}\frac{\partial^2}{\partial Z^2} + V_0(Z) + \hbar\omega\left(a^+ a + \frac{1}{2}\right) + V_1(Z)\, x\,, \qquad (6.40)$$

where Z and x are the coordinates of the atom and the oscillator, respectively. The linear coupling $V_1(Z)$ leads to a displacement of the surface oscillator. Now we assume that the motion of the atom is hardly influenced by the excitation of the surface oscillator. Furthermore we treat the atom as a classical particle that is subject to the force $F = -\partial V_0(Z)/\partial Z$. This is called the *trajectory approximation* since the atomic projectile moves on a fixed classical trajectory given by the equation of motion

$$M\ddot{Z} = -\frac{\partial V_0(Z)}{\partial Z}. \qquad (6.41)$$

The classical trajectory $Z(t)$ introduces a time-dependent force in the Hamiltonian for the oscillator, therefore it is called the *forced oscillator model*. The time-dependent Hamiltonian is given by

$$\begin{aligned} H_{\rm osc} &= \hbar\omega\left(a^+ a + \frac{1}{2}\right) + V_1(Z(t))\, x \\ &= \hbar\omega\left(a^+ a + \frac{1}{2}\right) + \lambda(t)\,(a^+ + a)\,, \end{aligned} \qquad (6.42)$$

where we have introduced the coupling parameter

$$\lambda(t) = V_1(Z(t))\sqrt{\frac{\hbar}{2m\omega}}. \qquad (6.43)$$

In the following we use the Heisenberg representation of quantum mechanics, i.e. we assume that the operators a^+ and a are time-dependent. a obeys the equation of motion

$$i\hbar\,\dot{a} = [a, H_{\rm osc}] = \hbar\omega\, a + \lambda(t) \qquad (6.44)$$

with the solution

$$a(t) = -\frac{i}{\hbar}\, e^{-i\omega t}\int_{-\infty}^{t} e^{i\omega t'}\,\lambda(t')dt' + e^{-i\omega t} a_0\,, \qquad (6.45)$$

where we have used the notation $a_0 = a(t = -\infty)$. Now we assume that before the scattering event no phonons have been excited, i.e. $a_0|\Psi(t = -\infty)\rangle = 0$. The expectation or mean value of excited phonons $\bar{n}$ after the scattering is obtained from

$$\begin{aligned} \bar{n} &= \langle\Psi(t = -\infty)|\, a^+(t = \infty) a(t = \infty)\, |\Psi(t = -\infty)\rangle \\ &= \left|\frac{1}{\hbar}\int_{-\infty}^{\infty} e^{i\omega t'}\,\lambda(t')dt'\right|^2, \end{aligned} \qquad (6.46)$$

which means that $\bar{n}$ is basically given by the square of the Fourier transform of the coupling $\lambda(t)$. In fact, for the forced oscillator the probability P_{ji} for a transition from an initial oscillator state i to the final state j can be explicitly derived from $\bar{n}$ [272]:

$$P_{ji} = \frac{i!}{j!} \, e^{-\bar{n}} \, \bar{n}^{j-i} \, [L_j^{j-i}(\bar{n})]^2 \,, \quad j \geq i \,, \tag{6.47}$$

where L_j^{j-i} is an associated Laguerre polynomial [273]. For an excitation from the ground state one obtains a Poisson distribution

$$P_{j0} = \frac{\bar{n}^j}{j!} \, e^{-\bar{n}} \,. \tag{6.48}$$

The probabilities (6.47) or (6.48), respectively, yield the energy distribution of excited phonons and hence also the energy transfer to the surface. This means that they correspond to the probability $P_E(\epsilon)$ entering the expression (6.32) for the sticking probability. The determination of the sticking probability within the trajectory approximation seems to be inconsistent since the sticking probability is derived from trajectories that correspond to non-sticking, namely scattering events. This approximation can be improved somewhat by considering backscattered trajectories with less kinetic energy and by introducing reduced mass corrections [209, 271].

In fact, a compact expression can be derived for the energy distribution in the scattering of an atom at a system of phonon oscillators with a Debye spectrum at a temperature T_s [274, 275]. Assuming a Morse potential for the potential $V_0(Z)$, this expression depends on a small set of parameters such as the potential well depth, the potential range, the mass of the surface oscillator and the surface Debye temperature. This model was used in order to reproduce the measured sticking probabilities of rare gas atoms on a Ru(001) surface at a temperature of $T_s = 6.5\,\mathrm{K}$ [271].

A comparison between the measured and calculated sticking probabilities for Ne, Ar, Kr and Xe on Ru(001) is shown in Fig. 6.9. The lighter the atoms, the smaller the sticking probability. At small energies, the sticking probabilities do not reach unity except for the heaviest rare gas atom Xe due to the quantum nature of the substrate phonons. Indeed, attempts to reproduce the measured sticking probabilities with purely classical methods have failed, at least for Ne and Ar [270, 271]. A classical treatment of the solid is only appropriate if the energy transfer to the surface is large compared to the Debye energy of the solid [276].

At even lower kinetic energies than reached in the experiment [271], the quantum nature of the adsorbing particles cannot be neglected any longer and the trajectory approximation cannot be applied any more. In fact, for short-range forces the matrix elements vanish for $E \to 0$ while the classical Fourier transform (6.46) remains finite [276, 277]. Therefore the quantum mechanical sticking probability also vanishes for $E \to 0$. However, in order to see this effect extremely small kinetic energies corresponding to a temperature

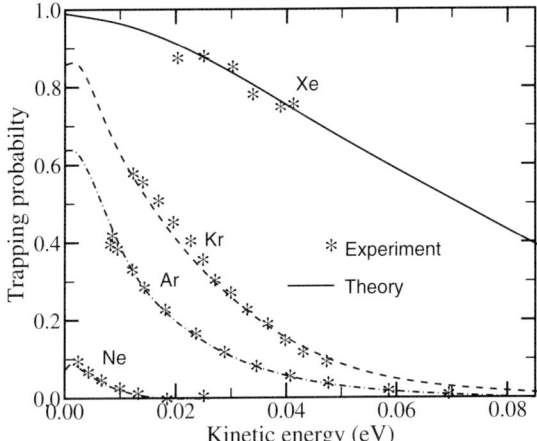

Fig. 6.9. Sticking probability of rare gas atoms on Ru(0001) at a surface temperature of $T_s = 6.5$ K. Stars (*): experiment; lines: theoretical results obtained with the forced oscillator model (after [271], not all measured data points are included)

below 0.1 K are required [276]. Nevertheless, this quantum phenomenon in the sticking at surfaces has been verified experimentally for the adsorption of atomic hydrogen on thick liquid ^{4}He films [278].

There is yet another effect that also leads to zero sticking at very low energies. There is a certain probability that a quantum particle is reflected at attractive parts of the potential. If the potential falls off asymptotically faster than $1/Z^2$, then the reflection amplitude R exhibits the universal behavior [279, 280]

$$|R| \underset{k \to 0}{\longrightarrow} 1 - bk\,, \tag{6.49}$$

where k is the wave number corresponding to the asymptotic kinetic energy $E = \hbar^2 k^2 / 2M$. This means that in the low energy limit the reflection probability $|R|^2$ goes to unity even if the particle does not reach a classical turning point. Such a *quantum reflection* has indeed been observed in the scattering of an ultracold beam of metastable neon atoms from silicon and glass surfaces [144]. In order to reproduce the measured reflectivities, an $1/Z^4$ dependence of the potential has to be assumed [144, 280] which indicates that the atoms are scattered at the long-range tail of the Casimir-van der Waals potential (see page 104).

6.6 Dissociative Adsorption and Associative Desorption

In the case of dissociative adsorption there is another channel for energy transfer, which is the conversion of the kinetic and internal energy of the

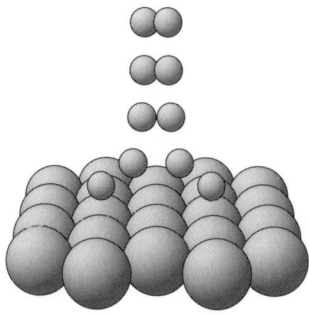

Fig. 6.10. Illustration of the dissociative adsorption process on a surface for a diatomic molecule

molecule into translational energy of the atomic fragments on the surface relative to each other. This process which is illustrated in Fig. 6.10 represents the fundamental difference to atomic or molecular adsorption. It is true that eventually the atomic fragments will also dissipate their kinetic energy and come to rest at the surface. However, especially in the case of light molecules like hydrogen dissociating on metal surfaces the energy transfer to the substrate is very small due to the large mass mismatch. Whether a molecule sticks on the surface or not is almost entirely determined by the bond-breaking process for which the energy transfer to the substrate can be neglected. This makes it possible to describe the dissociative adsorption process within low-dimensional potential energy surfaces neglecting the surface degrees of freedom if furthermore no substantial surface rearrangement upon adsorption occurs, as it is usually the case in the dissociative adsorption on close-packed metal surfaces.

The dynamics of the interaction of hydrogen with metal surfaces has been well-studied, both experimentally [281] and theoretically [233, 234, 282, 283]. The dynamics of hydrogen require a quantum mechanical description because of its light mass. Due to the high computational effort of quantum methods, for a long time the theoretical treatment was limited to studies within a reduced dimensionality. Only recently the first quantum studies were performed in which the full dimensionality of the hydrogen molecule was taken into account [236, 284].

As far as the activated adsorption is concerned, the interaction of hydrogen with copper surfaces has served as a model system [233, 238, 282, 284, 286–288]. It was also the first system for which high-dimensional potential energy surface were mapped out with DFT methods [235, 289, 290]. Based on an ab initio PES, six-dimensional wave-packet calculations of the dissociative adsorption of H_2/Cu(100) were performed [284, 285]. Figure 6.11 shows the calculated sticking probabilities for molecules that are initially non-rotating and either in the vibrational ground state or in the first excited state, respectively. These sticking probabilities are rather close to the experimental results which were derived from an analysis of both adsorption and desorption ex-

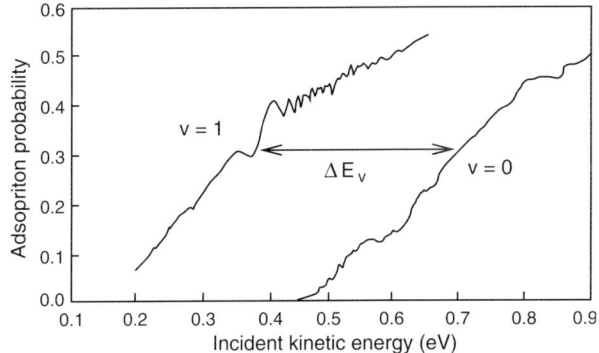

Fig. 6.11. Dissociative adsorption probability of H_2 on Cu(100) as a function of the incident kinetic energy determined by six-dimensional quantum wave-packet calculations for molecules initially in the vibrational ground state and first excited state, respectively. (After [285])

periments [291]. This indicates the reliability of both the DFT calculations determining the PES as well as of the quantum dynamics simulations.

The onset of the sticking probability at approximately 0.5 eV for H_2 molecules initially in the vibrational ground state is given by the minimum energy barrier including zero-point effects. The zero-point effects arise from the quantization of the molecular levels due to the localisation of the wave function in the degrees of freedom perpendicular to the reaction path at the minimum barrier position. The rise in the sticking probability is determined by the distribution of the barrier heights for dissociative adsorption in the multidimensional potential energy surface [238]. Thus sticking can be understood in terms of the region of the surface that classically is available to dissociation. This so-called *hole model* [292] is valid at high kinetic energies when the incoming particles are not significantly redirected by the shape of the potential energy surface.

As Fig. 6.11 demonstrates, in the system H_2/Cu the sticking probability is significantly enhanced if the impinging molecules are intially vibrationally excited. In order to quantify the effect the *vibrational efficacy* is introduced. It is defined as

$$\chi = \frac{\Delta E_v}{\hbar \omega_{\mathrm{vib}}}, \tag{6.50}$$

where ΔE_v is the energetic shift between the sticking curves for molecules in the vibrationally ground and first-excited state. In Fig. 6.11 we have indicated the energy shift which is of course not uniquely defined since the two sticking curves are not really parallel to each other. This shift is approximately 0.3 eV so that for the vibrational frequency of H_2, $\hbar \omega_{\mathrm{vib}} = 0.516$ eV, the vibrational efficacy is $\chi \approx 0.6$. This means that 60% of the vibrational energy is used to overcome the barrier for dissociative adsorption.

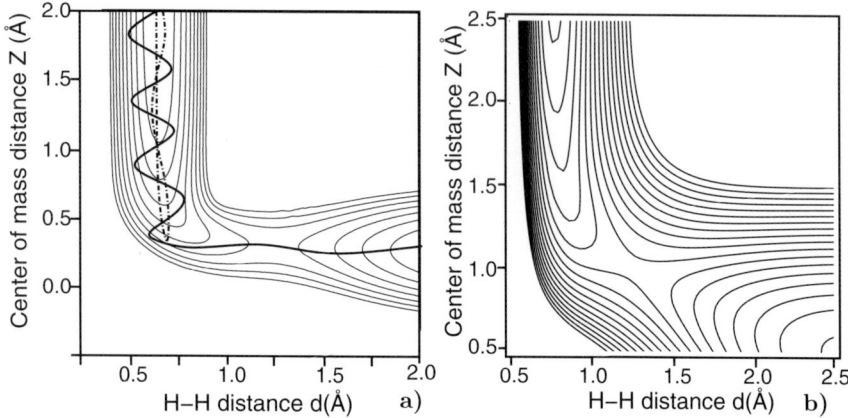

Fig. 6.12. Potential energy surfaces describing dissociative adsorption. The energy spacing between the contour lines is 0.1 eV. (**a**) Model potential with a late barrier. In addition, two typical trajectories are plotted illustrating the vibrationally enhanced dissociative adsorption process; (**b**) elbow plot of the minimum barrier in the system $H_2/Cu(100)$ determined by DFT-GGA calculations [76]

Vibrationally enhanced dissociation has been known for years in the gas phase dynamics [293]. The basic mechanism can be discussed within a two-dimensional elbow plot as shown in Fig. 6.12a. The plotted model PES corresponds to a so-called late barrier system which refers to the fact that the barrier is located after the curved region of the PES. Two typical trajectories are included in the elbow plot. A dissociation event corresponds to a trajectory that crosses the barrier and enters the exit channel. If initially non-vibrating molecules have a kinetic energy that is less than the barrier height, they are scattered back into the gas phase. However, if the molecule is already initially vibrating, i.e., if it is oscillating back and forth in the d-direction, then the vibrational energy can be very efficiently used "to make it around the curve" and enter the dissociation channel.

Potential energy surfaces such as the one plotted in Fig. 6.12a had been used to model the adsorption dynamics in the system H_2/Cu [294–296] before potential energy surfaces derived from electronic structure calculations became available [289, 290]. Figure 6.12b shows the elbow plot of the minimum barrier for $H_2/Cu(100)$. It is obvious that the curvature of the reaction path is much less than assumed in the model potential. Furthermore, the barrier is "earlier", i.e. it is not located after the curved region but rather in the curved region of the PES. Still the PES plotted in Fig. 6.12b produces the same vibrational efficacy of the PES of Fig. 6.12b. This is due to vibrationally adiabatic effects that will be explained in much more detail on p. 178.

While the system H_2/Cu serves as the benchmark system for *activated* dissociative adsorption on surfaces, H_2/Pd plays the same role for the non-activated adsorption [298]. Figure 6.13 compares the sticking probability for

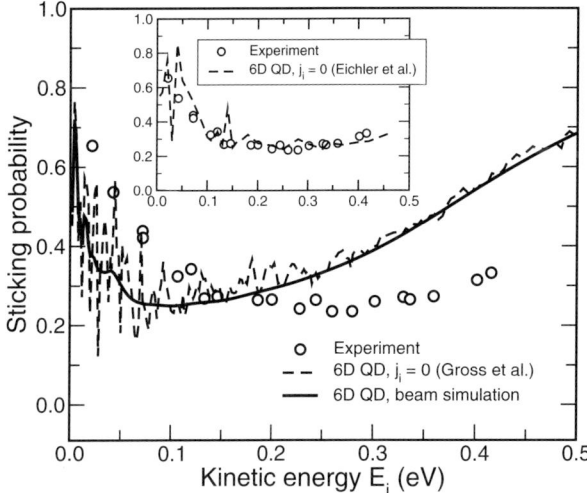

Fig. 6.13. Sticking probability of $H_2/Pd(100)$ as a function of the initial kinetic energy. Circles: experiment [193], dashed and solid line: theory according to H_2 intially in the ground state and with a thermal distribution appropriate for a molecular beam, respectively [236]. The inset shows the theoretical results using an improved ab initio potential energy surface [297]

$H_2/Pd(100)$ as a function of the kinetic energy obtained by molecular beam experiments [193] with the results of six-dimensional quantum calculations based on ab initio potential energy surfaces [236, 297]. The experiment shows an initial decrease of the sticking probability as a function of the kinetic energy while at larger kinetic energies the sticking probability slowly rises again.

The decrease of the sticking probability is typical for atomic or molecular adsorption (see Figs. 6.6 and 6.7). Consequently, the measured results were explained to be caused by the so-called *precursor mechanism* [193, 281]: before dissociation, the hydrogen molecule is assumed to be temporarily trapped in a molecular precursor state from which it then dissociates, and it is the trapping probability into the precursor state that determines the dependence of the sticking probability on the kinetic energy.

However, there is a large mass mismatch between the impinging hydrogen molecule and the palladium substrate. If one makes a simple hard-cube model analysis taking the mass of one palladium atom as the mass of the hard cube and assuming a typical precursor well depth of 0.25 eV, then the sticking probability is below one percent for kinetic energies above 85 meV. Furthermore, the calculated potential energy surface shows no evidence of a metastable prescursor state of H_2 at $Pd(100)$. Still the quantum results of the sticking probability [236] are in semi-quantitative agreement with the experiment. As the inset of Fig. 6.13 demonstrates, with an improved potential

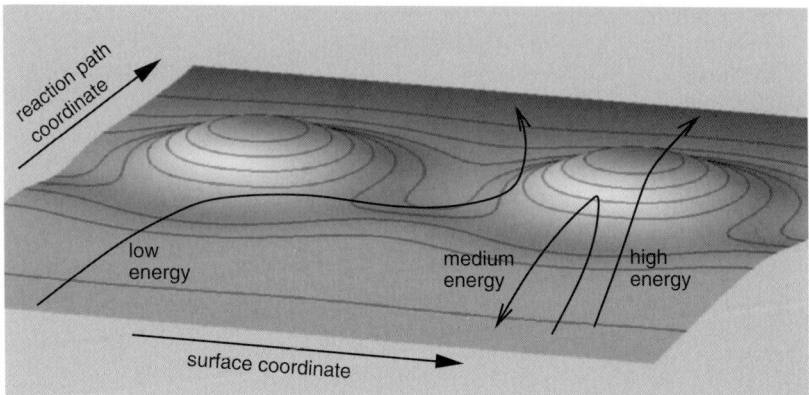

Fig. 6.14. Illustration of the steering effect. The potential energy surface is plotted as a function of the reaction path coordinate and one surface coordinate. The reaction path coordinate connects the molecule in the gas phase with the dissociated molecule on the surface. Three typical trajectories are included. By symmetry, the low and medium energy trajectories have the same initial conditions except for the initial kinetic energy

energy surface based on more ab initio points even quantitative agreement with the experiment can be achieved [297].

The advantage of a computer simulation compared to an experiment is that the simulation is performed under well-defined conditions and can be analyzed at any point of the simulation. This analysis showed that the initially decreasing sticking probability is caused by a dynamical process which had been proposed before [299] but whose efficiency had been grossly underestimated: dynamical steering. This process can only be understood if one takes the multi-dimensionality of the PES into account. The PES of $H_2/Pd(100)$ shows purely attractive paths towards dissociative adsorption, but the majority of reaction paths for different molecular orientations and impact points exhibits energetic barriers hindering the dissociation.

At very low kinetic energies the particles are so slow that they can be very efficiently steered to a favorable configuration for dissociation. This leads to a very high dissociation probability. Since this mechanism becomes less effective at higher kinetic energies, the reaction probability decreases. This scenario is illustrated in Fig. 6.14. A cut through the six-dimensional potential energy surface of $H_2/Pd(100)$ is plotted along the reaction path coordinate and one surface coordinate. The reaction path coordinate connects the molecule in the gas phase with the dissociated molecule on the surface. There is one purely attractive path in the center which corresponds to the dissociation at the hollow-bridge-hollow configuration indicated in Fig. 5.19a while the path directly over the maximum barrier in Fig. 6.14 (the "hilltop") represents the dissociation above the top site (Fig. 5.19b).

6.6 Dissociative Adsorption and Associative Desorption

Three typical trajectories are included in Fig. 6.14. The low and medium energy trajectories are related to each other by the symmetry along the surface coordinate. They are supposed to have the same initial conditions except for the initial kinetic energy. Both energies are too small to allow a direct crossing of the barrier the particles are directed at. However, at the low kinetic energy the forces acting on the incoming particle can redirect it so that it follows a path that leads to the purely attractive region of the PES. At the medium energy, of course the same forces act on the incoming particle. But now it is too fast to be steered significantly. It is reflected at the repulsive part of the potential and scattered back into the gas phase. This suppression of the steering effect for increasing kinetic energy leads to the initial decrease of the sticking probability in Fig. 6.13. If the energy is further increased, then the particle will eventually have enough kinetic energy to directly cross barriers, as the high-energy trajectory illustrates in Fig. 6.14. This leads to the rise of the sticking probability at high kinetic energies.

In general, the reactive trajectories are not always as simple as illustrated in Fig. 6.14. In particular in the low-energy regime, steered particles may not directly find a path through the barrier region. However, due the conversion of the initial kinetic energy into internal and lateral degrees of freedom, the particles may neither have enough kinetic energy to escape back into the gas phase. This leads to a dynamical trapping of the particles [300, 301]. In this state, the particles can bounce back and forth several times with respect to the surface [300] before their fate is decided. However, once a particle is dynamically trapped, the probability that it will eventually stick to the surface is much larger than the probability that it will scatter back into the gas phase. In quantum dynamics, the existence of these transient trapping states leads to the occurence of the peaked resonance structure in the sticking probability (see Fig. 6.13).

The most favorable path towards dissociative adsorption in the system $H_2/Pd(100)$ is purely attractive and has a rather small curvature (Fig. 5.19a). Therefore one would not expect any substantial influence of the vibrational state of H_2 on the sticking probability. Still the six-dimensional quantum calculations show a significantly larger dissociation probability for molecules in the first excited vibrational state $v = 1$ compared to the vibrational ground state [302] which is demonstrated in Fig. 6.15a. In fact, the energetic separation of the sticking curves for the two vibrational state is about 0.4 eV, as indicated in Fig. 6.15a, leading to a vibrational efficacy of $\chi \approx 0.75$ which is even larger than in the H_2/Cu system (see p. 173).

In order to clarify the nature of the vibrationally enhanced dissociation in the system $H_2/Pd(100)$, vibrationally adiabatic five-dimensional calculations have also been performed in which the molecules were kept in their initial vibrational state. Although the vibrational state is kept fixed, the vibrational frequency still changes along the reaction pathway. As Fig. 6.15a shows, the five-dimensional results are very close to the full dimensional 6-D results.

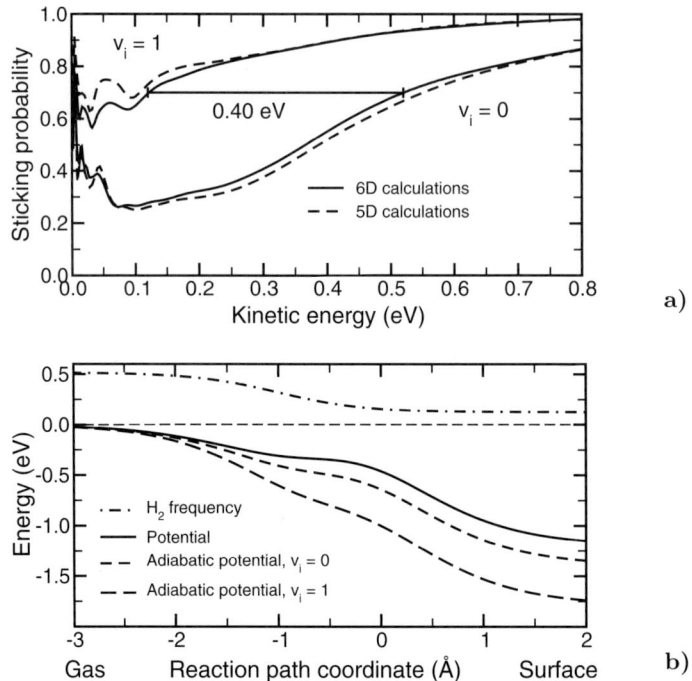

Fig. 6.15. Vibrational effects in the dissociation of $H_2/Pd(100)$. (**a**) Sticking probability for a H_2 beam initially in the vibrational ground state (*lower curves*) and the first excited state (*upper curves*). The results of six-dimensional and five-dimensional vibrationally adiabatic quantum calculations are shown. (**b**) H_2 vibrational frequency $\hbar\omega_{\text{vib}}$, potential and vibrational adiabatic potentials along the reaction path. (After [302])

This reflects two facts. First, the molecular vibrational state is a sufficiently good quantum number and is almost conserved during the scattering, i.e., the probability for transitions between different vibrational states during the scattering event is rather low. This can be understood from the fact that the vibrational motion corresponds to the fastest degree of freedom in this system so that the vibrational energy acts as an adiabatic invariant. And second, the curvature of the reaction path of the $H_2/Pd(100)$ PES is not crucial for the vibrational effects in this system because in the 5-D calculations no curvature is present in the Hamiltonian.

Consequently, the influence of the molecular vibrations on the dissociation can only be understood within an adiabatic picture. To see that, we introduce the "vibrationally adiabatic potentials" which are defined by

$$V_{\text{adia}}^{v_i}(s) = V_0(s) + (\hbar\omega(s) - \hbar\omega_{\text{vib}})\left(v_i + \frac{1}{2}\right), \qquad (6.51)$$

where s is the coordinate along the reaction path (see Fig. 5.19a). The vibrationally adiabatic potential is the relevant potential for the H_2 molecule moving on the PES in a fixed vibrational state. It takes the change of the vibrational frequency along the reaction path into account. In Fig. 6.15b) we have plotted the vibrational frequency together with the potential and the vibrational adiabatic potentials for $v_i = 0$ and $v_i = 1$ along the reaction path coordinate s. At $s = 0$, the point of maximum curvature along the reaction path, the vibrational frequency is strongly reduced from its gas phase value of $\hbar\omega = 516$ meV to about 150 meV. This leads to a lowering of the vibrationally adiabatic potential by 180 meV for $v_i = 0$ and by 550 meV for $v_i = 1$. Such a lowering does not only occur for the most favorable adsorption path, but also for other non-activated and activated pathways, i.e., for other impact sites in the surface unit cell and for other molecular orientations. There are now two points of view to describe the vibrational adiabatic effects. Either one says that the decrease of the vibrational frequency leads to an effective energy transfer from the vibrations to the translation which increases with the vibrational quantum number v, or one states that vibrationally excited molecules experience a potential energy surface with effectively lower barriers than molecules in the vibrational ground state. Both descriptions are equivalent and explain why vibrationally excited H_2 molecules have a higher dissociation probability on Pd(100). These adiabatic effects also contribute to the vibrational enhanced dissociation in the system H_2/Cu.

Information about the dissociation process at surfaces can not only be gained by studying the adsorption, but also by investigating the desorption of molecules. Adsorption and desorption are related to each other by time-reversal symmetry. Using the principle of detailed balance [258, 259], relative adsorption probabilities can be derived from the measurement of the state-resolved desorption flux. This has been used extensively to obtain the dependence of dissociative adsorption probabilities on the initial vibrational and rotational state (see, e.g., [291, 303]). However, it is important to note that detailed balance couples adsorption and desorption at a particular coverage and temperature. Adsorption experiments are usually done on a clean substrate at low temperatures while desorption fluxes are measured at high surface temperatures and high coverages.

Theoretically, the time-reversal symmetry of reaction probabilities can be directly employed to relate adsorption and desorption. We denote by $S_n(E_\perp, T_s)$ the state-specific sticking probabilities at a surface with temperature T_s as a function of the incident normal kinetic energy $E_\perp$. Here n stands for a multi-index that describes the initial vibrational, rotational and parallel momentum state of the molecule. The population D_n of the state n in desorption at a surface temperature of T_s is directly related to the sticking probability via

$$D_n(E_\perp, T_s) = \frac{1}{Z} S_n(E_\perp, T_s) \exp\left(-\frac{E_n + E_\perp}{k_B T_s}\right). \tag{6.52}$$

In the desorption distribution (6.52), $E_\perp$ is the kinetic energy perpendicular to the surface of the desorbing particles, E_n is the energy associated with

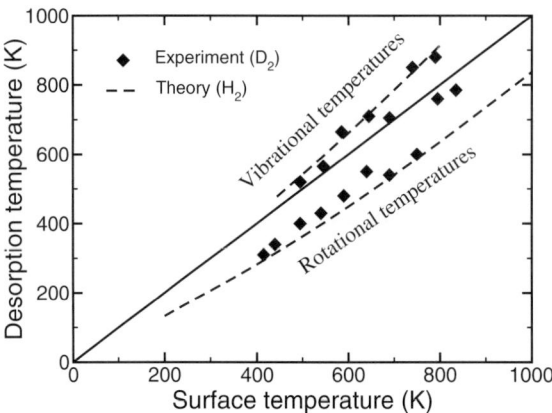

Fig. 6.16. Vibrational and rotational temperatures of hydrogen desorbing from Pd(100) as a function of the surface temperature. The experimental results have been determined by tunable vacuum ultraviolet laser ionization spectroscopy for D_2 while the theoretical results have been derived from six-dimensional quantum calculations for H_2 (After [304])

the internal state n, and Z is the partition sum that ensures the normalization of the distribution. In quantum dynamical simulations of the dissociative adsorption, often the substrate is kept fixed so that it does not participate dynamically in the adsorption/desorption process. Still desorption distributions can be derived by assuming that the substrate acts as a heat bath that determines the population distribution of the molecular states on the surface.

The vibrational distribution in desorption is often characterized by the so-called *vibrational temperature* as a function of the surface temperature. Classically it corresponds to the mean vibrational energy in desorption at the surface temperature T_s converted into a temperature by dividing it by the Boltzmann constant

$$T_{\rm vib}(T_s) = \langle E_{\rm vib}\rangle_{T_s}/k_B \,. \tag{6.53}$$

The vibrational energy quantum $\hbar\omega_{\rm vib}$ is usually large compared to $k_B T_s$. Hence only the vibrational ground and first excited state are significantly populated in desorption. Therefore the vibrational temperature is normally determined via

$$T_{\rm vib}(T_s) = \frac{\hbar\omega_{\rm vib}}{k_B \ln(D(v=0,T_s)/D(v=1,T_s))}, \tag{6.54}$$

where $D(v=n, T_s)$ is the population of the n-th vibrational level. It should be noted that the vibrational temperature does not correspond to a real temperature. It is rather a parameter characterizing the distribution. If the sticking probability is independent of the vibrational state, then the vibrational temperature equals the surface temperature. A sticking probability increasing with the vibrational quantum number leads to vibrational temperatures in desorption that are larger than the surface temperature. This is called *vibrational heating*. Conversely, if vibrational excitation suppressed the sticking probability, vibrational cooling would result.

The measured and calculated vibrational temperatures of hydrogen desorbing from Pd(100) are plotted in Fig. 6.16. The experimental results have been obtained by tunable vacuum ultraviolet laser ionization spectroscopy for D_2 [304]. Deuterium is often used in desorption experiments because the results are hardly influenced by background signals in contrast to the case of H_2 for which there is always an unavoidable background in the vacuum chambers. The calculations, on the other hand, are done for H_2 because of the much smaller computational effort for light hydrogen in quantum methods. Still both experiment and theory agree well as far as the vibrational heating is concerned.

Figure 6.16 also shows the rotational temperatures as a function of the surface temperature determined via

$$T_{\rm rot}(T_s) = \langle E_{\rm rot}\rangle_{T_s}/k_B \,. \tag{6.55}$$

In contrast to the vibrations, the rotational temperatures in desorption are below the value expected for thermal equilibrium. According to the principle of detailed balance, this rotational cooling in desorption should be reflected by rotational hindering in adsorption, i.e. a suppression of the sticking probability for rapidly rotating molecules.

This rotational hindering of the steering effect has actually been confirmed in the system H_2/Pd(111) [305, 306] where also steering dominates the dissociative adsorption at low kinetic energies. By seeding techniques the translational energy of a H_2 beam has been changed in a nozzle experiment without altering the rotational population of the beam. The rotationally hot beams showed a much smaller sticking probability than rotationally cold beams [305, 306]. The mechanism underlying the rotational hindering in dissociative adsorption is illustrated in Fig. 6.17. Hydrogen molecules can only dissociate on metal surfaces if their axis is parallel to the surface. A rapidly rotating molecule hitting the surface might in fact be in this favorable orientation for a short time, but it will rotate out of this orientation during the time it takes to complete the dissociation. In the upright position, the interaction between the molecule and the surface is purely repulsive. Thus rotating molecules have a higher probability to probe the repulsive regions of the potentials which leads to an enhanced reflectivity and consequently to a suppression of the sticking probability.

However, in state-resolved experiments it has been observed in the systems H_2/Cu and H_2/Pd that for higher rotational quantum numbers j the sticking probability increases again [303, 306]. This effect was confirmed in three-dimensional quantum calculations for H_2/Cu(111) [307]. It is caused by an adiabatic effect [307] similar to the one resulting from the the lowering of the vibrational frequency upon adsorption. Along the reaction path,

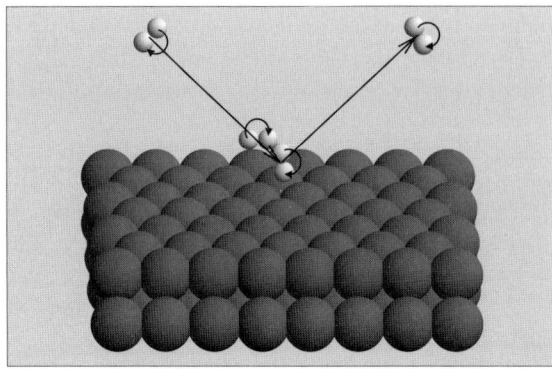

Fig. 6.17. Schematic trajectory illustrating the rotational hindering in dissociative adsorption. A rapidly rotating molecule impinging on the surface will rotate out of a favorable orientation for dissociation, in experiences a repulsive interaction and is scattered back into the gas phase

the bond length $r_e(s)$ and thus the moment of inertia $I = \mu r_e^2(s)$ increases. Consequently, the rotational energy

$$E_{\rm rot}(j) = \frac{\hbar^2 j(j+1)}{2I} = \frac{\hbar^2 j(j+1)}{2\mu r_e^2(s)} \tag{6.56}$$

associated with the rotational state j decreases. Assuming that the dissociation occures rotationally adiabatic, i.e., without any change in the rotational quantum state, this would correspond to an effective energy transfer from the rotation to the translation which increases with the rotational quantum number.

The rotational effects in the system H_2/Pd have been studied in detail in five-dimensional vibrational adiabatic quantum calculations [308]. As the potential energy surface, a simplified version of the ab initio PES for the system H_2/Pd(100) was used in order to allow a systematic variation of certain features of the potential. In Fig. 6.18, the sticking probabilities are plotted for a range of initial rotational states j as a function of the kinetic energy under normal incidence. As for the initial rotational state, in addition to the rotational quantum number j also the azimuthal quantum number m has to be specified. In the field of gas-surface dynamics m is usually defined with respect to the surface normal. Molecules rotating with $m = 0$ have their rotational axis preferentially oriented parallel to the surface. This means that they correspond to molecules rotating in the cartwheel fashion which is illustrated in Fig. 6.8. Consequently, azimuthal quantum numbers $m = j$ indicate molecules rotating in the helicopter fashion with their rotational axis preferentially perpendicular to the surface.

Figure 6.18a demonstrates that rotational motion with $(j = 2, m = 0)$ leads to a suppression of the sticking probability compared to non-rotating molecules, but a closer look reveals that there is no monotonous trend in

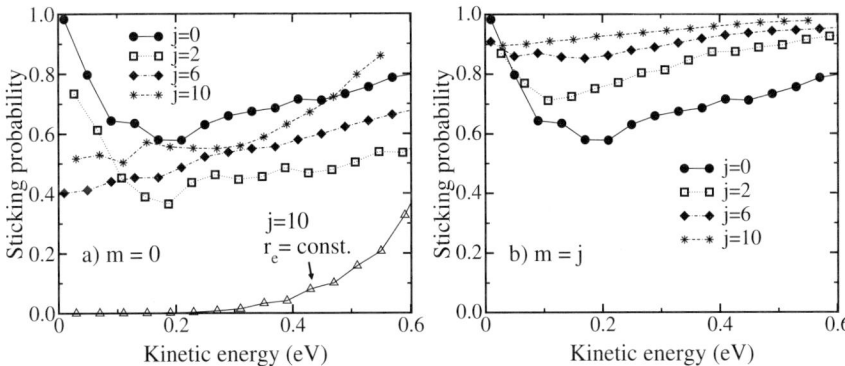

Fig. 6.18. Five-dimensional calculation of the sticking probability of $D_2/Pd(100)$ for a range of initial rotational states j as a function of kinetic energy under normal incidence. (**a**) Molecules initially rotating in the cartwheel fashion $m = 0$. For $j = 10$, addiationally the results for a fixed hydrogen bond length are plotted. (**b**) Molecules initially rotating in the helicopter fashion $m = j$

the dependence of the sticking probability on j. Over the whole considered energy range the sticking probability only decreases by going from $j = 0$ to $j = 2$, but for higher quantum numbers the trend is reversed. Increasing the rotational quantum number from $J = 6$ to $J = 10$ does not lead to a further suppression but rather to a rising sticking probability.

In order to confirm that the eventual increase in the sticking probability at higher j is indeed caused by the adiabatic rotational-translational energy transfer due to the lowering of the rotational energy quantum, the five-dimensional quantum calculations have been repeated using exactly the same potential energy surface except for the hydrogen bond length and consequently also the moment of inertia which were kept frozen at their gas phase values. The corresponding results for $j = 10$ are also included in Fig. 6.18a. And indeed, if the elongation of the molecular bond length along the reaction path is not taken into account in the Hamiltonian, the sticking probability is strongly suppressed for rapidly rotating molecules with $j = 10$, i.e., there is pure rotational hindering.

The sticking probabilities for molecules rotating in the helicopter fashion are plotted in Fig. 6.18b. Interestingly enough, no rotational hindering is apparent. The sticking probability is monotonously rising with increasing rotational quantum number j. In addition to the effective rotational-translational energy transfer, there is a steric effect promoting the dissociative adsorption. Helicopter molecules have their axis already parallel to the surface which is favorable for dissociation while cartwheel molecules have a high probability to hit the surface in the unfavorable upright orientation. In fact, the alignment of the molecular axis parallel to the surface is the better, the higher the rotational quantum number is. This counteracts the rotational hindering which is still present due to the azimuthal anisotropy of the potential en-

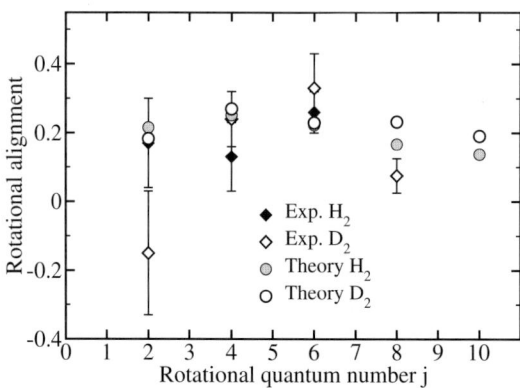

Fig. 6.19. Rotational alignment in the desorption of hydrogen from Pd(100) at a surface temperature of $T_s = 700$ K. Comparison of the experimental results for H_2 [309] and D_2 [310] with five-dimensional quantum calculations [308]

ergy surface. This steric effect of the rotational motion in adsorption has not been observed yet because it is difficult to align rotating hydrogen molecules in the gas phase. Again, the way out is to look at the time-reverse process, the associative desorption. Using state-specific laser techniques, it is possible to measure the *rotational alignment* of desorbing molecules. The alignment parameters $A_0^{(k)}$ contain the complete dynamical information about a reaction product. They correspond to the expectation values of the monopole, quadrupole and higher multipole moments of the angular momentum operators J:

$$A_0^0 = 1 \tag{6.57}$$

$$A_0^{(2)} = \left\langle \frac{3J_z^2 - J^2}{J^2} \right\rangle , \tag{6.58}$$

where J_z is the z-component of the angular momentum operator J. The values of the alignment parameter $A_0^{(2)}$ lie in the range

$$-1 \leq A_0^{(2)} \leq 2 \tag{6.59}$$

An alignment parameter of $A_0^{(2)} \leq 0$ means that the cartwheel rotations are preferentially populated in desorption while positive alignment $A_0^{(2)} > 0$ reflects a preferential population of helicopter states. A comparison between the measured rotational alignment for H_2 [309] and D_2 [310] with the five-dimensional quantum calculations [308] is shown in Fig. 6.19. Within the experimental uncertainty, there is a satisfactory agreement between theory and experiment. Except for the $j = 2$ experimental result of D_2, the rotational alignment parameters are positive reflecting the fact that the parallel orientation of the molecules is favorable for dissociation. In agreement with the experiment, the theoretical results show a non-monotonous behavior. The rotational alignment parameters show a maximum for $j = 6$ and decrease then again. This is caused by the same mechanism that is responsible for the non-monotonous behavior of the sticking probability of cartwheel molecules as a function of the rotational quantum number. The rotationally adiabatic effects reduce the difference between cartwheel and helicopter dissociation probabilities and thus lead to a smaller rotational alignment. This has again been confirmed by determining the rotational alignment with a fixed hydrogen bond length. In this case the alignment is a monotonously increasing function of j [308].

The role of the anisotropy of the PES is formally equivalent to the influence of the corrugation on the sticking probability. The energetic corrugation, i.e. the variation of the barrier height within the surface unit cell, leads to a suppression of the sticking probability for non-normal incidence [311, 312]. On the other hand, in the case of geometric corrugation, i.e. the variation of the barrier location within the surface unit cell, the section of the surface unit cell in which the incoming beam hits barriers locally in a perpendicular fashion is increased for non-normal incidence which enhances the sticking probability.

In the case of dissociative adsorption on surfaces, the higher barriers are often further away from the surface than the lower ones. These features are for example present in the calculated PES of H_2/Cu [289, 290]. For this so-called balanced corrugation [233], the opposing effects of energetic and geometric corrugation can cancel each other to a large extent [307]. This effect is the reason for the observed normal energy scaling of H_2/Cu [313] in spite of the strong corrugation of this system.

In contrast to close-packed metal surfaces, semiconductor surfaces show a strong surface rearrangement upon hydrogen adsorption. This is caused by the covalent nature of the bonding in semiconductors where an additional chemisorbed adsorbate strongly perturbs the bonding situation of the substrate. Hydrogen on silicon has become *the* model system for the study of adsorption on semiconductor surfaces [234, 314]. This system is not only interesting from a fundamental point of view, but also because of its technological relevance. Hydrogen desorption is the rate determining step in the growth of silicon wafers from the chemical vapour deposition (CVD) of silane.

This system provides a good example of the close fruitful collaboration between experiment and theory that is possible in surface science, but it also demonstrates that progress is not always achieved in a straightforward, but rather in an erratic way. Therefore the chronological progress in the understanding of this system will be sketched here. Desorption of hydrogen from Si(100) shows first-order kinetics [315, 316]. For associative desorption one normally expects second-order kinetics since two atoms have to find each other on the surface before they can desorb. The unusual first-order desorption kinetics has been explained by a *prepairing mechanism* [316, 317]: Desorbing molecules originate from the same dimer since it is energetically favorable for two hydrogen atoms to bind on the same dimer rather than on two independent dimers.

The interest in this system was even further increased by the so-called barrier puzzle: While the sticking coefficient of molecular hydrogen on Si surfaces is very small [318, 319] indicating a high barrier to adsorption [320], the low mean kinetic energy of desorbed molecules [321] suggests a small adsorption barrier. This puzzle was assumed to be caused by the strong surface rearrangement of Si upon hydrogen adsorption [320, 321]: The hydrogen molecules impinging on the Si substrate from the gas phase typically encounter a Si configuration which is unfavorable for dissociation, while desorbing hydrogen molecules leave the surface from a rearranged Si configuration with a low barrier. In the adsorption process such a surface rearrangement can only be achieved by thermal excitations of the lattice since due to the mass mismatch the silicon substrate atoms are too inert to change their configuration during the time the hydrogen molecule interacts with the surface.

It was immediately realized that the large influence of the dissociation barrier on the substrate configuration should cause a strong surface temperature dependence of the hydrogen dissociation probability on Si [320]. This proposed phonon assisted sticking motivated experimental studies of the temperature dependence of the sticking probability which indeed confirmed the theoretical predictions [319, 322, 323]. However, the microscopic details of the H_2 dissociation path in the coupling to Si substrate degrees of freedom remained unclear. Furthermore, the situation was confusing because of large quantitative discrepancies between different experiments [318, 323].

As for the ab initio electronic structure calculations, H_2/Si(100) is a system that initially was studied extensively by quantum chemical methods in which the extended substrate was modeled by finite clusters [324–326], such as the one shown in Fig. 3.5. Due to the localized nature of the covalent bonds in semiconductors it was believed that the cluster description for a Si surface might be appropriate. These cluster studies could not reproduce the experimentally observed activation energy for H_2 adsorption from Si(100) for a clean surface. Therefore defect-mediated desorption mechanisms had been proposed [324, 326, 327]. DFT calculations based on the slab approach, however, were in good agreement with experiment, as far as the desorption barrier was concerned [230, 328, 329]. There were speculations whether the difference between cluster and slab calculations was due to the different treatment of the electron exchange-correlation [330]. A later study showed [331] that one has to use rather large clusters to appropriately model the extended Si substrate. For example, the correct buckled structure of the Si(100) surface is only reproduced in cluster calculations if more than one surface dimer is included in the cluster.

There are two possible dissociation pathways for hydrogen on clean Si(100): the *intradimer* pathway where the hydrogen atoms of the dissociating molecule end up on both ends of a dimer, and the *interdimer* pathway

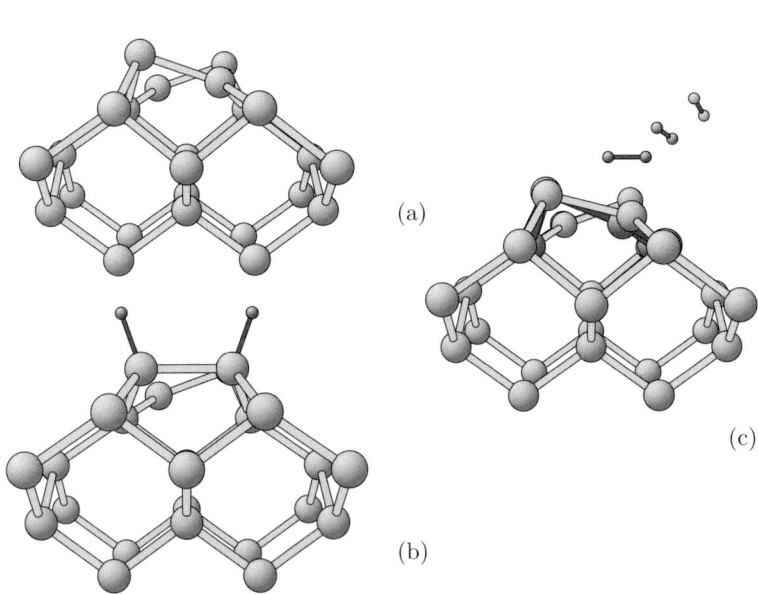

Fig. 6.20. Structural changes of the Si(100) surface upon hydrogen adsorption determined with DFT-GGA calculations [230, 328, 329]. (**a**) (2 × 2) structure for the clean surface with anti-buckled dimers. (**b**) Symmetric surface dimer upon the dissociative intradimer adsorption of a hydrogen molecules. (**c**) Snapshots of an ab initio molecular dynamics simulation of a hydrogen molecule desorbing from Si(100). The resulting surface relaxation is indicated by the darker Si atoms [230]

where the H-H bond is oriented perpendicular to the Si dimers and the hydrogen atoms adsorb at two neighboring dimers. An earlier study suggested that the adsorption barrier of the interdimer pathway is approximately 0.3 eV higher than the intradimer barrier [332]. Therefore, most DFT slab studies first focused on the intradimer pathway [230, 328, 329] which is illustrated in Fig. 6.20. Upon adsorption of H_2 on a Si dimer, the buckling of the dimer (Fig. 6.20a) is lifted and the dimer becomes symmetric in the monohydride phase (Fig. 6.20b). This strong surface rearrangement was considered as a possible candidate responsible for the barrier puzzle [328, 329]. Ab initio molecular dynamics simulations were performed in order to determine the energy distribution of hydrogen molecules desorbing from Si(100) [230]. Snapshots of one of the forty calculated trajectories are shown in Fig. 6.20c. The dark Si atoms correspond to the relaxation of the Si lattice after the desorption event. Approximately 0.1 eV of the potential energy at the transition state is transferred to vibrations of the Si lattice. The simulations reproduced the vibrational heating and the rotational cooling observed in the desorption experiments [314]. However, the kinetic energy in desorption was still much larger in the ab initio molecular dynamics runs than in the experiment [321]. This is due to the fact that the elastic energy of the surface frozen in the transition state configuration is only about 0.15 eV [329] which is too little in order to explain the barrier puzzle.

Later it turned out that it is not sufficient to focus on the H_2 dissociation on clean Si(100). Instead it is important to consider the influence of the hydrogen coverage on the adsorption/desorption barriers [333, 334]. The progress in the understanding has been achieved in a close collaboration between experiment and electronic structure calculations. It was realized that it is very important to determine the exact surface structure. At surface imperfections like steps the reactivity of a surface can be extremely altered. Indeed it was found experimentally on vicinal Si(100) surfaces that the sticking coefficient at steps is up to six orders of magnitude higher than on the flat terraces [335]. This finding was supported by DFT studies which showed that non-activated dissociation of H_2 on the so-called rebonded D_B steps on Si(100) is possible [335, 336], while on the flat Si(100) terraces the dissociative adsorption is hindered by a barrier of 0.4 eV [230].

Indeed adsorbates can have a similar effect on the dissociation probability as steps since the electronic structure of the dangling bonds is perturbed in a similar way by both steps and adsorbates [333]. Recent scanning tunneling microscope (STM) experiments demonstrated that predosing the Si(100) surface by *atomic* hydrogen creates active sites at which the H_2 adsorption is considerably facilitated [337]. Actually the predosing of atomic hydrogen makes the adsorption of H_2 in an interdimer configuration possible. This renewed the interest in the theoretical study of the interdimer pathway. The interdimer pathway was revisited by DFT-GGA calculations [333] which in fact found that its barrier is *smaller* than the barrier along the intradimer pathway. The discrepancy to the former calculations [332] was attributed to the different transition state geometry considered. The DFT-GGA calculations further confirmed that on hydrogen-precovered Si(100) highly reactive sites exist at which H_2 can spontaneously dissociate. Hence now a consistent picture of the adsorption/desorption of H_2/Si(100) has emerged. From hydrogen-covered Si(100) at full coverage, H_2 molecules can desorb without being accelerated towards the gas phase which explains the low kinetic energy measured in desorption experiments. In adsorption experiments, on the other hand, this dissociation path without a barrier is not present at clean Si(100) leading to the small observed sticking probability. At intermediate coverages, both activated as well as non-activated adsorption paths are present leading to a crossover from activated dissociation dynamics to non-activated dissociation dynamics.

Exercises

6.1 Verlet Algorithm

a) Show that the velocity Verlet algorithm

$$r_i(t+h) = r_i(t) + h\, v_i(t) + \frac{h^2}{2}\frac{F_i(t)}{m}$$
$$v_i(t+h) = v_i(t) + h\,\frac{F_i(t+h) + F_i(t)}{2m} \qquad (6.60)$$

is mathematically equivalent to the Verlet algorithm

$$r_i(t+h) = 2r_i(t) - r_i(t-h) + \frac{h^2}{2}\frac{F_i(t)}{m}. \qquad (6.61)$$

b) Now consider a diatomic molecule interacting with a frozen substrate described using center-of-mass coordinates. The relative coordinates of the molecule are given in spherical coordinates (r, θ, ϕ). Try to formulate the Verlet algorithm using spherical coordinates. Why does this not work?

6.2 Isotope Effect in Classical and Quantum Dynamics

a) Show that for a given interaction potential, the sticking probability as a function of the kinetic energy only depends on the mass ratio between the considered atoms in classical mechanics. In particular, this means that for a system of atoms with equal masses the sticking probability does not depend on the mass of the atoms [338], i.e. there cannot be an isotope effects in such systems in classical dynamics.
Hint: Consider the Lagragian of the system and perform appropriate transformations.

b) Why is this not true in quantum mechanics?

6.3 Energy Transfer in Collision on Surfaces

Consider the elastic binary collision between an atom (mass m) and a substrate atom (mass M) as a model for the interaction of a gas phase atom with a surface. μ denotes the mass ratio $\mu = m/M$.

a) Show that the energy transfer $\Delta = E_i - E_f$ to a substrate atom is given by

$$\Delta = \frac{4\mu}{(1+\mu)^2}\, E_i, \qquad (6.62)$$

if the substrate atom is initially at rest.

b) In order to take the surface temperature T_s into account, we assume that the surface atom moves with an initial velocity v_s before the collision. Show that now an energy

$$\Delta = \frac{4\mu}{(1+\mu)^2}\left(E_i - \frac{1}{2}k_B T_s\right) \tag{6.63}$$

is transferred to the substrate.

Hint: Average the energy transfer to the substrate over a velocity distribution with $\langle v_s \rangle = 0$ and $\langle v_s^2 \rangle = k_B T_s / M$

c) In the hard-cube model [265, 266], a weighted Maxwellian velocity distribution

$$P_c(v_c) = \frac{\alpha}{\sqrt{\pi}v_g}(v_g - v_c)\exp\{-\alpha^2 v_c^2\} \tag{6.64}$$

is used for the velocity of the cube which represents the substrate. Proove that this velocity distribution yields an energy transfer of

$$\Delta = \frac{4\mu}{(1+\mu)^2}\left\{E_i - \left(1 - \frac{\mu}{2}\right)k_B T_s\right\}. \tag{6.65}$$

6.4 Hard-Cube Model

Verify that the trapping probability

$$P_{\text{trap}}(v_g) = \int_{-\infty}^{v_{\text{lim}}} P_c(v_c) dv_c. \tag{6.66}$$

in the hard-cube model is given by

$$P_{\text{trap}}(v_g) = \frac{1}{2} + \frac{1}{2}\text{erf}(\alpha v_{\text{lim}}) + \frac{\exp\{-\alpha^2 v_{\text{lim}}^2\}}{2\sqrt{\pi}\alpha v_{\text{well}}} \tag{6.67}$$

with

$$v_{\text{lim}} = \frac{\mu+1}{2}\sqrt{\frac{2E_{\text{ad}}}{m}} + \frac{\mu-1}{2}\sqrt{v_g^2 + \frac{2E_{\text{ad}}}{m}}, \tag{6.68}$$

where $P_c(v_c)$ is the weighted Maxwellian velocity distribution (6.64).

7. Kinetic Modelling of Processes on Surfaces

Many processes on surfaces such as diffusion, desorption etc. are hindered by large energetic barriers. On a microscopic time scale, these processes occur very rarely. Many unsuccessful attempts are performed before eventually the corresponding barrier is crossed. The time between two successful events can easily be in the order of nanoseconds or even longer. In any microscopic molecular dynamics simulation all unsuccessful events are explicitly included. Since the time scale of MD runs is typically limited to picoseconds, the simulation of these rare events is prohibited. Besides, such a simulation would mean a waste of computer time because a lot of useless information would be gathered.

Therefore for the simulation of these processes a kinetic approach is necessary in which the single processes are described by the corresponding rates. In this chapter we will first show how the rates can be determined from microscopic information via transition state theory. We will then show how processes such as diffusion and growth can be described either by rate equations or by kinetic Monte Carlo simulations. This allows to extend the information gained from microscopic electronic structure calculations to simulations on mesoscopic or even macroscopic time and length scales which will be illustrated in detail.

7.1 Determination of Rates

Experimentally it is well established that the rate of many processes as a function of temperature follows the Arrhenius behavior

$$k = k_0 \, \exp\left(-\frac{E_a}{k_\mathrm{B} T}\right). \tag{7.1}$$

In (7.1), E_a is the apparent activation energy which is usually interpreted as the minimum barrier hindering the particular process. From a microscopic point of view, it is desirable to derive an expression for a rate from the properties of the underlying potential energy surface. This can in fact be done using transition state theory (TST) [339].

Consider a potential $V(x)$ along the reaction path coordinate x connecting to locally stable states (Fig. 7.1). At $x = x_\mathrm{T} = 0$ the transition state

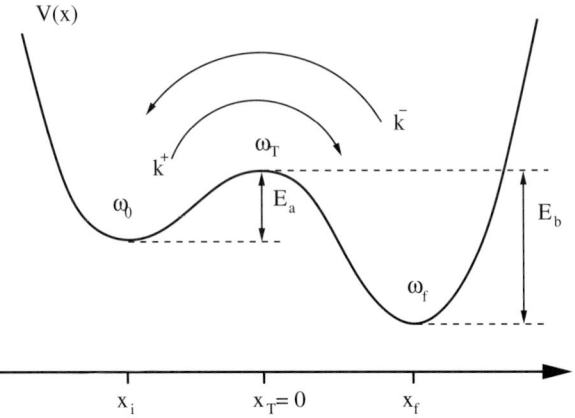

Fig. 7.1. Schematic representation of escape processes between to locally stable states with forward rate k^+ and backward rate k^-

is located. The barrier for the forward escape is given by E_a while the corresponding barrier for the backward process is E_b. In the following we are mainly concerned with the derivation of the forward rate k^+. The frequency of particles vibrating in the initial well around x_i is given by ω_0. The transition state is characterized by an imaginary frequency ω_T.

There are two basic assumptions underlying transition state theory. First, it is assumed that the moving particles are sufficiently strongly coupled to a heat bath so that there is local thermodynamic equilibrium along the whole reaction path. Secondly, the transition state corresponds to a point of no return which means that any trajectory passing through the transition state will not recross it. This last assumption is inherently coupled to classical mechanics because in quantum mechanics it does not make sense to speak of single trajectories with no recrossings. Hence transition-state theory is fundamentally a classical mechanical theory, although the concept can be generalized to consider the leading quantum corrections within semiclassical quantum theory [339].

Using the basic assumptions mentioned above, the equilibrium average of the one-way forward flux at the transition state can be expressed as

$$k^+_{\text{TST}} = \frac{\langle \delta(x)\, \dot{x}\, \Theta(\dot{x}) \rangle}{\langle \Theta(-x) \rangle} . \tag{7.2}$$

Here the average $\langle \ldots \rangle$ denotes the thermal expectation value. Θ is the step function defined by

$$\Theta(x) = \begin{cases} 1, & \text{for } x > 0 \\ 0, & \text{for } x < 0 \end{cases} . \tag{7.3}$$

Hence $\langle \Theta(-x) \rangle = \bar{n}_i$ corresponds to the equilibrium population for $x < 0$, i.e., of the initial state i in the left well of Fig. 7.1. It is important to note that the rate (7.2) always gives an upper bound for the true rate, i.e. $k_{\mathrm{TST}}^+ \geq k$, since recrossing of reactive trajectories are neglected.

In order to derive the rate expression within TST, we start with the simple case of an one-dimensional system according to Fig. 7.1. A particle of mass m moves in the potential $V(x)$ with two local minima. Using (7.2), the transition-state forward rate from state i to state f is given by

$$k_{\mathrm{TST}} = Z_0^{-1} \frac{1}{2\pi\hbar} \int dq\, dp\, \delta(q)\, \dot{q}\, \theta(\dot{q})\, \exp(-\beta H(q,p)), \tag{7.4}$$

where we have identified the variable q with the reaction-path coordinate x. $\beta = 1/k_{\mathrm{B}}T$ is the inverse temperature and Z_0 denotes the partition sum in the initial well

$$Z_0 = \frac{1}{2\pi\hbar} \int_{q<0} dq\, dp\, \exp(-\beta H(q,p)). \tag{7.5}$$

The integral over the momentum coordinate p in (7.4) can easily be evaluated to yield

$$\int_{-\infty}^{\infty} dp\, \dot{q}\, \theta(\dot{q})\, \exp\left(-\beta \frac{p^2}{2m}\right) = \int_0^{\infty} dp\, \frac{p}{m} \exp\left(-\frac{p^2}{2mk_{\mathrm{B}}T}\right)$$

$$= k_{\mathrm{B}}T, \tag{7.6}$$

while the integral over the coordinate q simply gives

$$\int dq\, \delta(q)\, \exp\left(-\frac{V(q)}{k_{\mathrm{B}}T}\right) = \exp\left(-\frac{E_a}{k_{\mathrm{B}}T}\right). \tag{7.7}$$

Inserting the expressions for the integrals into (7.4), we get for the reaction rate

$$k_{\mathrm{TST}} = \frac{k_{\mathrm{B}}T}{h} \frac{1}{Z_0} \exp\left(-\frac{E_a}{k_{\mathrm{B}}T}\right). \tag{7.8}$$

In the harmonic approximation, the partition sum in the initial well is given by $Z_0 = k_{\mathrm{B}}T/\hbar\omega_0$, i.e., we obtain the following rate

$$k_{\mathrm{TST}} = \frac{\omega_0}{2\pi} \exp\left(-\frac{E_a}{k_{\mathrm{B}}T}\right). \tag{7.9}$$

Here we already see that the prefactor before the exponential in (7.9) corresponds to an attempt frequency that yields the number of attempts that the particle tries to get over the barrier. The Boltzmann factor then gives the thermal probability that the particle has enough energy to cross the transition state.

This formalism can relatively easily be extended to the multi-dimensional case starting from the flux expression (7.2), in particular for the case of a nonlinear coordinate ($i = 0$) coupled to N vibrational degrees of freedom.

In the harmonic approximation, which is valid for $\hbar\omega \ll k_BT$, the partition sums Z_0 and Z_{TS} in the initial well and at the transition state are simply given by

$$Z_0 = \prod_{i=0}^{N}\left\{\frac{k_BT}{\hbar\omega_i^{(0)}}\right\}, \quad Z_{TS} = \prod_{i=1}^{N}\left\{\frac{k_BT}{\hbar\omega_i^{TS}}\right\}, \tag{7.10}$$

where $\omega_i^{(0)}$ and ω_i^{TS} are the vibrational frequencies in the initial well and the transition state, respectively. We note in passing that often the transition state is denoted by ($\neq$), i.e. $Z_{TS} \equiv Z^{\neq}$. The transition rate can now be expressed as [339]

$$\begin{aligned}k_{TST} &= \frac{k_BT}{h}\frac{Z^{TS}}{Z_0}\exp\left(-\frac{E_a}{k_BT}\right)\\&= \frac{1}{2\pi}\frac{\prod_{i=0}^{N}\omega_i^{(0)}}{\prod_{i=1}^{N}\omega_i^{TS}}\exp\left(-\frac{E_a}{k_BT}\right).\end{aligned} \tag{7.11}$$

This result for the TST rate can be reformulated employing the Helmholtz free energy F via the substitution

$$Z = \exp[-(E-TS)/(k_BT)] = \exp[-F/(k_BT)] \tag{7.12}$$

Inserting this expression into (7.11), we arrive at

$$k_{TST} = \frac{k_BT}{h}\exp\left(\frac{\Delta S}{k_B}\right)\exp\left(-\frac{E_a}{k_BT}\right) = k_0\exp\left(-\frac{E_a}{k_BT}\right), \tag{7.13}$$

where $\Delta S = S_{TS} - S_0$ is the entropy change and S_{TS} and S_0 are the entropy of the $2N$-dimensional phase space at the transition state and the $2(N+1)$-dimensional phase space in the initial well. Hence we have derived the Arrhenius expression (7.1) from transition state theory. Note that the prefactor k_0 which is usually assumed to be temperature-independent has in fact a linear dependence on the temperature according to (7.13). However, since the temperature dependence is dominated by the exponential term, it is still often justified to neglect the temperature dependence of the prefactor.

In principle, all variables in (7.11) can be evaluated from electronic strucuture calculations since they are all related to the potential energy surface. However, usually the determination of eigenmode-frequencies is computationally very demanding since it involves the evaluation and diagonalization of a Hesse matrix. Hence often the prefactor in (7.11) is just estimated and only barrier heights are computed from first principles.

Furthermore, it is not trivial to locate the transition state. There is no uniquely defined way of finding the minimum energy path between a given initial and final state. One way to determine this path is to simply map out the relevant potential energy surface in great detail, but this can be computationally very demanding. However, there are robust methods that can be used to find a minimum energy path. In the *nudged elastic band method* [340,341], first the energy and the forces of the system are determined along a string usually interpolating linearly between initial and final state. Neighboring points along the string are connected by springs in order to guarantee a continuous path. Then an optimization algorithm is performed which involves force projections of both the true forces and the spring forces. Thus the string of points is dragged closer and closer to the minimum energy path until the transition state is located. Thus also nonintuitive transition state geometries might be detected.

7.2 Diffusion

A single particle on a surface can jump laterally along the surface from one stable adsorption site to the next. This process is a typical process that is driven by thermal fluctuations. Each adsorbate performs thermal vibrations around the equilibrium site in the adsorption well. The rate for a jump to a next nearest neighbor site is then given by an Arrhenius expression

$$k_j = k_0 \exp\left(-\frac{E_a}{k_B T_s}\right), \tag{7.14}$$

where E_a is the energetic barrier to the next-nearest neighbor site and T_s is the surface temperature. To obtain a more detailed understanding of the so-called self-diffusion process, we introduce the probability $P(\bm{R}, t)$ that the lattice site $\bm{R}$ on the surface is occupied a time t. The probability that the atom is still at site $\bm{R}$ at time $t + \Delta t$ can be expressed as

$$P(\bm{R}, t + \Delta t) = \sum_{\bm{R'}} W(\bm{R}, \bm{R'}, \Delta t)\, P(\bm{R'}, t). \tag{7.15}$$

Here $W(\bm{R}, \bm{R'}, \Delta t)$ describes the conditional probability that the atom is at site $\bm{R}$ at time $t + \Delta t$ given that it was at site $\bm{R'}$ at time t. Now we assume that only nearest-neighbor (n.n.) jumps can occur and that the time Δt is so short that at most one jump happens during this time. Since k_j is the overall rate that a jump to any of the nearest neighbors occurs, the rate of a jump to a particular nearest neighbor is given by k_j/N where N is the number of nearest neighbors. Then the probability of a jump to this nearest neighbor site within the time Δt is simply $\Delta t \cdot k_j/N$. Thus we can write the probability W as

$$W(\bm{R}, \bm{R'}, \Delta t) = \begin{cases} k_j \Delta t/N, & \text{if } \bm{R}, \bm{R'} \text{ n.n.}, \\ 1 - \sum_{n.n.} k_j \Delta t/N, & \text{if } \bm{R} = \bm{R'}, \\ 0 & \text{else}, \end{cases} \tag{7.16}$$

where we have used the fact that W is a probability, i.e. $\sum_{\bm{R}} W(\bm{R}, \bm{R'}, \Delta t) = 1$.

If we insert (7.16) in (7.15), we obtain

$$P(\mathbf{R}, t+\Delta t) = \sum_{n.n.} \frac{k_j \Delta t}{N} P(\mathbf{R'}, t) + \left(1 - \sum_{n.n.} \frac{k_j \Delta t}{N}\right) P(\mathbf{R}, t). \quad (7.17)$$

Now we substract $P(\mathbf{R}, t)$ from both sides of (7.17), divide by Δt, and take the limit $\Delta t \to 0$. This leads to the differential equation

$$\frac{\partial P(\mathbf{R}, t)}{\partial t} = \frac{k_j}{N} \sum_{n.n.} [P(\mathbf{R'}, t) - P(\mathbf{R}, t)]. \quad (7.18)$$

An equation such as (7.18) is called a *master equation*. In general, master equations give the time dependence of probability distributions of physical observables. Equation (7.18) is valid for diffusion via nearest-neighbor jumps in arbitrary environments. Let us now assume that the jumps are confined to a two-dimensional square lattice with lattice constant a. The number of nearest neighbors is 4. If we now perform a Taylor expansion of $P(\mathbf{R'}, t)$ at the nearest-neighbor sites, e.g.

$$P(\mathbf{R} \pm a\hat{e}_x, t) = P(\mathbf{R}, t) \pm a \frac{\partial P(\mathbf{R}, t)}{\partial x} + \frac{a^2}{2} \frac{\partial^2 P(\mathbf{R}, t)}{\partial x^2} \pm \dots, \quad (7.19)$$

where $\hat{e}_x$ is the unit vector in x-direction, we obtain the well-known diffusion equation

$$\frac{\partial P(\mathbf{R}, t)}{\partial t} = \frac{k_j a^2}{4} \nabla^2 P(\mathbf{R}, t). \quad (7.20)$$

The factor

$$D_s = \frac{k_j a^2}{4} \quad (7.21)$$

is called the *self-diffusion* or *tracer-diffusion coefficient*. It is often denoted by D^*. This coeffient also enters the mean square displacement of the particle on the surface. Let us assume that the adatom was at the origin $\mathbf{R} = 0$ at time $t = 0$. After the time t the particle has performed $t \cdot k_j$ jumps, each with a square displacement of a^2. Hence the mean square displacement is given by

$$\langle \mathbf{R}^2(t) \rangle = t \, k_j \, a^2$$
$$= 4 \, D_s \, t. \quad (7.22)$$

In fact, (7.22) is often used to define the self-diffusion coefficient D_s. It is important to note that D_s differs from the chemical diffusion coefficient D_c. Consider an ensemble of particles on the surface. If one assumes that the number of particles stays constant with time, their particle distribution function $n(\mathbf{R}, t)$ obeys a conservation law:

$$\frac{\partial n(\mathbf{R}, t)}{\partial t} + \nabla \cdot \mathbf{j}(\mathbf{R}, t) = 0, \quad (7.23)$$

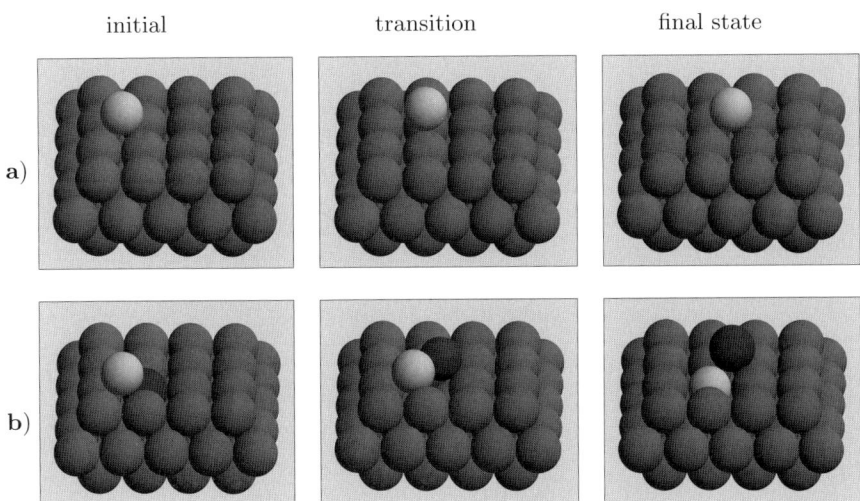

Fig. 7.2. Diffusion mechanisms on a fcc(100) surface. (**a**) hopping diffusion, (**b**) exchange diffusion

where $j(\boldsymbol{R},t)$ is the particle current. Now we assume that the particle current is driven by the non-uniformity of the density distribution. The simplest assumption is a linear dependence

$$j = -D_c \, \nabla n(\boldsymbol{R},t) \,. \tag{7.24}$$

This equation is known a *Fick's law*. Substituting Fick's law into the conservation law yields Fick's second law, the diffusion equation

$$\frac{\partial n(\boldsymbol{R},t)}{\partial t} = \nabla \cdot (D_c \nabla n(\boldsymbol{R},t)) \,. \tag{7.25}$$

The chemical diffusion coefficient does not describe the motion of a single atom due to thermal fluctuations, but the transport of a large number of atoms due to a gradient in the particle density. In general, D_c depends on the particle density. Only in the case of vanishing particle densities when D_c does not depend on the density any more, D_s and D_c become equal, as a comparison of (7.20) and (7.25) confirms.

The usual mode of surface diffusion is assumed to proceed via hops between adjacent equilibrium adsorption sites (see Fig. 7.2a). This is indeed true for a number of close-packed metal surfaces. However, for self-diffusion on Pt(100) [342] and Ir(100) [343] another diffusion mechanism has been experimentally observed. On these surfaces the adatoms diffuse along the [001] and [010] directions by displacing a neighboring surface atom. This exchange mechanism is illustrated in Fig. 7.2b.

An explanation for the driving force of the exchange diffusion has been given by Feibelman based on LDA-DFT calculations for the diffusion of Al adatoms on Al(100) [344]. He realized that it is important to regard a diffusion process as a motion associated with the making and breaking of chemical bonds rather than as a hard sphere rolling over a corrugated plane. For the hopping diffusion along the [110] direction, the atom at the transition state is only twofold coordinated resulting in a rather large barrier of 0.65 eV. On the other hand, for the exchange mechanism the transition state is threefold coordinated which is especially favorable for the trivalent aluminum where it leads to a reduction of the diffusion barrier to 0.20 eV [344]. The coordination argument is not restricted to trivalent systems. It has been shown that the exchange mechanism in the late $5d$ metals Ir, Pt and Au can be explained by the unusually high surface stress which is a consequence of relativistic effects [345, 346]. This high stress pulls the two atoms of the exchange transition state closer to the surface which lowers the energy barrier. As far as other fcc(100) metal surfaces are concerned, exchange diffusion has only been found experimentally [347] and confirmed theoretically by DFT calculations [348] for Ni(100) and predicted for strained Ag(100) [346].

7.3 Kinetic Lattice Gas Model

The derivation of the self-diffusion coefficient (7.15)–(7.21) actually presented a special case of the application of a much more general model that allows the description of processes on the surface such as diffusion, adsorption, desorption or growth on mesoscopic length and time scales: the *kinetic lattice gas model* [349]. The prerequisites for its application are that the geometry of the surface remains unperturbed and that adsorption occurs at well-defined sites. Then the substrate can be divided into N_c cells where each cell corresponds to a possible adsorption site. In principle these cells can also describe different layers. The cells can either be occupied by an adsorbate or empty. This means that the whole microscopic motion of the atom around its equilibrium position is entirely neglected. Let us denote by $\{\boldsymbol{R}_i\}$ the set of occupied cells. We further introduce the occupation numbers n_i which are either 0 or 1 depending on whether the adsorption site in cell $\boldsymbol{R}_i$ is empty or occupied. Then the energetics of the system are described by the lattice gas Hamiltonian

$$H(\{\boldsymbol{R}\}) = \sum_{\boldsymbol{R}_i} E(\boldsymbol{R}_i)\, n_i + \frac{1}{2} \sum_{\boldsymbol{R}_i} \sum_{\boldsymbol{R}_j} V_2(\boldsymbol{R}_i, \boldsymbol{R}_j)\, n_i n_j$$
$$+ \frac{1}{6} \sum_{\boldsymbol{R}_i} \sum_{\boldsymbol{R}_j} \sum_{\boldsymbol{R}_k} V_3(\boldsymbol{R}_i, \boldsymbol{R}_j, \boldsymbol{R}_k)\, n_i n_j n_k + \ldots , \qquad (7.26)$$

where $E(\boldsymbol{R}_i)$ is the single-particle Helmholtz free energy of the atom at $\boldsymbol{R}_i$, and V_2 and V_3 are the two-particle and three-particle interactions, respectively. Usually only two-particle interactions are included, but for certain problems also so-called *triples* have to be included.

Time-dependent phenomena on a mesoscopic or macroscopic time and length scale are best described in an approach based on nonequilibrium statistical mechanics [259] using time-dependent distribution functions. Usually one assumes that the variables at any moment determine the further development of the system in time, i.e. one uses the Markov approximation. One then introduces the function $P(\{\boldsymbol{R}\}, t)$ which gives the probability that the system is in the state $\{\boldsymbol{R}\} = (n_1, n_2, \ldots, n_{N_c})$ at time t. The time-evolution of this probability function is determined by the master equation

$$\frac{dP(\{\boldsymbol{R}\}, t)}{dt} = \sum_{\{\boldsymbol{R'}\}} [W(\{\boldsymbol{R}\}, \{\boldsymbol{R'}\}) \, P(\{\boldsymbol{R'}\}, t) - W(\{\boldsymbol{R'}\}, \{\boldsymbol{R}\}) \, P(\{\boldsymbol{R}\}, t)] \,. \quad (7.27)$$

Here $W(\{\boldsymbol{R'}\}, \{\boldsymbol{R}\})$ describes the transition probability per unit time, i.e. it corresponds to the rate that the system changes from state $\{\boldsymbol{R}\}$ to $\{\boldsymbol{R'}\}$. Equation (7.27) corresponds to the generalization of (7.18). The transition rates $W(\{\boldsymbol{R'}\}, \{\boldsymbol{R}\})$ are in fact not entirely independent. They must satisfy the *principle of detailed balance* [259] which is related to the time-reversal symmetry or microscopic reversibility. It ensures that forward and backward rates weighted by the Boltzmann factor are the same and is expressed as

$$W(\{\boldsymbol{R}\}, \{\boldsymbol{R'}\}) \, \exp\left(-\beta \left[H(\{\boldsymbol{R'}\}) - \sum_{\{\boldsymbol{R'}_j\}} n'_j \, \mu(\boldsymbol{R'}_j)\right]\right)$$

$$= W(\{\boldsymbol{R'}\}, \{\boldsymbol{R}\}) \, \exp\left(-\beta \left[H(\{\boldsymbol{R}\}) - \sum_{\{\boldsymbol{R}_i\}} n_i \, \mu(\boldsymbol{R}_i)\right]\right) \,. \quad (7.28)$$

Here $\beta = 1/k_\mathrm{B} T$ is the inverse temperature and $\mu(\boldsymbol{R}_i)$ is the equilibrium chemical potential of the atom at site $\boldsymbol{R}_i$. The change from state $\{\boldsymbol{R}\}$ to $\{\boldsymbol{R'}\}$ in general involves a number of different microscopic processes. These are usually assumed to be independent from each other so the transition rate $W(\{\boldsymbol{R'}\}, \{\boldsymbol{R}\})$ is expressed as a sum over individual microscopic processes. For example, the rate can be written as a sum of adsorption-desorption and diffusion terms

$$W(\{\boldsymbol{R}\}, \{\boldsymbol{R'}\}) = W_{ad-des}(\{\boldsymbol{R}\}, \{\boldsymbol{R'}\}) + W_{diff}(\{\boldsymbol{R}\}, \{\boldsymbol{R'}\}) \quad (7.29)$$

Except for the detailed balance relation there is no constraint within the lattice gas model on the rates. They have either to be guessed or fitted or derived from electronic structure calculations.

7.4 Kinetic Modelling of Adsorption and Desorption

In Sects. 6.5 and 6.6 we have already treated the dynamics of atomic and molecular adsorption on surfaces and of the time-reverse process, desorption.

In the examples that were given, the sticking probabilities were always calculated for initially adsorbate-free surfaces. However, in a typical adsorption scenario, the surface might be initially uncovered, but after the first atoms or molecules have been adsorbed, the sticking probability will be modified by the presence of the adsorbates. The determination of sticking probabilities as a function of the coverage is usually not tractable with dynamical methods since huge simulation systems are required. Therefore a kinetic description is necessary which, however, can still use input from electronic structure calculations.

In the following we consider an adsorbate on the surface for which the interaction energy is described by the lattice gas Hamiltonian (7.26). The change of the adsorbate coverage is due to adsorption processes that increase the coverage and desorption processes which decrease the coverage. The time rate of the change is therefore given by [350]

$$\frac{d\theta}{dt} = R_{\text{ad}} - R_{\text{des}}, \tag{7.30}$$

where R_{ad} and R_{des} give the rate of adsorption and desorption, respectively. The flux F of particles impinging from the gas phase at pressure P and temperature T on the surface unit cell with area a_s is given by

$$F = \frac{P a_s \lambda_{\text{th}}}{h}, \tag{7.31}$$

where

$$\lambda_{\text{th}} = \frac{h}{\sqrt{2\pi m k_B T}} \tag{7.32}$$

is the thermal wavelength of a molecule of mass m. Introducing the coverage and temperature dependent sticking probability $S(\theta, T)$, the rate of adsorption can be expressed as

$$R_{\text{ad}} = \frac{S(\theta, T) P a_s}{\sqrt{2\pi m k_B T}}. \tag{7.33}$$

In order to make contact with the kinetic lattice gas model, we have to specify the transition probabilities entering the master equation (7.27). For the sake of simplicity we consider only nearest neighbor interactions between sites $\boldsymbol{R}_i$ and $\boldsymbol{R}_{i+a}$. The adsorption-desorption term becomes [351]

$$\begin{aligned} W_{\text{ad-des}}(\{\boldsymbol{R'}\}, \{\boldsymbol{R}\}) = \\ W_0 \sum_i \bigg\{ (1 - n_i) \bigg(1 + A_1 \sum_a n_{i+a} + A_2 \sum_{a,a'} n_{i+a} n_{i+a'} + \ldots \bigg) \\ + D_0 n_i \bigg(1 + D_1 \sum_a n_{i+a} + D_2 \sum_{a,a'} n_{i+a} n_{i+a'} + \ldots \bigg) \bigg\} \\ \times \delta(n'_i, 1 - n_i) \prod_{j \neq i} \delta(n'_j, n_j), \end{aligned} \tag{7.34}$$

where the A_i refer to adsorption processes and the D_i to desorption processes. The Kronecker delta for sites $\boldsymbol{R}_i \neq \boldsymbol{R}_j$ excludes multiple transitions so that only one transition occurs at any given time. To study the time evolution, the coverage is defined as

$$\theta(t) = N_c^{-1} \sum_i \sum_{\{\boldsymbol{R}\}} n_i\, P(\{\boldsymbol{R}\}, t) = N_c^{-1} \sum_i \langle n_i \rangle. \tag{7.35}$$

The time evolution of the coverage is then obtained by multiplying the master equation (7.27) with n_i and summing over all sites and configurations. By comparison with the phenomenological ansatz (7.33), the rate W_0 in (7.34) can be identified as

$$W_0 = \frac{S_0(T) P a_s}{\sqrt{2\pi m k_\mathrm{B} T}}, \tag{7.36}$$

where $S_0(T)$ is the temperature-dependent sticking coefficient for zero coverage. The coverage-dependent sticking coefficient is then given by a product of $S_0(T)$ with a sum that yields the coverage dependence,

$$S(\theta, T) = S_0(T) \left\{ (1-\theta) + \sum_{\{\boldsymbol{R}\}} \left(A_1 \sum_a \langle (1-n_i) n_{i+a} \rangle \right.\right.$$

$$\left.\left. + A_2 \sum_{a,a'} \langle (1-n_i) n_{i+a} n_{i+a'} \rangle + \ldots \right) \right\}. \tag{7.37}$$

It is important to note that within the lattice gas model not all the parameters can be uniquely specified. Detailed balance yields only one condition for each pair A_i and D_i [350]. For example, for the first pair of parameters this condition is

$$1 + A_1 = (1 + D_1) \exp(-V_{nn}/k_\mathrm{B} T), \tag{7.38}$$

where V_{nn} is the interaction energy between adsorbates on nearest neighbor sites. Furthermore, the sticking probability $S_0(T)$ can only be determined in a dynamical calculation since it depends on the energy transfer to the substrate degrees of freedom such as phonons or electron-hole pairs (see Sects. 6.5 and 8.3). The functional relation between A_i and D_i and the sticking probability $S_0(T)$ must therefore be postulated *ad hoc* or derived from a microscopic theory that takes the necessary couplings to the substrate explicitly into account.

Still the relative sticking coefficient $S(\theta, T)/S_0(T)$ can be evaluated in closed form within the kinetic lattice gas model under the assumption that the temperature is so low that the surface diffusion is too slow to establish equilibrium during the adsorption process [350]. Then the adsorbate remains disordered and the correlation functions appearing in (7.37) can be factorized to give products of only θ and $(1-\theta)$. For strong nearest neighbor repulsion,

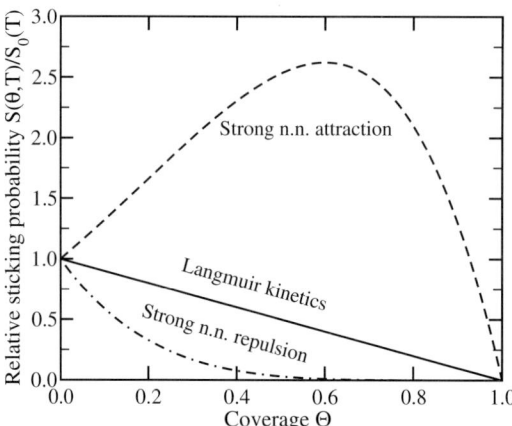

Fig. 7.3. Coverage dependence of the relative sticking probability $S(\theta, T)/S_0(T)$ on a disordered adsorbate (low T) on a square lattice for strong nearest neighbor (n.n.) attraction, repulsion and pure site exclusion (Langmuir kinetics) determined within the kinetic lattice gas model [350]

i.e., $V_{nn} \gg k_\mathrm{B} T$, one obtains a simple expression for the relative sticking coefficient,

$$S(\theta, T)/S_0(T) = (1 - \theta)^{z+1} , \qquad (7.39)$$

where z is the number of nearest neighbors in the surface. For strong nearest neighbor attraction $-V_{nn} \gg k_\mathrm{B} T$, on the other hand, and symmetric lateral interaction for adsorption and desorption, i.e. $A_i = -D_i$, the sticking probability is given by

$$S(\theta, T)/S_0(T) = (1 - \theta)(1 + \theta)^z . \qquad (7.40)$$

The symmetric effect of the lateral interactions for adsorption and desorption can be made reasonable by considering that lateral repulsion will suppress adsorption and aid desorption, whereas for lateral attraction it is the other way around.

We have plotted the coverage dependent sticking probabilities in these two limiting cases on a square lattice in Fig. 7.3. It is obvious how significantly the lateral interactions influence the sticking probabilities. In addition, we have included the sticking probability when the adsorption is independent of the neighboring sites. This so-called *Langmuir kinetics* is obtained by setting all $A_i = 0$ in (7.37). The sticking probability is then simply given by the availability of empty sites, i.e., $S(\theta, T)/S_0(T) = (1-\theta)$. Langmuir kinetics has only been observed in a few systems [350]; in most systems the dependence of the sticking probability is modified by lateral interactions, multilayer growth or the existence of precursors. The influence of precursors and longer-range lateral interactions on adsorption and desorption properties can be build

7.4 Kinetic Modelling of Adsorption and Desorption

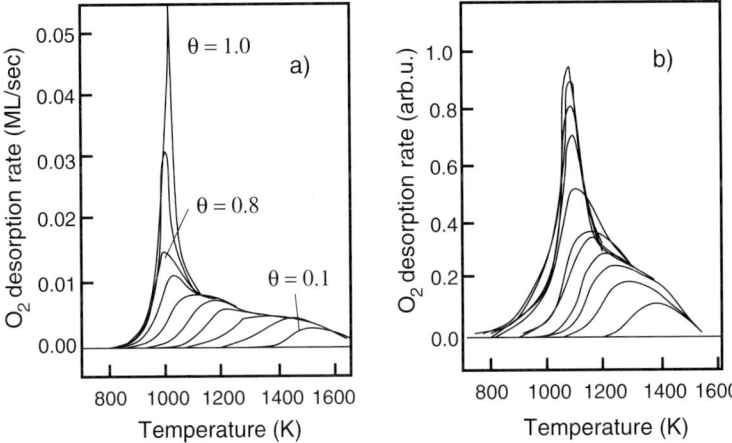

Fig. 7.4. TPD spectra for the associative desorption of oxygen from Ru at a heating rate of 6 K/s for initial coverages up to one monolayer. (**a**) Theoretical results derived from ab initio results using the kinetic lattice gas model [352], (**b**) experimental results [353]

into the kinetic lattice gas formalism without any principal difficulty, the equations just become more complex [350, 351].

Since adsorption and desorption are related through detailed balance, the same mechanisms must be operative for both processes. If the surface diffusion is much faster than adsorption and desorption, it can be assumed that there is always a quasi-equilibrium on the surface. In such a situation, the desorption rate can be derived from non-equilibrium thermodynamics [349]. It can be written as a product of the sticking probability and the activity, i.e., the chemical potential of the adsorbate:

$$R_{\text{des}} = S(\theta, T) \frac{a_s}{\lambda_{\text{th}}^2} \frac{k_B T}{h} Z_{\text{int}} \exp(\beta \mu). \tag{7.41}$$

Here Z_{int} is the internal partition function of the adsorbate in the gas phase and μ is its chemical potential.

Traditionally, in the description of the adsorption and desorption kinetics, the parameters of the model have been adjusted in order to reproduce experimental data. Nevertheless, the most significant interaction parameters entering the lattice gas Hamiltonian are accessible by total-energy calculations which has been demonstrated for the system O/Ru(0001) [352, 354]. Based on a series of DFT-GGA calculations for different adsorption structures, the parameters for first, second and third-neighbor two-body interaction energies and nearest-neighbor three-body interactions have been derived (see Exercise 7.2). Since oxygen desorbs associatively, the expression for the desorption rate slightly differs from the general expression (7.41) for atomic desorption. Furthermore, the chemical potentials of the gas phase and the adsorbate are the same in equilibrium. This allows to break up the term $\exp(\beta\mu)$ into different factors. In total, one obtains for the desorption rate [352]

$$R_{\text{des}} = 2S(\theta,T)\, \frac{a_s}{\lambda_{\text{th}}^2}\, \frac{k_B T}{h}\, \frac{Z_{\text{int}}}{q_3^2}\, \frac{\theta^2}{(1-\theta)^2}\, \exp(-2\beta|V_0|)\exp(\beta\mu_{\text{lat}})\,, \quad (7.42)$$

where q_3 is the partition function of the oxygen atoms on the surface, $2|V_0|$ is the energy required to desorb two atoms from the substrate and associate them in the gas phase, and μ_{lat} is the contribution of the lateral interaction to the chemical potential of the adsorbate. The thermodynamic information about the system such as for example the chemical potential has been determined from the lattice gas Hamiltonian by the so-called transfer matrix technique [349]. The coverage-dependent sticking probability $S(\theta,T)$ of O_2 on Ru(0001) entering (7.42), however, had to be parametrized and adjusted to experimental data.

The calculated desorption rate as a function of the substrate temperature for different initial coverages is shown in Fig. 7.4a. In panel b, the corresponding experimental results [353] are plotted. The spectra are obtained using the method of temperature programmed desorption (TPD). The surface temperature is increased linearly with time, i.e. $T = T_0 + \alpha t$, and the desorption flux is determined as a function of the temperature. The maxima in the TPD spectra at the temperature T_{max} can be related to the desorption energy E_{des}. For a simple n-th order desorption process, for which the desorption rate can be written as

$$\frac{d\theta}{dt} = R_{\text{des}} = R_0\, \theta^n\, \exp(-\beta E_{\text{des}})\,, \quad (7.43)$$

the relation between E_{des} and T_{max} reads (see Exercise 7.3 [101])

$$\ln\left(\frac{T_{\text{max}} R_0\, \theta^{n-1}}{\alpha}\right) = \frac{E_{\text{des}}}{k_B T_{\text{max}}} + \ln\left(\frac{E_{\text{des}}}{k_B T_{\text{max}}}\right). \quad (7.44)$$

However, often TPD spectra are distorted due to lateral interactions or precursor and dynamical effects, or additional peaks show up. This makes the derivation of interaction energies from the spectra ambiguous. Therefore the calculation of realistic theoretical TPD spectra derived from first principles is very helpful for the interpretation of experimental spectra.

The comparison of the calculated and the measured TPD spectra in Fig. 7.4 demonstrates that there is a satisfactory agreement between both. At low initial coverages, in the theory the oxygen desorbs at temperatures that are about 100 K higher than in the experiment. This has been traced back to the overbinding present in the DFT calculation. Except for that, all features of the experimental TPD spectra are nicely reproduced. The peak maximum shifts to lower temperatures for higher initial coverages. This is a consequence of the repulsive interaction between the adsorbed oxygen atoms which lowers the binding energy and thus leads to an onset of desorption at lower temperatures. The maximum for higher initial coverages is due to the

(1 × 1) oxygen monolayer on Ru(0001). The very steep leading edge is caused by the rapidly decreasing sticking coefficient for higher coverages. This leads to a delay of the desorption of the first oxygen molecules to higher temperatures over a very narrow temperature range. The shoulders at 1100 K and 1300 K in the spectra can be related to the formation of ordered (2 × 1) and (2 × 2) structures [352].

The influence of the trio interactions on the TPD spectra has also been investigated. It turns out that neglecting them leads to broader spectra for higher coverages which reduces the agreement with experiment. Thus the trio interactions are indeed important for a reliable description of the desorption kinetics in the system O/Ru(0001).

7.5 Growth

In this chapter so far we have addressed the diffusion on the surface and adsorption/desorption processes. We have not focused on the changes of the surface structure caused by these processes. Now consider a flux of particles impinging on a surface. If more particles stick on the surface per unit time than desorb again, adlayers will start to grow on the surface. This is a typical non-equilibrium situation because in thermal equilibrium the impinging flux would be balanced by the flux of scattered and desorbing particles and there would be no net growth of the surface.

If the temperature is sufficiently high, the resulting surface structures in growth can still be classified by thermodynamic stability arguments. Experimentally it is well established to distinguish between three different growth modes [100], the so-called *Frank–van der Merwe* (FV), *Volmer–Weber* (VW) and *Stranski–Krastanov* (SK) growth modes. These growth modes are illustrated in Fig. 7.5. The FW growth mode corresponds to a layer-by-layer growth while in the VW growth mode the adsorbate grows in a three-dimensional fashion by building small crystallites. The SK growth mode is an intermediate case where a few monolayers adsorb in a layer-by-layer fashion before three-dimensional growth starts.

In Fig. 7.5, the free surface energy of the substrate γ_s and of the adsorbate overlayer γ_a and the substrate-adsorbate interface energy γ_i are depicted. If the sum of the adsorbate surface energy plus the interface energy is smaller than the substrate surface energy, i.e. $\gamma_a + \gamma_i < \gamma_s$, then it is energetically favorable if the adsorbate layer covers the whole substrate (FV growth). This two-dimensional growth scenario is also called *wetting*. On the other hand, if $\gamma_s < \gamma_a + \gamma_i$, then it is energetically more favorable if not the whole substrate is covered by an overlayer which leads to three-dimensional growth (VK growth). In the intermediate SK case, the energy difference $\Delta\gamma = \gamma_a + \gamma_i - \gamma_s$ changes sign at a critical layer thickness leading to a transition from two-dimensional growth to three-dimensional growth.

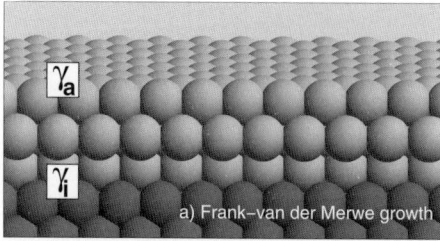
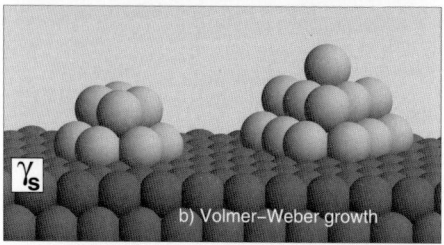
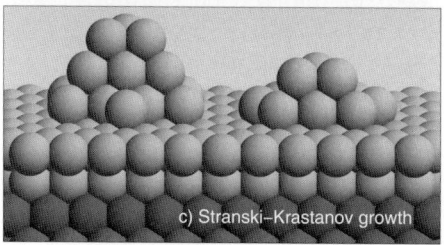

Fig. 7.5. Illustration of three distinct growth modes, Frank–van der Merwe, Volmer–Weber and Stranski–Krastanov growth. γ_s, γ_a and γ_i denote the free surface energy of the substrate, the adsorbate overlayer and the substrate-adsorbate interface energy, respectively

This argumentation is entirely based on macroscopic properties of the surface and the interface and neglects any microscopic details of the growth process. For a microscopic understanding, however, one has to take into account that even in the case of two-dimensional growth a whole layer is not grown at one time. Before a new layer is completed, the uppermost layer will exhibit a rough structure with single adatoms, islands, steps and kinks. The morphology of the resulting surface structure depends sensitively on the mobility of the adatoms which have deposited from the gas phase. If this mobility is high, the situation is close to thermal equilibrium and the atoms can travel far enough on the surface to find the most favorable adsorption sites. These are usually located at steps and kinks. The adatoms thus travel along the surface until they attach to a step or kink, and through this attachment steps advance along the surface. This growth mode is called *step flow*. During the step flow process the macroscopic morphology of the surface is not modified, and the resulting structure corresponds to a flat film.

If the mobility is not that high, the adatoms do not necessarily find the most favorable adsorption sites. Then the growth mode is determined by the kinetics of the transport and diffusion processes on the surface. Therefore it is important to consider the microscopic diffusion processes and the proba-

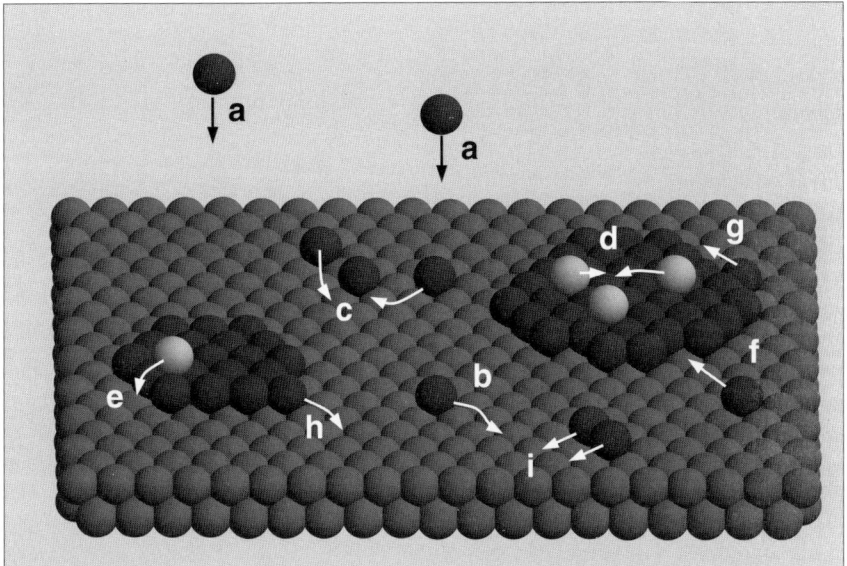

Fig. 7.6. Atomistic processes during growth: a) deposition, b) diffusion on terraces, c) nucleation of an island, d) nucleation of a second-layer island, e) diffusion to a lower terrace, f) attachment to an island, g) diffusion along a step edge, h) detachment from an island, i) diffusion of a dimer (or a larger island). (After [345])

bilities with which they occur. In Fig. 7.6 the atomistic processes during film growth are illustrated. After the deposition of the atoms (a) the atoms can diffuse on the surface (b). Some atoms may meet to form the small nucleus of an island, either on a flat terrace (c) or on top of an island (d). If the atom was deposited on top of an existing island, it might move down to the lower terrace (e). An atom on the terrace can directly attach to an existing island (f). Once it is attached, it might diffuse along the step edge (g) or it might again detach from the island (h). Finally, a dimer or larger islands can move as a whole (i).

Before we proceed, we should consider the barrier for diffusion across a step. Consider the situation depicted in Fig. 7.7 At the upper step edge, a diffusing atom is only bound to a small number of neighbors. This low coordination leads to a barrier for diffusion across the step E_S that is larger than the diffusion barrier E_T on the terraces. Consequently there is an additional step-edge barrier

$$E_{ES} = E_S - E_T, \qquad (7.45)$$

the so-called Ehrlich–Schwoebel barrier [355,356], which hinders the diffusion across a step. The magnitude of this barrier in fact decides whether two-dimensional or three-dimensional growth occurs in the kinetic regime. For a large Ehrlich–Schwoebel barrier, adatoms that have landed on top of an

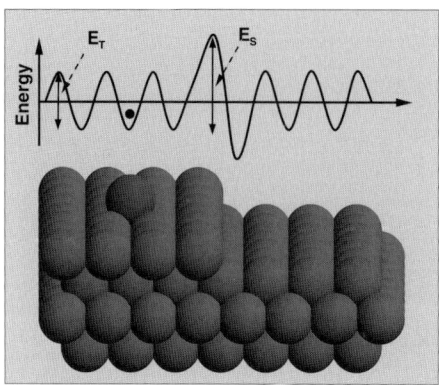

Fig. 7.7. Schematic representation of the Ehrlich–Schwoebel barrier for an adatom moving over a step edge. Due to the lower coordination for adatoms above a step edge atom barrier for diffusion E_S is larger than the diffusion barrier E_T on the terraces. This leads to an additional barrier for the diffusion over the step, the Ehrlich–Schwoebel barrier $E_{ES} = E_S - E_T$

existing island cannot move to the lower terrace. They stay on the island so that the islands become three-dimensional. On the other hand, if this barrier is small, adatoms are able to move down to the lower terrace and attach to the edge of the island so that the islands grow and eventually coalesce leading to two-dimensional growth. We will later see that the Ehrlich–Schwoebel barrier can be quite small for metals. This is particularly true if diffusion across the steps proceeds via the exchange mechanism because then the low coordination during the diffusion process is avoided.

There is another important aspect illustrated in Fig. 7.7. The energy well for adatoms attached to the steps is much larger than for adatoms adsorbed on the flat terraces. Again this can be understood by the higher coordination of the adatoms at the steps. One consequence of this higher binding energy is that the barrier for moving up a step is even much larger than for moving down a step. This is the reason why we did not include this process in Fig. 7.6 because adatoms usually do not move to upper terraces. Just a note of caution should be added here. Note that going up or down has nothing to do with gravity here. Gravitational forces are entirely negligible for all our considerations here.

Many qualitative and even some quantitative aspects of growth processes at surface can already be derived from mean-field nucleation theory using phenomenological rate equations [357]. A central concept is the critical cluster size i. Neglecting any effects of the shape of the islands, the critical cluster size is defined as the size i at which islands are just not stable, i.e. all clusters of size $j > i$ are stable. Stability is defined dynamically here, i.e. a stable cluster grows more rapidly than it decays during deposition.

In order to derive the basic equation of nucleation theory [358], we assume that only single atoms are mobile. We define n_x as the density of stable clusters or islands:

$$n_x = \sum_{j=i+1}^{\infty} n_j. \tag{7.46}$$

By considering all possible processes, we can derive the rate equations for the densities n_j of clusters of size j. For single atoms the rate equation is

$$\frac{dn_1}{dt} = F - \sigma_i D n_i n_1 - \sigma_x D n_x n_1 - F\theta. \tag{7.47}$$

The number of single atoms on the surface increases through the flux F of deposited particles. On the other hand, a single atom can diffuse with diffusion coefficient D to a cluster of critical size n_i leading to the nucleation of a new stable cluster. This is described by the nucleation term $\sigma_i D n_1 n_i$. Furthermore, an atom can attach to an existing stable cluster at the diffusion capture rate $\sigma_x D n_x$. Finally, the deposited atom can directly attach to a stable cluster with a rate $F\theta$ where θ is the coverage of stable clusters. Reevaporation of atoms back into the gas phase is neglected in (7.47). This approximation is justified if the growth temperature is not too high.

We assume that there is local thermodynamic equilibrium between subcritical clusters. This can be expressed as

$$\frac{dn_j}{dt} = 0, \quad 2 \leq j \leq i. \tag{7.48}$$

As far as the stable clusters are concerned, their number can increase by nucleation processes given by the nucleation term $\sigma_i D n_1 n_i$. On the other hand, stable cluster can coalesce which reduces the number of stable clusters by one. The rate of this process is given by $2n_x(d\theta)/(dt)$. Hence the rate equation for n_x becomes

$$\frac{dn_x}{dt} = \sigma_i D n_1 n_i - 2n_x \frac{d\theta}{dt}. \tag{7.49}$$

Solving the coupled rate equations (7.47)–(7.49) is not trivial [358]. Assuming steady-state conditions, i.e. $dn_1/dt = 0$, and local thermodynamic equilibrium between n_i and n_1, the island density of stable clusters is given by

$$n_x = f(\theta, i) \left(\frac{D}{F}\right)^{-\frac{i}{i+2}} \exp\left(\frac{E_i}{(i+2)k_B T}\right), \tag{7.50}$$

where E_i is the binding energy of a i-sized cluster. Equation (7.50) is often expressed as a scaling relation connecting the island density with the flux and the diffusion coefficient

$$n_x \propto \left(\frac{D}{F}\right)^{-\frac{i}{i+2}}. \tag{7.51}$$

This scaling relation can be used to extract microscopic parameters from experimental results. It the island density n_x is measured as a function of the flux F, the critical size i can be derived. The temperature dependence of n_x for a known critical size i yields the diffusion barrier E_a and the prefactor D_0 entering the diffusion coefficient.

Simple scaling theory often works rather well, but it is only valid for compact islands in the coverage regime of saturation. Furthermore, as a mean-field theory it cannot properly describe island size distributions and coalescence effects. Therefore a microscopic approach is needed that explicitly includes the randomness of the size and shape distribution of islands, but still allows the simulation of systems having mesoscopic time and size scales. Kinetic Monte Carlo (KMC) simulations can in fact meet all these requirements. In a kinetic Monte Carlo simulation, first of all a list of all relevant processes that are allowed for the various atoms present on the surface has to be made. Furthermore, a rate with which each process occurs has to be specified. These rates can be derived from electronic-structure calculations via transition state theory (see Sect. 7.1).

A typical kinetic Monte Carlo algorithmus proceeds as follows [341, 345]. For a given configuration, all possible processes and their corresponding rates k_i are determined. Then the sum of all the rates $R = \sum k_i$ is formed. A random number ρ_1 in the range (0,1] is picked and the process with rate k_l which satisfies

$$\sum_{i=0}^{l-1} k_i \leq \rho_1 R < \sum_{i=0}^{l} k_i \tag{7.52}$$

is performed. The average time until the next event will occur is given by $\tau = 1/R$. However, it is more realistic to allow for fluctuations in the time intervall. Hence another random number ρ_2 between 0 and 1 is chosen and the simulation time is updated by $t = t + \Delta t$ with $\Delta t = -\ln(\rho_2)/R$. After the process l has been performed, the configuration of the system has changed. Hence there might be other processes possible for this new configuration. Therefore the list of all possible processes has to be updated, and the procedure described above starts again.

In any reliable kinetic Monte Carlo simulation it is crucial to take into account *all* relevant microscopic processes. Unfortunately, there is no scheme that guarantees that one indeed has considered all important transitions. One has to try to find all relevant processes by either clever search algorithms or based on experience and intuition. Hence an otherwise sound KMC simulation can still yield unrealistic results if some important microscopic mechanism has been overlooked.

In this elementary introduction, we focus on the growth processes at fcc(111) surfaces. Before we proceed, it is important to note that on fcc(111) surfaces two different kinds of close-packed steps exist that exhibit different facets. This is illustrated in Fig. 7.8. The close-packed steps running along the

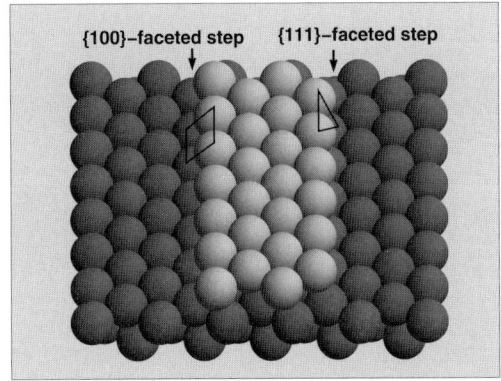

Fig. 7.8. Orientation of the close-packed steps on fcc(111) surfaces

[110] direction can either consist of {100} microfacets or of {111} microfacets. Sometimes these steps are also called A and B steps, respectively. These steps have slightly different formation energies on Al(111): $E_{\{100\}} = 0.248$, $E_{\{111\}} = 0.232\,(\mathrm{eV/atom})$ [359].

At high surface temperatures, the shape of the island should be close to their equilibrium structure. Neglecting entropy effects, the edges of equilibrium islands are determined by the step formation energies in analogy to the Wulff construction for the shape of three-dimensional crystallites (see p. 67). On Al(111), this results in hexagonally shaped islands with close-packed steps along the [110] directions. Due to the difference in the formation energies of the two kinds of close-packed steps, the hexagons are slightly distorted. The equilibrium shape determined by the Wulff construction yields a length ratio of $L_{\{100\}} : L_{\{111\}} = 4 : 5$ [359].

At lower surface temperatures, the island size distribution and their shape occuring in growth no longer follows from equilibrium considerations. In fact, the size and shape is determined by the diffusion processes which are dominant in a particular temperature range. This is beautifully demonstrated by a kinetic Monte Carlo study of the growth of Al(111) based on microcopic input from LDA-DFT total-energy calculations [360].

As already mentioned, the first step in a kinetic Monte Carlo algorithm is the microscopic determination of all relevant processes. Table 7.1 provides a list of the diffusion mechanisms considered in the KMC study. The diffusion barriers have been evaluated by LDA calculations [359]. On Al(111), hopping self-diffusion is rather fast, hindered by a barrier of only 0.04 eV. Along the {111} faceted close-packed steps, diffusion occurs via the exchange mechanism. The exchange mechanism is favored because hopping would go through a transition state with only two neighbors. Along the {100} faceted step, on the other hand, an Al atom has four nearest neighbors at the transition state for hopping diffusion. This coordination leads to a diffusion barrier of 0.32 eV

Table 7.1. Diffusion mechanisms and barriers for the self-diffusion of Al/Al(111) surface. (From [359])

adatom diffusion	mechanism	E_a (eV)
flat Al(111)	hopping	0.04
parallel to {111}-faceted step	exchange	0.42
parallel to {100}-faceted step	hopping	0.32
descent down {111}-faceted step	exchange	0.06
descent down {100}-faceted step	exchange	0.08
corner jump (bridge)	hopping	0.17
corner jump (atop)	hopping	0.28

that is 0.1 eV smaller than the barrier for diffusion along the {111} faceted steps.

The barriers for diffusion across the steps in the descending direction are only slightly larger than for diffusion on flat Al(111) which means that the Ehrlich–Schwoebel barrier (7.45) is rather small. This low barrier is a consequence of the exchange mechanism. Adatoms landing on top of existing islands will rather easily move to lower terraces where they will attach to the step edges. Hence a three-dimensional growth of Al on Al(111) is rather improbable. Furthermore, the barriers for diffusion around the corners are included in Table 7.1.

In order to estimate the different diffusion coefficients, the prefactors D_0 are needed. Since these prefactors are related to an analysis of the relevant phase space or, equivalently, of the normal modes perpendicular to the diffusion path, they require a high computational effort. For the diffusion processes on Al(111), the prefactors had not been evaluated. Therefore they had to be guessed. Since hopping and exchange mechanisms correspond to microscopically rather different processes, it is reasonable to assume that they have different prefactors. The prefactors used for the main processes in the KMC simulation are listed in Table 7.2. They were estimated as weighted averages of data found in the literature for (111) surfaces of metallic system [360]; hence they can be regarded as an educated guess.

Using the diffusion processes listed in Table 7.1, kinetic Monte Carlo simulations of the growth of Al/Al(111) using a (600 × 600) array have been

Table 7.2. Assumed prefactors for the main diffusion mechanisms of Al/Al(111) surface. (From [360])

adatom diffusion	mechanism	Prefactor (cm^2/s)
flat Al(111)	hopping	2×10^{-4}
parallel to {111}-faceted step	exchange	5×10^{-2}
parallel to {100}-faceted step	hopping	5×10^{-4}

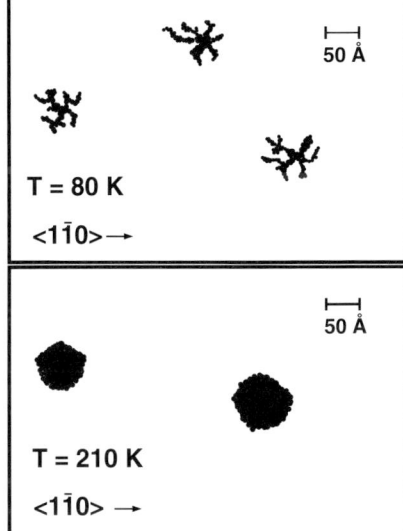
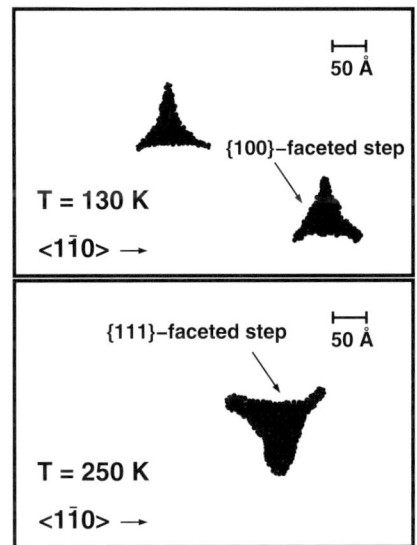

Fig. 7.9. Kinetic Monte Carlo simulation of the growth of Al(111) for four different temperatures. The plotted simulation area correponds to 1/8 of the total simulation area. The deposition flux is 0.08 ML/s and the coverage is $\theta = 0.08$ ML. (After [360])

performed. Note that no detachment from any island is taken into account in the simulations. This means that irreversible attachment of the adatoms to form dimers is assumed or, in other words, the critical cluster size is taken to be $i = 1$. The results of the KMC simulations for different surface temperatures are shown in Fig. 7.9 where approximately 1/8 of the simulation area is plotted. The deposition flux corresponds to 0.08 monolayer per second (ML/s) which is realistic, but still larger than typical deposition fluxes realized in experiment. The high flux is used in the simulation in order to keep the computational effort still tractable. The coverage reached in the simulation is 0.08 ML which means that the total simulation time is one second.

First of all it is evident in Fig. 7.9 that the shape of the resulting islands strongly depends on the temperature regime. At low temperatures ($T = 80$ K) the islands are very irregular. The mechanism responsible for the irregular shapes is called diffusion limited aggregation [361, 362] which corresponds to a hit and stick mechanism. At these low temperature the adatoms irreversibly stick to the site of first attachment to an existing island without any further mobility along the edges. The randomness of the site of the attachment then leads to the fractal growth patterns.

If the temperature is increased, the islands become triangular bounded by {100} faceted steps at $T = 130$ K. At $T = 210$ K the islands approximately have a hexagonal shape and then become triangular again at $T = 250$ K, but with a different orientation caused by the termination of the island by {111} faceted steps.

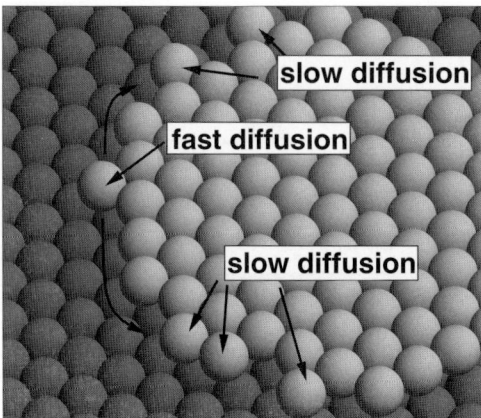

Fig. 7.10. Illustration of the growth direction of islands on a fcc(111) surface. Adatoms will preferentially remain at edges with slow diffusion or, equivalently, a low mobility. These edges will therefore advance, their length will shorten due to the finite island size and these step edges will eventually disappear

The orientation of the islands in this temperature regime is in fact a consequence of the adatom mobilities along the steps. The basic mechanism is illustrated in Fig. 7.10. Consider an island with edges that have quite different diffusion coefficients. Adatoms at an edge with a high diffusion coefficient will be very mobile while adatoms at the edge with a low diffusion coefficient will stay much longer at a specific site. Consequently, the step edges with a low lateral mobility for adatoms will advance. Because of the finite islands size, the advancing steps will become shorter until they eventually disappear.

However, this does still not explain why the Al islands in Fig. 7.9 change their orientation for temperatures of 130–250 K. According to the resulting island shapes, for $T = 130$ K the diffusion along the {100}-facetted steps should be faster than along the {111}-facetted steps while for $T = 250$ K it should be the other way around. In order to understand this phenomenon, it is important to consider the temperature dependence of the diffusion coefficient $D = D_0 \exp(-E_a/k_B T)$. Note that at high temperatures when the exponent is small, the diffusion coefficient is dominated by the prefactor D_0 while for low temperature the diffusion barrier E_a is crucial for the mobility of the adatoms.

As Table 7.1 demonstrates, the barrier for diffusion along the {100}-facetted steps is 0.1 eV smaller than along the {111}-facetted steps, which means that $E_a^{\{100\}} < E_a^{\{111\}}$. Consequently, at the lower temperature of $T = 130$ K, where the diffusion coefficient is dominated by the diffusion barrier, the adatoms are less mobile along the {111}-facetted steps so that they disappear and the triangular islands become bounded by the {100}-facetted steps. On the other hand, the prefactor for diffusion along the {111}-facetted steps is by two orders of magnitude larger than along the {100}-facetted steps. At the higher temperature of $T = 250\,\text{K}$, the diffusion along the {100}-facetted steps is therefore slower and the triangular shape of the islands becomes inverted. The temperature of $T = 210\,\text{K}$ corresponds to an intermediate case for which the diffusion coefficient along both step edges is similar resulting in a hexagonally shaped islands.

This explanation, however, has been questioned [363]. It has been argued that instead of the diffusion along the step the barrier for diffusion around the corner is crucial for the evolution of the island shape. The corner barrier for diffusion from the {111}-facetted step to the {100}-facetted step differs from the barrier for the reverse motion. If this is taken into account, the diffusion along the steps has almost no influence on the island shape any more. This demonstrates how severe the neglect of some crucial processes in KMC simulations can be. Still this issue is not fully settled yet since both models yield the same result.

The simulations of the epitaxial growth of Al on Al(111) provide a detailed picture of the important microscopic processes determining the resulting structures. Unfortunately, due to experimental problems the island shapes in the two-dimensional growth of Al on Al(111) have not been measured yet so that the results of the simulations are not confirmed.

The growth of Pt on Pt(111), on the other hand, has been studied in detail by scanning tunneling microscopy [364–366]. In fact, the mechanism for the inversion of the triangular shapes had already been proposed before the simulation of the growth of Al on Al(111) were performed, based on a STM study of the growth of Pt on Pt(111) [364]. At submonolayer coverages this system shows exactly the same qualitative trend as far as the shape of the islands as a function of the temperature is concerned, only at a higher temperature range of 350–650 K.

The homoepitaxial growth of Pt on Pt(111) has also been the subject of a series of LDA-DFT studies by Feibelman [367–369]. Some of the results of these studies were seemingly at variance with the experimental results. For example, at $T > 700\,\text{K}$, i.e., near equilibrium, it was found that the {111}-facetted edges of the distorted hexagonal Pt islands are longer by about 50% than the {100}-facetted edges, corresponding to a formation energy that is 13% lower [364]. The LDA-DFT calculations, on the other hand, yielded step formation energies of $0.46\,\text{eV/atom}$ for the {111}-facetted edge and $0.47\,\text{eV/atom}$ for the {100}-facetted edge, i.e., almost equal formation energies [367].

Furthermore, experiments had shown that Pt grows on Pt(111) in a three-dimensional fashion [365]. This was at variance with the DFT calculations [368] which produced an Ehrlich-Schwoebel barrier for the downward self-diffusion across the {100}-facetted steps in the exchange mechanism of only $0.02\,\text{eV}$. Such a small barrier would always favor two-dimensional instead of three-dimensional growth.

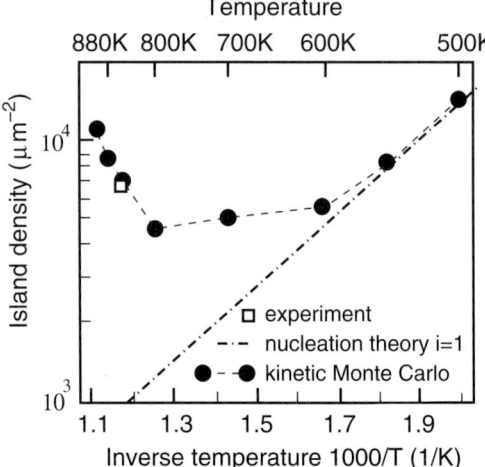

Fig. 7.11. Island density on GaAs(100) as a function of the inverse growth temperature. The filled circles correspond to the results of kinetic Monte Carlo simulations [370] whereas the dashed line shows the predictions of nucleation theory with a critical island size $i = 1$. One experimental data point [371] is shown as an open box

These apparent contradictions were resolved when the experiments were repeated under improved conditions [366]. These new experiments showed that island shapes, island densities and step-edge barriers are strongly affected by minute CO coverages as low as 10^{-3} monolayers. In fact, under clean conditions three-dimensional growth is suppressed and the two-dimensional islands are only bounded by {111}-facetted steps in the entire temperature range from about 350–650 K.

The description of growth processes on semiconductor surfaces is much more involved than on metal surfaces because of the complex surface reconstructions that can occur (see Sect. 4.3). Nevertheless, the island nucleation and growth of compound semiconductor has already been addressed by kinetic Monte Carlo simulations on the basis of rates derived from DFT calculations [370]. In order to study the island growth on the $\beta 2(2 \times 4)$ reconstruction of GaAs(100) (Fig. 4.15c), the barriers for 32 microscopically different Ga diffusion processes and As_2 adsorption/desorption have been determined. The derived rates span a time scale from picoseconds to milliseconds. The simulation cell had been chosen to be 160×320 sites which corresponds to $64 \, \text{nm} \times 128 \, \text{nm}$.

The kinetic Monte Carlo simulations reveal that the island growth at $T = 700$ K starts in the trenches of the $\beta 2(2\times 4)$ reconstruction of GaAs(100) (right part of the surface unit cell shown in Fig. 4.15c). From there, growth proceeds along the trenches (for an animation, see [370]). The island density has been determined in the temperature range from 500 up to 900 K. The results are compared to predictions of the nucleation theory with a critical island size of

$i = 1$ in Fig. 7.11. At lower temperatures up to $T = 600\,\mathrm{K}$ nucleation theory and kinetic Monte Carlo simulations agree, however, at higher temperatures the results start to deviate, and for $T > 800\,\mathrm{K}$ the island density even rises again in the KMC simulation, a result that cannot be understood at all within nucleation theory. At such a high temperature, $T = 850\,\mathrm{K}$, the islands density has been determined experimentally [371]. This STM results actually agrees well with the kinetic Monte Carlo simulations.

A detailed analysis of the simulation revealed that at these elevated temperatures the Ga-As-As-Ga$_2$ complexes forming the island nuclei in the trenches are no longer stable, which means that they can dissolve again. Only if an existing island already extends into a new layer, it will continue to grow. However, due to detachment processes of Ga atoms from the Ga-As-As-Ga$_2$ complexes, the density of mobile Ga adatoms becomes higher. This causes an increased nucleation rate of new islands and thus leads to the observed rise in the island densities. This analysis demonstrates the importance of microscopic considerations for the understanding of macroscopic growth phenomena.

7.6 Reaction Kinetics on Surfaces

A catalytic process on a surface often consists of a number of different microscopic reaction steps. First of all the reactants have to adsorb and possibly dissociate on the surface, then the wanted product has to be formed on the surface and desorb again. Furthermore, there can be a large number of intermediate steps. As an example, I present a proposed scheme for the ammonia (NH$_3$) synthesis

$$N_2 + * \rightarrow N_2^{(\mathrm{ads})}, \tag{7.53}$$

$$N_2^{(\mathrm{ads})} + * \rightarrow 2N^{(\mathrm{ads})}, \tag{7.54}$$

$$H_2 + 2* \rightarrow H_2^{(\mathrm{ads})}, \tag{7.55}$$

$$N^{(\mathrm{ads})} + H^{(\mathrm{ads})} \rightarrow (NH)^{(\mathrm{ads})} + *, \tag{7.56}$$

$$(NH)^{(\mathrm{ads})} + H^{(\mathrm{ads})} \rightarrow (NH_2)^{(\mathrm{ads})} + *, \tag{7.57}$$

$$(NH_2)^{(\mathrm{ads})} + H^{(\mathrm{ads})} \rightarrow (NH_3)^{(\mathrm{ads})}, \tag{7.58}$$

$$(NH_3)^{(\mathrm{ads})} \rightarrow NH_3 + *, \tag{7.59}$$

where $*$ denotes a free surface site and $X_n^{(\mathrm{ads})}$ is an adsorbed X_n species. The dissociation of N$_2$, step (7.54), is actually the rate limiting step for the NH$_3$ synthesis.

The determination of the reaction paths and barriers of such a large number of processes from electronic structure calculations is still computationally very demanding if not impossible. Even if the potential energy surface of all reaction steps were known, a dynamical simulation of such a complex process would not be feasible since many of the microscopic reaction steps correspond to rare events on the scale of typical simulation times. Again, only a kinetic modelling makes the theoretical determination of catalytic reaction rates possible.

The reaction rates entering the kinetic description of reactions such as the ammonia synthesis (7.53)–(7.59) are usually taken from experiment. These experiments are mostly performed under ultrahigh vacuum (UHV) conditions on single-crystal surfaces. Catalytic reactors, on the other hand, run at pressures that are many orders of magnitude higher. This can influence the reaction rates and microscopic mechanisms when the high pressure induces changes in the surface coverages, composition and structure. In addition to the structure gap addressed in Sects. 5.8 and 5.9 there is therefore also the *pressure gap* between surface science and heterogeneous catalysis.

The ammonia synthesis seems to be a system were the pressure gap is not relevant. Using data from UHV studies of the N_2, H_2 and NH_3 adsorption on clean and K-precovered Fe single crystal surfaces, kinetic simulations correctly predicted the NH_3 synthesis rate for an iron-based industrial NH_3 catalyst at 1–300 atm and 375–500°C [372]. This suggests that the insight gained from well-defined chemisorption systems can be transferred to the more complex catalytic systems.

However, there are also systems that show a pronounced pressure gap. Under UHV conditions, Ru surfaces are very inactive, as far as the catalytic CO oxidation is concerned. However, at high pressures the situation is reversed, and the catalytic properties of Ru are superior compared to other metals [139]. The reason for this peculiar behavior remained unclear for a long time. Now the formation of RuO_2 at high pressures is considered to be the source of the high catalytic activity [373].

One of the most fascinating phenomenon with respect to reactions at surfaces is the evolution of spatiotemporal self-organization pattern formed in the course of a catalytic reaction [375, 376]. The model system in this respect is the catalytic oxidation of CO on Pt(110). Figure 7.12 corresponds to an photoemission electron microscopy (PEEM) image of a Pt(110) surface exposed to O_2 and CO [374,377]. The contrast imaged in the PEEM is caused by local differences of the work function associated with varying adsorbate concentrations. The work function change $\Delta\phi$ is proportional to the oxygen coverage θ_O which in turn is directly related to the reaction rate of CO oxidation. The image shows three different spirals with different rotation periods and wavelengths.

In order to understand the origin of this pattern formation, the microscopic steps of the CO oxidation on Pt(110) have to be considered. The overall reaction $2CO + O_2 \rightarrow 2CO_2$ proceeds through the Langmuir–Hinshelwood mechanism which corresponds to the recombinative desorption of the surface species $CO^{(ads)}$ and $O^{(ads)}$. These are formed by the adsorption of O_2 and CO from the gas phase. $CO^{(ads)}$ is bound relatively weakly to the surface and

Fig. 7.12. Photoemission electron microscopy (PEEM) image of a Pt(110) surface exposed to 4×10^{-4} mbar O_2 and 4.3×10^{-5} mbar CO at a temperature of $T = 448$ K. (After [374])

may desorb as well as diffuse on the surface while oxygen is so strongly bound that it is assumed to remain at its adsorption site. Furthermore, depending on the CO coverage, the Pt(110) surface undergoes reversible structural changes between a 1×2 missing-row reconstruction and a 1×1 phase. In total, the scheme for the CO oxidation is given by

$$\mathrm{CO} + * \leftrightarrow \mathrm{CO}^{(\mathrm{ads})}\,, \tag{7.60}$$

$$\mathrm{O}_2 + * \to 2\mathrm{O}^{(\mathrm{ads})}\,, \tag{7.61}$$

$$\mathrm{O}^{(\mathrm{ads})} + \mathrm{CO}^{(\mathrm{ads})} \to 2* + \mathrm{CO}_2\,, \tag{7.62}$$

$$1 \times 2 \leftrightarrow 1 \times 1\,. \tag{7.63}$$

The adsorption energy of CO on the 1×1 phase is larger than on the 1×2 phase which induces a $1 \times 2 \to 1 \times 1$ transformation for CO coverages larger than $\theta_{\mathrm{CO}} = 0.2$ [378]. On the other hand, the 1×1 phase exhibits a larger sticking probability for oxygen than the 1×2 phase. For a specific external parameter range (temperature and partial pressures of the reactants), the following scenario occurs: CO adsorbs on the 1×2 phase which eventually causes the $1 \times 2 \to 1 \times 1$ surface transition. On the 1×1 phase, oxygen adsorbs more easily and reacts with the adsorbed CO to CO_2 which desorbs. This lowers the CO concentration again until the surface switches back to the 1×2 phase which is less reactive with respect to O_2 adsorption, and the cycle can start again. The strongly nonlinear behavior of this oscillating reaction leads to a wide range of spatiotemporal patterns such as propagating wave fronts, spiral waves, solitary-type waves, standing waves or chaotic structures denoted *chemical turbulence*.

For a mathematical modeling of this spatiotemporal behaviour, the local CO and O coverage are denoted by $u = \theta_{CO}$ and $v = \theta_O$, respectively. In addition, w is the fraction of the surface area consisting of the 1×1 phase. Using these three variables, the following kinetic model has been proposed for the CO oxidation on Pt(110) [378]

$$\dot{u} = s_{CO}\, p_{CO} - k_2 u - k_3 uv + D\nabla^2 u, \tag{7.64}$$
$$\dot{v} = s_{O_2}\, p_{O_2} - k_3 uv, \tag{7.65}$$
$$\dot{w} = k_5 \left[f(u) - w \right]. \tag{7.66}$$

The rate constants for CO desorption and CO_2 reactive combination are given by k_2 and k_3, respectively. D is the surface diffusion coefficient of adsorbed CO. The sticking coefficients s_{CO} and s_O of CO and O are a function of the coverages. O adsorption is assumed to be prohibited on CO covered sites and to be dependent on the surface structure while CO adsorption is taken to be independent of the O coverage and surface structure. Hence for s_{CO} the simple form

$$s_{CO} = k_1 \left(1 - u^3\right) \tag{7.67}$$

has been assumed while s_O is much more complicated:

$$s_{O_2} = k_4 [s_1 w + s_2(1-w)] (1 - u - v)^2. \tag{7.68}$$

A modified simpler version of (7.64) and (7.65) can in fact be solved analytically and yields a temporal oscillatory structure (see Exercise 7.1). However, a complex set of coupled differential equations such as (7.64)–(7.68) can only be solved numerically [379]. Since u and v are strictly anticorrelated [380], one adsorbate variable can be eliminated. Finally one ends up with a nonlinear set of equations which corresponds to a two-variable model with so-called *delayed inhibitor production* because the phase transition of the Pt(110) surface sets only in at a CO coverage of 0.2 ML.

Depending on the CO partial pressure and the temperature, the solution of the kinetic model yields oscillatory, excitable or bistable regimes. Nonlinear phenomena such as traveling solitary pulses [381] or spiral waves [380] can indeed been reproduced within such a model. In a defect-free region, spirals result from a broken plane wave. Then the open ends of the wave start to curl and form rotating waves. It turns out that the rotation period is mainly determined by the rate k_5 (see (7.66)) which is therefore used as an adjustable parameter in the simulations. At a defect-free surface there should be only spirals with almost the same rotation periods. However, as Fig. 7.12 demonstrates, in the experiment spirals with quite different rotation periods are found [374]. These can be simulated in the kinetic model by introducing artificial nonexcitable regions that are larger than the core size of free spirals ($\geq 1\,\mu\text{m}$, depending on the temperature). Then the spirals are pinned to these defects with rotation periods that depend almost linearly on the defect radius [374]. This size is large compared to microscopic length scales. Hence these spirals must be caused by patches of enlarged roughness or accumulated impurities that change the kinetic parameters mesoscopically from the values of a perfect Pt(110) surface.

Most of the kinetic parameters entering (7.64)–(7.68) are derived empirically or have to be guessed. Furthermore, due to the simplified structure of the kinetic equations, it is not guaranteed that the solution is unique, i.e. the same spatiotemporal pattern might be reproduced with another set of kinetic equations and parameters. Still these kinetic simulations yield valuable qualitative insight into the underlying reaction mechanisms, as demonstrated above. The determination of the crucial parameters from first principles seems to be out of reach at the moment, but it might well become possible by a combination of kinetic lattice gas models and ab initio total-energy calculations.

Exercises

7.1 Rate Equations

Consider the system of coupled rate equations

$$\frac{dx}{dt} = \alpha_1 x - \beta_1 xy \tag{7.69}$$

and

$$\frac{dy}{dt} = -\alpha_2 y + \beta_2 xy \tag{7.70}$$

with $\alpha_i, \beta_i > 0$, $i = 1, 2$. These equations can be considered as a modified version of (7.64) and (7.65) describing the temporal evolution of two different species.

a) Solve the set of coupled differential equations analytically.
Hint: show that

$$H(x, y) = -\alpha_2 \ln(x) + \beta_2 x - \alpha_1 \ln(y) + \beta_2 x \tag{7.71}$$

is constant along the trajectory $(x(t), y(t))$. Such a function is called a first integral of the system.

b) Sketch the solution as a curve in the xy plane for $\alpha_1 = \alpha_2 = \beta_1 = \beta_2 = 1$ for different values of the $H(x, y) = c > 0$. Give an interpretation of the resulting curves in terms of a population analysis.

7.2 Lattice Gas Hamiltonian from First Principles

a) The lattice gas Hamiltonian of one species of atoms adsorbed on a square lattice considering first and second-neighbor two-body interactions is given by

$$H = E_s \sum_i n_i + \frac{1}{2}\left(V_{1n} \sum_{i,a} n_i n_{i+a} + V_{2n} \sum_{i,b} n_i n_{i+b} \right). \quad (7.72)$$

The indices a and b indicate first and second-neighbor distances. How can the parameters E_s, V_{1n} and V_{2n} be determined from first-principles calculations? Specify the adsorbate structures and the set of linearly coupled equations necessary to derive the parameters for $T = 0$, i.e., if no distinction between binding energy and Helmholtz free energy has to be made.

Hint: For the case of a hexagonal substrate, see [352].

b) Now consider additionally third-neighbor interactions $V_{3n} \sum_{i,c} n_i n_{i+c}$ in (7.72). How can the parameters of the lattice gas Hamiltonian be evaluated from total-energy calculations in this case?

c) Three-body terms account for modifications of the interaction between two adsorbed atoms when a third adatom is adsorbed close by. Considering only so-called trio interactions between three nearest neighbors, we get an additional term in the lattice gas Hamiltonian $\sum_{i,a,a'} V_{3n}(i,a,a') n_i n_{i+a} n_{i+a'}$, where the interaction energy $V_{3n}(i,a,a')$ depends on the specific configuration of the trio. Which trio configurations exist on a square lattice? Repeat the derivation of the lattice gas parameters including first, second and third-neighbor two-body interactions and trio three-body interactions.

7.3 Temperature Programmed Desorption (TPD)

In TPD experiments, the surface temperature is increased linearly, i.e. $T = T_0 + \alpha t$, and the desorption flux is monitored as a function of the temperature. Under the assumption that the rate for n-th order desorption can be written as

$$\frac{d\theta}{dt} = R_{\text{des}} = R_0\, \theta^n \exp(-\beta E_{\text{des}}), \quad (7.73)$$

where θ is the coverage, proove (7.44), i.e. show that the maxima in the TPD spectra are given by

$$\ln(T_{max} R_0\, \theta^{n-1}/\alpha) = \frac{E_{\text{des}}}{k_B T_{max}} + \ln\left(\frac{E_{\text{des}}}{k_B T_{max}}\right). \quad (7.74)$$

7.4 Random Deposition

Random deposition is one of the simplest growth models [382]: Particles are deposited randomly on the surface. They stay where they are landed so that all columns grow independently, as illustrated in the figure.

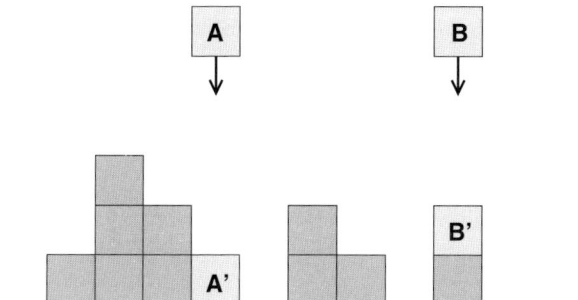

a) There are no correlations between the columns. Hence the height $h(i,t)$ of every column grows independently by one with a probability $p = 1/L$ where L is the number of lattice sites. Determine the probability $P(h, N)$ that a column has height h after the deposition of N particles.

b) Let the time be defined by the mean number of deposited layers, $t = N/L$. Show that the average height

$$\bar{h}(t) = \langle h \rangle = \sum_{h=1}^{N} h P(h, N) \tag{7.75}$$

grows linearly in time.

c) The *surface width* that characterizes the *roughness* of the surface is defined by the root mean square fluctuation in the height,

$$w(t) = \sqrt{\langle (h - \langle h \rangle)^2 \rangle} = \sqrt{\frac{1}{L} \sum_{i=1}^{L} [h(i,t) - \bar{h}(t)]^2}. \tag{7.76}$$

Determine the growth exponent β that is given by $w(t) \propto t^\beta$.

8. Electronically Non-adiabatic Processes

So far we have almost entirely dealt with surface structures and processes that correspond to the electronic ground state. This means that we have assumed that the Born–Oppenheimer approximation is justified. Although there are many important processes at surfaces that involve electronic transitions [225, 383, 384], the status of the theoretical treatment of processes with electronically excited states is not very satisfactory. Many factors still hamper the development of quantitative models incorporating electronic excitations. Neither the determination of the electronically excited states nor the calculation of coupling matrix elements between these excited states is trivial. But even if the excited states and the coupling between them is known, the simulation of the reaction dynamics with electronic transitions still represents a challenge. In the next sections I will illustrate why the treatment of excited states is so complicated, but I will also show that there are some promising approaches to overcome the problems. In addition, concepts to treat reaction dynamics with electronic transitions will be discussed.

8.1 Determination of Electronically Excited States

In Sect. 4.3, we already saw that for example the GW approximation allows an accurate determination of the electronic band structure including excited electronic states. However, no total energies can be derived from the GW approximation so that the evaluation of excited state potential energy surfaces is not possible.

In principle, energies of excited states can be determined by quantum chemistry methods. This has in fact been done successfully for the description of electronically nonadiabatic processes at surfaces, as will be shown in Sect. 8.5. Still, quantum chemistry methods are limited to finite systems of a rather small size. Density functional theory, which is so successful for electronic ground state properties of extended system, can not be directly used for electronically excited states since it is in principle an electronic ground-state theory.

Still DFT can be extended to allow the determination of excited states energies, namely in the form of the time-dependent density-functional theory (TDDFT) [385, 386]. It rests on the *Runge–Gross theorem* which is the analogue to the Hohenberg–Kohn theorem of time-independent density-functional theory. The Runge–Gross theorem states

> The densities $n(r,t)$ and $n'(r,t)$ evolving from a common initial state $\Psi_0 = \Psi(t_0)$ under the influence of two potentials $v(r,t)$ and $v'(r,t)$ are always different provided that the potentials differ by more than a purely time-dependent, i.e. r-independent function $v(r,t) \neq v'(r,t) + c(t)$.

This means that there is an one-to-one mapping between time-dependent potentials and time-dependent densities. However, it is important to note that the functional depends on the initial conditions, i.e. on $\Psi_0 = \Psi(t_0)$. Only if the initial state corresponds to the electronic ground state, the functional is well-defined since it depends on the density $n(r,t)$ alone.

The proof is a little bit more complex than the one for the Hohenberg–Kohn theorem. Still it is not too complicated [386]. Here I only sketch the main ideas of the lines of reasoning. First one shows by using the quantum mechanical equation of motion that the current densities

$$j(r,t) = \langle \Psi(t)|\hat{j}_p|\Psi(t)\rangle \tag{8.1}$$

and

$$j'(r,t) = \langle \Psi'(t)|\hat{j}_p|\Psi'(t)\rangle \tag{8.2}$$

are different for different potentials v and v'. The current densities are related to the density by the continuity equation. By using

$$\frac{\partial}{\partial t}(n(r,t) - n'(r,t)) = -\nabla \cdot (j(r,t) - j'(r,t)), \tag{8.3}$$

the one-to-one mapping between time-dependent potentials and densities can be proven.

Still it does not seem to be obvious why time-dependent DFT should lead to the determination of electronic excitation energies. To see this, we have to use linear response theory. We consider an electronic system subject to an external potential of the form

$$v_{\text{ext}}(r,t) = \begin{cases} v_0(r) & ; \quad t \leq t_0 \\ v_0(r) + v_1(r,t) & ; \quad t > t_0 \end{cases}, \tag{8.4}$$

In perturbation theory, the change of the density due to the external potential $v_1(r,t)$ is expanded in powers of v_1, i.e. the density is written as

$$n(r,t) = n_0(r) + n_1(r,t) + n_2(r,t) + \ldots, \tag{8.5}$$

where the single terms correspond to the different orders of v_1. The first order or linear response to the perturbation $v_1(r,t)$ is given by

$$n_1(r,t) = \int dt' \int d^3r' \, \chi(r,t,r',t') v_1(r',t'), \tag{8.6}$$

8.1 Determination of Electronically Excited States

where the density-density response function χ is defined as

$$\chi(\mathbf{r}, t, \mathbf{r}', t') = \left. \frac{\delta n[v_{\text{ext}}](\mathbf{r}, t)}{\delta v_{\text{ext}}(\mathbf{r}', t')} \right|_{v_0}. \tag{8.7}$$

The Runge–Gross theorem is not only valid for interacting particles, but also for non-interacting particles moving in an external potential $v_s(\mathbf{r}, t)$. The density-density response function of non-interacting particles with unperturbed density n_0 corresponds to the Kohn–Sham response function and is given by

$$\chi_s(\mathbf{r}, t, \mathbf{r}', t') = \left. \frac{\delta n[v_s](\mathbf{r}, t)}{\delta v_s(\mathbf{r}', t')} \right|_{v_s[n_0]}. \tag{8.8}$$

This linear response formalism can be used in order to determine polarizabilities not only of atoms and molecules [387], but also for large systems. For the fullerene molecule C_{60}, e.g., the results have been quite accurate, however, for polyacetylene chains the conventional exchange-correlation functionals fail [388]. I do not want to address this subject any further, but rather focus on another application of time-dependent DFT, the calculation of excitation energies [386, 389]. The main idea rests on the fact that the frequency-dependent linear response of a finite system, i.e. the Fourier transform of (8.6), has discrete poles at the excitation energies $\Omega_j = E_j - E_0$ of the unperturbed system. By using the functional chain rule, it can be shown that the noninteracting and interacting response functions are related by a Dyson-type equation [386, 389]. This leads to an integral equation for the frequency-dependent linear response $n_1(\mathbf{r}, \omega)$

$$\int d^3x \, K(\mathbf{x}, \mathbf{r}, \omega) \, n_1(\mathbf{x}, \omega) = \int d^3r' \, \chi_s(\mathbf{r}, \mathbf{r}', \omega) \, v_1(\mathbf{r}', \omega), \tag{8.9}$$

where the Kernel $K(\mathbf{x}, \mathbf{r}, \omega)$ is given by

$$K(\mathbf{x}, \mathbf{r}, \omega) = \delta(\mathbf{r} - \mathbf{x}) - \int d^3r' \, \chi_s(\mathbf{r}, \mathbf{r}', \omega)$$
$$\times \left(\frac{1}{|\mathbf{r}' - \mathbf{x}|} + f_{\text{xc}}[n_0](\mathbf{r}', \mathbf{x}, \omega) \right), \tag{8.10}$$

with the Fourier transform $f_{\text{xc}}[n_0](\mathbf{r}', \mathbf{x}, \omega)$ of the so-called time-dependent exchange-correlation kernel

$$f_{\text{xc}}[n_0](\mathbf{r}, t, \mathbf{r}', t') = \left. \frac{\delta v_{\text{xc}}[n](\mathbf{r}, t)}{\delta n(\mathbf{r}', t')} \right|_{n_0}. \tag{8.11}$$

Now one uses the fact that the Kohn–Sham excitation energies ω_j are in general not identical with the true excitation energies Ω_j. Hence the right hand side of (8.9) remains finite for $\omega \to \Omega_j$. On the other hand, the exact density response n_1 diverges for $\omega \to \Omega_j$. In order that the integral operator acting on n_1 on the left hand side of (8.9) yields a finite result, the eigenvalues of this integral operator have to vanish. This is equivalent to the statement

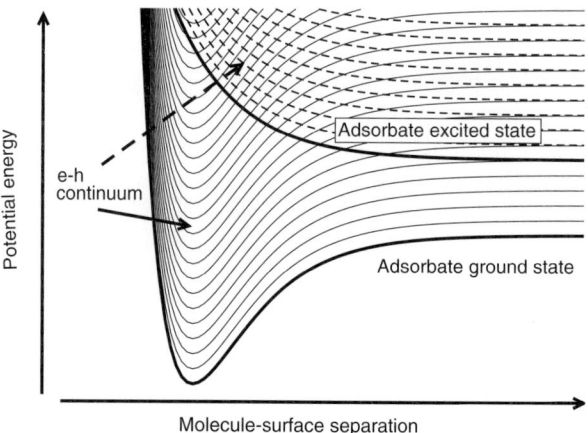

Fig. 8.1. Illustration of electronically excited states in the interaction of atoms or molecules with surfaces

that the true excitation energies Ω are characterized as those frequencies where the eigenvalues $\lambda(\omega)$ of

$$\int d^3x \int d^3r' \; \chi_s(\boldsymbol{r},\boldsymbol{r}',\omega) \left(\frac{1}{|\boldsymbol{r}'-\boldsymbol{x}|} + f_{xc}[n_0](\boldsymbol{r}',\boldsymbol{x},\omega) \right) g(\boldsymbol{r}',\omega)$$
$$= \lambda(\omega) \; g(\boldsymbol{r},\omega) \tag{8.12}$$

satisfy $\lambda(\Omega) = 1$. This can easily be shown by performing the integration over the delta function in (8.10). The solution of (8.12) is still not trivial. For practical purposes, further approximations have to be made. Often one uses the adiabatic local density approximation for the exchange-correlation functional which not only assumes locality in space, as in standard LDA calculations, but also locality in time.

8.2 Electronic Excitation Mechanisms at Surfaces

If we consider electronic excitations in the interaction of atoms or molecules with surfaces, we have to distinguish between delocalized excited states of the surface and localized excitations at the adsorbate or the adsorbate-surface bond. These two different kinds of excitation modes are illustrated in Fig. 8.1 where the potential energy surfaces for the interaction of a molecule with a surface are plotted.

First of all, the interaction of a molecule with a surface can lead to the excitation of a electron-hole (e-h) pair in the surface. In particular in metals, the electronic states show a continuous spectrum. More importantly, they are rather delocalized. Hence the local interaction of a molecule with a metal surface will hardly be influenced by the excitation of an electron-hole pair.

In Fig. 8.1, this is schematically illustrated by a multitude of shifted ground state potentials. Still, the excitation of an electron-hole pair corresponds to an energy transfer process to the surface that leads to dissipation effects. Thus the influence of the excitation of e–h pairs on the molecule-surface dynamics should be described within a friction-dissipation formalism.

The situation is entirely different in the case of an electronic excitation of the molecule or atom interacting with the surface. Then the shape of the excited-state potential might be entirely different from the ground-state potential. In a theoretical description of such a process usually the potential of the excited state must be explicitly considered. These two scenarios can be considered as the adiabatic and the diabatic limit of electronic transitions at surfaces. We will first discuss the adiabatic limit that can be modelled within a friction formalism.

8.3 Electronic Friction Effects

The determination of the role of electron-hole pairs in the scattering and sticking of molecules at surfaces is rather cumbersome. There are hardly any reliable studies where the influence of the e-h pairs has been investigated from first-principles. Hence there is also no accepted viewpoint over the importance of the e–h pairs. It seems that whenever there are some unclear results in a sticking or scattering experiment, e–h pairs are made responsible. Equivalently, the validity of Born–Oppenheimer molecular dynamics simulations is often questioned because of the neglect of e–h pair excitations.

Using a thin polycristalline Ag film deposited on n-type Si(111) as a Schottky diode device, the nonadiabatically generated electron-hole pairs upon both atomic and molecular chemisorption could be detected [390, 391]. A strong correlation between the adsorption energy and the measured *chemi-current* has been observed. For NO adsorption on Ag (adsorption energy $\sim 1\,\mathrm{eV}$), it has been estimated that one quarter of the adsorption energy was dissipated to electron-hole pairs. Adsorption-induced electron hole-pair creation has also been found for other metal substrates, such as Au, Pt, Pd, Cu, Ni and Fe, and even for semiconductors such as GaAs and Ge [390,392].

The understanding of this nonadiabatic dissipation channel is still rather incomplete. There are, however, theoretical studies that used ab initio input in order to assess the effects of electronic excitations in adsorption and reaction processes on surfaces. An approach based on time-dependent density functional theory has been used in order to estimate the electron-hole pair excitation due to an atom incident on a metal surface [393,394]. The method consists of three independent steps. First, a conventional Kohn-Sham DFT calculations is performed in order to determine the ground state potential energy surface for various configurations of the incident atom. Then, the resulting Kohn-Sham states are used in the framework of time-dependent DFT in order to obtain a position dependent friction coefficient. Finally, this friction coefficient enters a forced oscillator model (see p. 169) in which the probability density of electron-hole pair excitations caused by the classical motion of the incident atom is estimated.

This formalism has been applied [394] to address the chemicurrent measured in experiments of the adsorption of hydrogen atoms on copper surfaces [395]. Satisfactory agreement between theory and experiment has been obtained [394]. However, only one single trajectory for the hydrogen impinging on the top site has been used in the forced oscillator description so that the effect of corrugation has been entirely neglected. In the so-called *molecular dynamics with electronic friction* method [396], the effects of e–h pair excitation have been incorporated in an otherwise classical molecular dynamics simulation which allows to evaluate many trajectories. In this approach, the energy transfer between nuclear degrees of freedom and the electron bath of the surface is also modelled with a friction term, but additionally temperature-dependent fluctuating forces are included.

In detail, the equation of motion for the nuclear adsorbate degrees of freedom is

$$M_I \ddot{\boldsymbol{R}}_I = -\frac{\partial V}{\partial \boldsymbol{R}_I} - \sum_J K_{IJ} \dot{\boldsymbol{R}}_J + \boldsymbol{S}_I(t). \tag{8.13}$$

Here $\boldsymbol{S}_I(t)$ is a stochastic fluctuating force satisfying the *fluctuation-dissipation theorem*

$$\langle \boldsymbol{S}_I(t) \boldsymbol{S}_J(t') \rangle = k_\mathrm{B} T K_{IJ} \delta(t - t'). \tag{8.14}$$

K_{IJ} is the friction matrix in the adsorbate degrees of freedom which depends on the position of the adsorbate. Several assumptions enter the derivation of this methods. First of all, it is assumed that the coupling is weak. Furthermore, the metal density of states at the Fermi level should be smooth so that the important coupling matrix elements become energy-independent. Then the energy-independent friction kernel can be written as

$$K_{IJ} = \pi \hbar \; \mathrm{Tr}\{P(\varepsilon_F) \boldsymbol{G}_{\boldsymbol{R}_I} P(\varepsilon_F) \boldsymbol{G}_{\boldsymbol{R}_J}\}, \tag{8.15}$$

where $\boldsymbol{G}_{\boldsymbol{R}_I}$ corresponds to the derivative of the Hamiltonian and overlap matrices H and S with respect to the nuclear coordinates $\boldsymbol{R}_I$

$$\boldsymbol{G}_{\boldsymbol{R}_I} = \frac{\partial}{\partial \boldsymbol{R}_I} H - \varepsilon_F \frac{\partial}{\partial \boldsymbol{R}_I} S, \tag{8.16}$$

and where P is a split-time local density of states:

$$P(\varepsilon, t, t') = \sum_i c_i(t) c_i(t') \delta(\varepsilon - \varepsilon_i). \tag{8.17}$$

The friction kernel K_{IJ} has been evaluated for CO/Cu(100) by Hartree–Fock cluster calculations using single excitations and parametrized in a form suitable for molecular dynamics simulations. The interaction potential of CO/Cu(100) in the nuclear degrees of freedom was derived empirically.

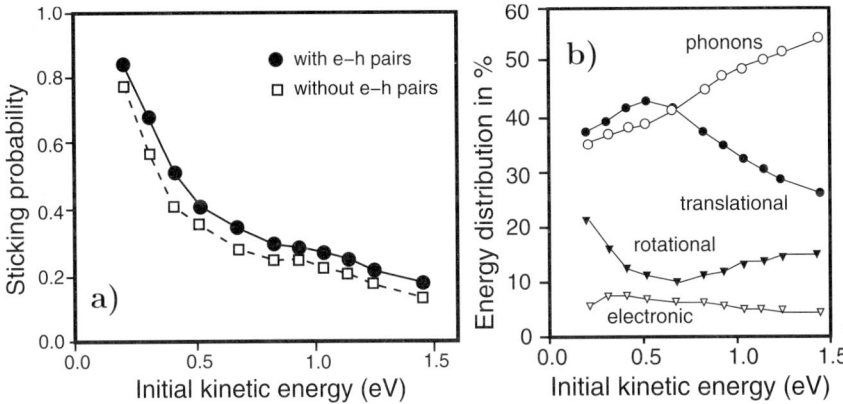

Fig. 8.2. Role of e–h pairs in the scattering and sticking of CO/Cu(111) at a surface temperature of $T_s = 100$ K; (**a**) sticking probability for CO/Cu(111) under normal incidence calculated without and with electronic friction, (**b**) Energy distribution of CO molecules scattered under normal incidence from Cu(111) in percent of the initial kinetic energy. (After [397])

By averaging over molecular dynamics trajectories with 108 surface atoms in the periodic surface unit cell and stochastic boundary conditions representing interactions with the bulk, the sticking probability of CO/Cu(100) was evaluated. Figure 8.2a) compares the sticking probability as a function of the kinetic energy with and without the consideration of e–h excitations. First of all it is evident that the sticking probability shows the typical monotonously decreasing behavior. The incorporation of e–h leads to an additional channel for energy transfer to the surface. Hence the sticking probability should increase by taking into accound e–h pair excitations. This is confirmed by the calculations. However, the effect is rather small. This means that e–h pair excitation plays only a minor role as a dissipation channel in the sticking and scattering of CO/Cu(100).

In order to quantify the energy transfer to the e–h pairs, the energy distribution for directly scattered molecules was determined (Fig. 8.2b). Less than 10% of the incident kinetic energy is transferred to e–h pairs in a direct scattering process which is less than observed for NO/Ag [390]. The main energy loss channel for CO/Cu(100) is the excitation of surface phonons. The electronic friction does not influence the equilibrium surface diffusion either, it only affects the transient mobility of molecules following adsorption [397]. The relative unimportance of electronic friction can be related to the different time scales of electronic and nuclear motion which already entered the derivation of the Born–Oppenheimer approximation. Even if it takes infinitesimal energies to excite e–h pairs as in the case of metal surfaces, still their excitation probability is small compared to the excitation of surface phonons.

It is important to note, however, that it is not appropriate to naively generalize the results for the CO/Cu(100) system to other systems. Copper has almost no d-band density of states at the Fermi level, furthermore CO has a closed shell electronic configuration. For other substrate materials and molecules the coupling between surface e–h pairs and impinging molecules might be much stronger. For example, the observed stronger nonadiabatic dissipation effects in the system NO/Ag [390] might be caused by the unpaired electron in NO. There is certainly plenty of room for further investigations.

8.4 Reaction Dynamics with Electronic Transitions

In order to couple the direct determination of excited states by electronic structure methods with a dynamical simulation, a full quantum treatment of the system would be desirable. However, this is computationally not feasible. In any case, for atoms heavier than hydrogen or helium a classical description of the reaction dynamics is usually sufficient. Hence a mixed quantum-classical dynamics method is appropriate in which a multi-dimensional classical treatment of the atoms is combined with a quantum description of the electronic degrees of freedom. The crucial issue in mixed quantum-classical dynamics is the self-consistent feedback between the classical and the quantum subsystems. There are two standard approaches that incorporate self-consistency, *mean-field* and *surface-hopping* methods [398].

These methods are based on the separation of the kinetic energy of the classical particles from the total Hamiltonian

$$H = T_{\mathbf{R}} + H_{\text{el}}(\mathbf{r}, \mathbf{R}) , \tag{8.18}$$

where $\mathbf{R}$ are the classical and $\mathbf{r}$ the quantum degrees of freedom. The time evolution of the quantum wave function is then given by the time-dependent Schrödinger equation using the electronic Hamiltonian H_{el}

$$i\hbar \frac{\partial}{\partial t} \psi(\mathbf{r}, \mathbf{R}, t) = H_{\text{el}}(\mathbf{r}, \mathbf{R}(t)) \, \psi(\mathbf{r}, \mathbf{R}, t) , \tag{8.19}$$

where the coordinates $\mathbf{R}$ of the classical degrees of freedom enter as parameters. In both the mean-field and the surface-hopping methods, the quantum particles are subject to a Hamiltonian that varies in time due to the motion of the classical particles. On the other hand, the quantum state of the system determines the forces that act on the classical particles. Thus the self-consistent feedback cycle between quantum and classical particles is realized.

The difference between the two methods lies in the treatment of the back-response of the classical system to quantum transitions. In the mean-field method, the motion of the classical particles is determined by a single effective potential that corresponds to an average over quantum states

$$M_I \frac{d^2}{dt^2} \mathbf{R}_I = -\nabla_{\mathbf{R}_I} \langle \psi(\mathbf{r}, \mathbf{R}) | H_{\text{el}}(\mathbf{r}, \mathbf{R}) | \psi(\mathbf{r}, \mathbf{R}) \rangle , \tag{8.20}$$

which, using the Hellmann–Feynman theorem, can be transformed to

$$M_I \frac{d^2}{dt^2} R_I = -\langle \psi(r,R) | \nabla_{R_I} H_{\mathrm{el}}(r,R) | \psi(r,R) \rangle. \tag{8.21}$$

The mean-field approach properly conserves the total energy, furthermore it does not depend on the choice of the quantum representation since the wave function can be directly obtained by the numerical propagation of the wave packet using (8.19). However, this approach violates microscopic reversibility, and it is subject to the deficiency of all mean-field methods: the classical path is mainly determined by the major channel trajectory so that branching and correlation effects in the time-evolution are not appropriately accounted for.

A proper treatment of the correlation between quantum and classical motion requires a distinct classical path for each quantum state. This is in fact fulfilled in the surface-hopping method. In this approach, the wave function is expanded in terms of a set of basis functions

$$\psi(r,R,t) = \sum_j c_j(t) \phi_j(r,R). \tag{8.22}$$

With respect to this basis, matrix elements of the electronic Hamiltonian are constructed

$$V_{ij} = \langle \phi_i(r,R) | H_{\mathrm{el}}(r,R) | \phi_j(r,R) \rangle. \tag{8.23}$$

Furthermore, the *nonadiabatic coupling vector* is defined as

$$d_{ij} = \langle \phi_i(r,R) | \nabla_R | \phi_j(r,R) \rangle, \tag{8.24}$$

where the gradient is taken with respect to all atomic coordinates R. Inserting the wave function (8.22) into (8.19), one obtains [399] (see Exercise 8.2)

$$i\hbar \dot{c}_k = \sum_j c_j (V_{kj} - i\hbar \dot{R} \cdot d_{kj}). \tag{8.25}$$

In an adiabatic representation, the matrix elements V_{kj} would be zero while in a diabatic representation the nonadiabatic coupling vector vanishes.

In any surface-hopping method, the classical particles move on the potential energy of one particular quantum state

$$M_I \frac{d^2}{dt^2} R_I = -\nabla_{R_I} \langle \phi_i(r,R) | H_{\mathrm{el}}(r,R) | \phi_i(r,R) \rangle. \tag{8.26}$$

At the same time, the set of coupled differential equations (8.25) is solved in order to obtain the amplitudes c_j of each electronic quantum state. What is left, is the specification of a rule for the switches between the different potential energy surfaces. This rule can in fact not be uniquely defined so that many different algorithms exist (see, e.g., [400]). One particularly elegant method is the so-called *fewest-switches algorithm* [399] which is a variationally-based hopping algorithm that guarantees the correct population $|c_j(t)|$ of each state in an ensemble of many calculated trajectories with the minimum number of hops (see Exercise 8.3).

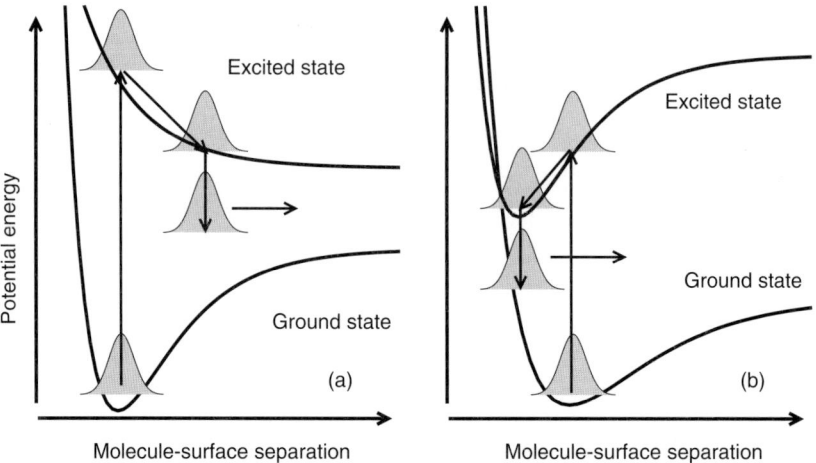

Fig. 8.3. Illustration of Menzel–Gomer–Redhead (MGR) (**a**) and the Antoniewicz (**b**) scenario to model desorption induced by electronic transtions (DIET)

There are also caveats of surface hopping algorithms. They are not independent of the quantum representation [399, 401], and there is some ambiguity in the velocity adjustment if $V_{kk}(\boldsymbol{R}) \neq V_{ll}(\boldsymbol{R})$ at the position of the switch between state k and l. In addition, they are computationally more demanding than mean-field methods, and in fact there are cases for which the mean-field method is more accurate [398].

8.5 Electronic Transitions

Reactions at surfaces with electronic transitions, in particular photochemical reactions, have attracted considerable attention from experimentalists [402]. Ultraviolet (UV) or visible phonons can induce dissociation or reactions on surfaces. A very common process is the *desorption induced by electronic transitions* (DIET) which encompasses photon-stimulated desorption (PSD) as well as electron-stimulated desorption (ESD). DIET processes are usually described using two different scenarios which are sketched in Fig. 8.3. Both correspond to one-dimensional two-electronic-state models.

In the *Menzel–Gomer–Redhead* (MGR) [403, 404] model (Fig. 8.3a) the molecule is excited to a repulsive potential energy surface. The repulsive excited potential accelerates the molecule away from the surface before it is transferred back to the ground-state potential in a *Franck–Condon transition*, i.e. without any change in position and momentum of the nuclei. Depending on how long the molecule has been accelerated on the repulsive potential, it can have gained enough kinetic energy to overcome the desorption barrier. This model was originally developed in order to provide an explanation for the rather large isotope effects observed is ESD experiments. For a given lifetime on the excited potential energy surface, the lighter isotope is stronger accelerated and consequently gains more energy than the heavier isotope; therefore is also shows a much larger desorption probability. To be more specific, if the total desorption probability σ_i of an isotope with mass M_i can be expressed as

$$\sigma_i = \sigma_{\mathrm{ex}}\, P_i\,, \tag{8.27}$$

where σ_{ex} is a primary excitation probability and P_i the escape probability for isotope i, then the isotope effect in the MGR model can be expressed as [405] (see Exercise 8.4)

$$\frac{\sigma_1}{\sigma_2} = \left(\frac{1}{P_1}\right)^{\sqrt{(m_2/m_1)}-1}. \tag{8.28}$$

However, often the excited molecule corresponds to an ionic resonance which is subject to image forces. Hence the excited complex is stronger bound and located closer to the surface. Antoniewicz showed [406] that in such a scenario still DIET processes can occur (Fig. 8.3b). The excited molecule is first accelerated towards the surface and then bounces off. Again, the excitation/deexcitation channel causes an enery transfer into the center-of-mass motion of the adsorbate which can lead to desorption.

A large number of one-dimensional, two-state dynamical simulations of DIET processes have been performed, pioneered by Gadzuk [407, 408], with the desorption of NO from Pt(111) serving as a prototype system [226]. The desorption process has been modelled using jumping wave packets: After initial excitation the wave packet is propagated for a certain residence time τ_r on the excited state potential and then transferred back to the ground state. Then the final fate of the wave packet on the ground state is determined by continuing the propagation and evaluating desorption probabilities and velocity distributions. Such simulations have also been performed for two-dimensional potentials, for example to model the photo-induced desorption of NH_3 from Cu(111) [409, 410] where a significant excitation of the NH_3 vibrational umbrella mode has been detected [411]. Also purely quantum mechanical simulations of DIET processes have been performed [412].

Almost all of the simulations of DIET processes have been based on empirical model potentials due to the problems associated with the first-principles determination of excited state potentials. There are exceptions, however. In order to address the laser-induced desorption of NO from NiO(100) configuration interaction (CI) calculations have been performed. One specific charge transfer state in which one electron was transferred from the NiO(100) surface to the NO molecule was used as a representative electronically excited state [413]. This charge transfer PES of NO^- on NiO(100) which is plotted in Fig. 8.4a has been determined as a function of the NO center of mass distance from the surface and the polar orientation of the molecule with respect to

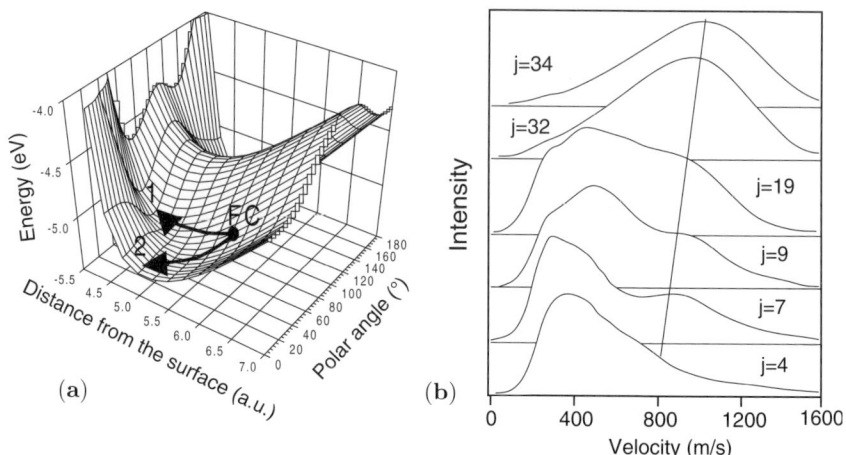

Fig. 8.4. Theoretical description of the photoinduced desorption of NO/NiO(100). (a) Charge transfer PES of NO/NiO(100) as a function of the NO center of mass distance from the surface and the polar orientation of the molecule α. FC denotes the Franck–Condon point at which the wave packet propagation is started. (b) Calculated velocity distribution of NO molecules as a function of the rotational quantum number j. (After [413])

the surface normal. In the CI calculations a NiO_5^{8-} cluster was embedded in a semi-infinite Madelung potential of $\pm 2e$ point charges for the simulation of the NiO(100) surface. In addition, the ground state potential was determined. The calculated ground state potential did in fact not reproduce the experimentally found binding energy of 0.52 eV [414]. This is a well-known problem of cluster calculations which often give good adsorption geometries and frequencies but poor adsorption energies [24]. The ab initio minimum has therefore been scaled to fit the experimental data [413].

The laser-induced desorption of NO from NiO(100) has been simulated using the jumping wave packet technique in three dimensions. The molecular distance from the surface and the polar and azimuthal orientation of the molecule were considered explicitly. The laser-induced electronic excitation was modelled by putting the three-dimensional ground state wave function of the electronic ground state onto the electronically excited PES. Thereby the center of the wave function is located at the Franck-Condon point FC shown on Fig. 8.4. The photoinduced desorption on NO/NiO(100) proceeds according to the Antoniewicz scenario [406] since the minimum of the excited state potential is closer to the surface than in the ground state potential. The wave packet has been propagated on the excited state potential for a number of different lifetimes. The coupling between adsorbate and substrate leading to the deexcitation of the charge transfer state is still unknown. Hence the mean residence time, the so-called *resonance* time τ_r, has entered as an adjustable parameter in the dynamics simulation. A value of $\tau_r = 24$ fs has been chosen which yields a desorption probability of 3.3% in agreement with typical experimental data [413, 415].

Since the wave packet calculations are performed within a restricted dimensionality, quantitative agreement with the experiment [415] cannot be the ultimate goal of this theoretical study. More important is to gain a qualitative understanding of experimental trends. And indeed this study provides such a qualitative concept. The excited state potential has a complicated shape as a function of the polar orientation of the molecule. Two possible pathways are sketched in Fig. 8.4 along which partial wave packets can propagate. This bifurcation of the wave packet causes a bimodality in the velocity distribution of desorbing molecules. This is demonstrated in Fig. 8.4b where the calculated velocity distribution of desorbing NO molecules as function of the rotational quantum number j is plotted. Furthermore, a strong correlation between rotational state and kinetic energy of the desorbing molecules is apparent. These features of the velocity distribution have also been found in the experiment [415].

In addition to the DIET mechanism, a transient high concentration of excited carriers makes multiple excitations in the adsorbate/substrate complex possible leading to a nonlinearly enhanced reaction probability in so-called *desorption induced by multiple electronic transitions* (DIMET). DIMET processes mediated by substrate excitations can be described in the open-system density-matrix approach by supplementing the Lindblad functional (6.23) with a term that accounts for the substrate-mediated excitation of the adsorbate [225]. DIMET of NO from metal surfaces, in particular Pt(111), has been treated by this approach using empirical potentials [416, 417]. These studies confirmed the high desorption yield in DIMET processes which increases nonlinearly with the peak temperature of the substrate carriers. However, already two-dimensional DIMET processes described by the density-matrix technique are computationally extremely demanding so that no quantitative agreement with real experiments can be achieved.

Not only lasers can induce the reactions of atoms and molecules on surfaces, they can also be triggered by the scanning tunneling microscope. In a pioneering work, the ability to control the hydrogen desorption from a Si surface by current injection from the STM tip was demonstrated [418]. The STM current is assumed to cause a $\sigma \rightarrow \sigma^*$ excitation of the Si-H bonds which weakens the bond and eventually leads to the hydrogen desorption.

This STM-induced desorption of hydrogen from the H-terminated Si(111)–(1×1) was addressed by first-principles molecular dynamics simulations [419] using a scheme based upon time-dependent density functional theory [420]. In order to reduce the computational cost, no electronic transitions were allowed. The time-evolution of the n-th electronic state was determined by integrating the time-dependent Kohn–Sham equation

$$i\hbar \frac{\partial \psi_n}{\partial t} = H_{\text{KS}} \psi_n \tag{8.29}$$

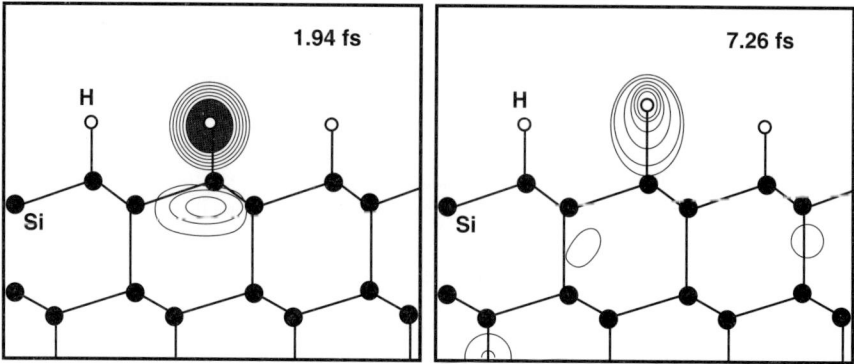

Fig. 8.5. Contour plots of the time evolution of a Si–H σ hole state determined by TDDFT calculations [419] using GGA. The contour plots were recorded 1.94 fs and 7.26 fs after the initial $\sigma \to \sigma^*$ excitation

for each state n separately and coupling it with classical equations of motions for the ions. Hence the ionic motion was restricted to one particular excited Born–Oppenheimer energy surface. Cluster and periodic slab calculations were performed in order to simulate the excitation process. A $\sigma \to \sigma^*$ excitation was prepared as the initial state and the time-evolution of the excitation and the location of the hydrogen atom was determined. The $\sigma \to \sigma^*$ creates both an electron in the conduction band plus an hole in the valence band.

It turned out that in the cluster models the localization of the electronic excitation was over-estimated due to the limited size of the clusters. Slab models are more appropriate to model the electronic excitations. However, in order to induce a localized $\sigma \to \sigma^*$ that is describe by one excited eigenstate of the Kohn–Sham Hamiltonian, one Si–H bond had to be deliberately elongated by 0.05 Å.

Figure 8.5 shows the time evolution of a Si–H σ hole state determined by the TDDFT calculations [419] using GGA. It is apparent that after 7 fs the initially strongly localized hole has become delocalized. An analysis of the electronic structure revealed that no direct $\sigma^* \to \sigma$ recombination has occured but that the σ hole rather relaxed to extended bulk states. Still, no breaking of the Si–H bond caused by the electronic excitation was observed. Thus the simulations provided no explanation for the STM-induced H desorption observed in the experiment [418]. However, it is fair to say that the first-principles treatment of reactions at surfaces with electronic excitations is still in its infancy. The computational schemes developed so far are still incomplete. For example, the TDDFT calculations could be coupled with a surface hopping method (see p. 233) in order to describe nonadiabatic transitions. Such an approach is certainly worth the effort since it allows the ab initio treatment of an important class of reactions at surfaces.

Exercises

8.1 Induced Potential in the Time-dependent Local Density Approximation

Show that the electric potential $\phi_{\text{ind}}(\mathbf{r}, \omega)$ induced by an external electric potential $\phi_{\text{ext}}(\mathbf{r}, \omega)$ in the time-dependent local density approximation is given by [421]

$$\phi_{\text{ind}}(\mathbf{r}, \omega) = \int d^3r' \, K(\mathbf{r}, \mathbf{r}') \, n_1(\mathbf{r}', \omega) \tag{8.30}$$

with the kernel

$$K(\mathbf{r}, \mathbf{r}') = \frac{1}{|\mathbf{r} - \mathbf{r}'|} + \left.\frac{\partial V_{\text{xc}}[n]}{\partial n}\right|_{n_0(\mathbf{r})} \delta(\mathbf{r} - \mathbf{r}'), \tag{8.31}$$

where $n_1(\mathbf{r}, \omega)$ is the frequency-dependent linear response:

$$n_1(\mathbf{r}, \omega) = \int d^3r' \, \chi(\mathbf{r}, \mathbf{r}', \omega) \phi_{\text{ext}}(\mathbf{r}', \omega). \tag{8.32}$$

Hint: Approximate the induced potential by a sum of electrostatic and exchange-correlation terms.

8.2 Surface Hopping

Consider a time-dependent electronic Hamiltonian $H_{\text{el}}(\mathbf{r}, \mathbf{R}(t))$. The electronic wave function is expanded in some suitable set of basis functions

$$\psi(\mathbf{r}, \mathbf{R}, t) = \sum_j c_j(t) \phi_j(\mathbf{r}, \mathbf{R}). \tag{8.33}$$

In this basis, the matrix elements of the electronic Hamiltonian are defined as

$$V_{ij} = \langle \phi_i(\mathbf{r}, \mathbf{R}) | H_{\text{el}}(\mathbf{r}, \mathbf{R}) | \phi_j(\mathbf{r}, \mathbf{R}) \rangle. \tag{8.34}$$

Show that inserting the wave function into the time-dependent Schrödinger equation yields the following set of coupled differential equations for the coefficients c_k

$$i\hbar \dot{c}_k = \sum_j c_j \left(V_{kj} - i\hbar \dot{\mathbf{R}} \cdot \langle \phi_k | \nabla_{\mathbf{R}} \phi_j \rangle \right). \tag{8.35}$$

8.3 Fewest Switches Algorithm

In the following we will use the density matrix notation for the occupation of the electronic states:

$$a_{kj} = c_k c_j^*. \tag{8.36}$$

a) Show that in this notation (8.35) becomes [399]

$$i\hbar \dot{a}_{kj} = \sum_l \left\{ a_{lj}(V_{kl} - i\hbar \dot{\mathbf{R}} \cdot \mathbf{d}_{kl}) - a_{kl}(V_{lj} - i\hbar \dot{\mathbf{R}} \cdot \mathbf{d}_{lj}) \right\}, \quad (8.37)$$

where $\mathbf{d}_{ij} = \langle \phi_i | \nabla_{\mathbf{R}} \phi_j \rangle$ is the nonadiabatic coupling vector. Show that the populations satisfy

$$\dot{a}_{kk} = \sum_{l \neq k} b_{kl} \quad (8.38)$$

with

$$b_{kl} = \frac{2}{\hbar} \text{Im}\, (a_{kl}^* V_{kl}) - 2\text{Re}\, (a_{kl}^* \dot{\mathbf{R}} \cdot \mathbf{d}_{kl}). \quad (8.39)$$

b) The fewest switches algorithm [399] for a two state system is defined as follows:
Consider a trajectory which is in state 1 at integration step i. Both the trajectory and (8.37) are integrated one time interval Δt to step $i+1$. Now a random number x between 0 and 1 is drawn. A state switch from state 1 to 2 will be performed if

$$\frac{\Delta t\, b_{21}}{a_{11}} > x. \quad (8.40)$$

Similarly, if the system is in state 2 it will switch to state 1 if

$$\frac{\Delta t\, b_{21}}{a_{22}} > x. \quad (8.41)$$

Show that this algorithm satisfies the *fewest switches criterion*, i.e., prove that it minimizes the number of state switches subject to the constraint that the correct statistical populations of states 1 and 2 is maintained at all times.

8.4 Isotope Effect in the MGR Model

Consider the DIET process in the Menzel–Gomer–Redhead scenario (Fig. 8.3(a)). The total desorption probability σ_i of an isotope with mass M_i is expressed as

$$\sigma_i = \sigma_{ex}\, P_i, \quad (8.42)$$

where σ_{ex} is a primary excitation probability and P_i the escape probability for isotope i. Show that ratio in the desorption probabilities between two isotopes is given by

$$\frac{\sigma_1}{\sigma_2} = \left(\frac{1}{P_1}\right)^{\sqrt{(m_2/m_1)} - 1}. \quad (8.43)$$

Hint: Assume that the repulsive potential curve of the excited state can be taken to be linear over the relevant range and that the lifetime of the excited state τ is independent of the distance from the surface.

9. Perspectives

Traditionally, surface science studies have focused on the investigation of low-index single crystal surfaces and the interaction of atoms and simple molecules with them. As shown in the previous chapters, the focus in experimental as well as theoretical surface science shifts more and more to surfaces with well-defined defects such as for example the steps of vicinal surfaces. However, even more complex systems have become the subject of microscopic studies [422]. In this final chapter I present some examples of such microscopic studies based on ab initio electronic structure calculations. The examples do not only show the state of the art but they also indicate the promising perspective and future directions of theoretical research in surface science.

9.1 Solid-liquid Interface

All the examples presented so far have been concerned with solid surfaces either in vacuum or interacting with gas particles, i.e. we did not consider any liquids in contact with a solid surface. Still, the solid-liquid interface is of significant importance in the fields of electrochemistry and electrocatalysis which deal with reactions of molecule at this interface [423]. Such reactions are of enormous technological relevance, for example with respect to the development of more efficient fuel cells. In theoretical surface science, however, the solid-liquid interface has hardly been studied microscopically yet because of the difficulties in the reliable description of both the liquid or electrolyte and the solid surface. Often electronic structure studies addressing electrochemical systems omit the description of the electrolyte with the hope that the influence of it on, e.g., adsorption and reaction properties of molecules is negligible [424].

In principle, it is not particularly difficult to include a liquid in periodic supercell calculations. Instead of leaving the region between the slabs empty, it can well be filled with the liquid. This is illustrated in Fig. 9.1. Indeed there are first supercell calculations that have addressed surface energies and structures [425,426] and even reactions at the solid-liquid interface. A particularly impressive study addressed the deprotonation of acetic acid (CH_3COOH) over Pd(111) [427]. In this study, DFT calculations were performed in order

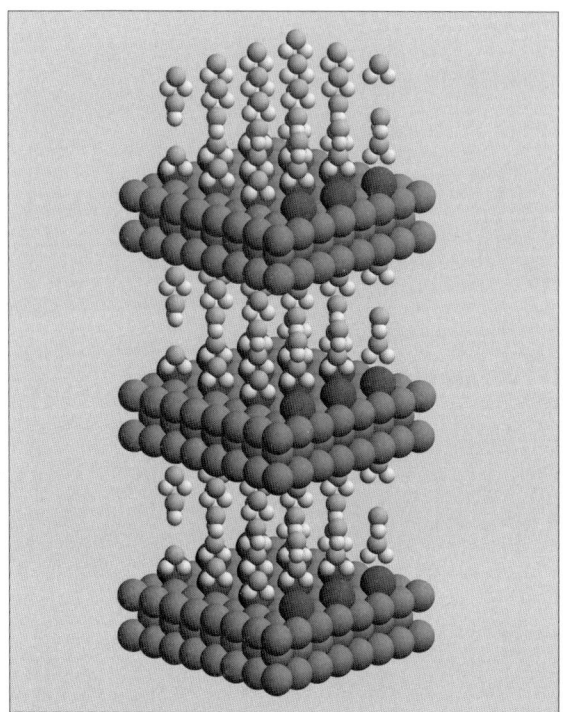

Fig. 9.1. Illustration of the description of the solid-liquid interface within the supercell appproach in periodic electronic structure calculations. The region between the adsorbate covered slabs is filled with water molecules

to examine how solvating molecules influence the bond-breaking and bond-making process at metal surfaces. The dissociation of acetic acid into the acetate anion and a proton is highly endothermic in the gas phase. However, the DFT calculations found that this dissociation is almost thermoneutral in the presence of water molecules [427]. The dissociation is facilitated in an aqueous environment since the fragments which are highly unstable when formed in the gas phase become stabilized by the solvation. The Pd(111) surface also catalyzes the deprotonation of acetic acid and strongly binds the acetate intermediates, but the dissociative adsorption is more endothermic than the dissociation in water without the metal surface. The dissociation is even less favorable at Pd(111) if water is present at the surface because the solvating water molecules weaken the interaction of the acetic acid with the Pd(111) surface.

In the field of electrochemistry, the knowledge about reaction steps and mechanisms at the liquid-solid interface is still rather limited. Therefore there is a strong need for studies like the one just presented which will open the way to a microscopic description and analysis of electrocatalytic reactions.

9.2 Nanostructured Surfaces

The last years have seen a tremendous interest in the so-called nanoscience and nanotechnology. Small particles or clusters with sizes in the nanometer range show strongly modified electronical, optical and chemical properties, compared to bulk materials. The research on nanosize particles has been fueled by the hope that the modified properties can be used to build new or better devices or chemical reactors [428].

The theoretical treatment of nanosize particles by electronic structure theory methods represents a great challenge. Due to the large number of symmetrically different atoms in nanostructures, the numerical effort required to treat these structures is enormous. On the other hand, there is definitely a need for the microscopic description of nanoparticles because the knowledge of the underlying mechanism leading to the modified nature of the particles is still rather limited. It is often not clear whether the specific properties are caused by the reduced dimension of the particles ("quantum size effects") or by the large surface area of the nanocluster where furthermore often many defects are present.

One of the most remarkable modifications of the properties of a material by going from the bulk to nanoscale particles has been found with respect to the catalytic activity of gold. While gold as a bulk material is chemically inert [173], nanoscale gold particles supported on various oxids show a surprisingly large catalytic activity, especially for the low-temperature oxidation of CO [429, 430].

In a colloboration between experiment and electronic structure theory, the CO oxidation catalyzed by size-selected Au_n clusters with $n \leq 20$ supported on defect-poor and defect rich MgO(100) films has been investigated [431]. The experiments revealed that the gold clusters deposited on defect-rich MgO-films have a dramatically increased activity compared to clusters deposited on defect-poor films at temperature between 200 and 350 K. Furthermore, the Au_8 cluster was found to be the smallest catalytically active particle.

In order to explore the microscopic mechanisms underlying the observed behavior, LDA-DFT calculations have been performed [431]. Between 27 and 107 substrate atoms have been embedded into a lattice of about $2000 \pm 2\,\mathrm{e}$ point charges at the positions of the MgO lattice. In order to model the defect-rich substrate, an oxygen vacancy was introduced at the MgO(100) surface which is called a colour center or F-center (from german "Farbzentrum") because of its optical properties. The equilbrium shape of a Au_8 cluster adsorbed on the defect-free MgO surface and on the F-center was determined and reaction paths of the CO oxidation catalyzed by the Au_8 cluster were explored. Figure 9.2 shows a side view of the Au_8 cluster adsorbed on the F-center. The structure corresponds to a deformed close-packed stacking. A sizable charge transfer of 0.5 e from the MgO(100) surface to the gold octamer has been found.

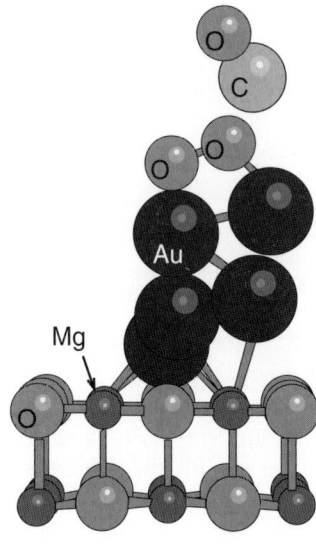

Fig. 9.2. CO oxidation on an Au_8 cluster adsorbed on a MgO(100) surface containing an oxygen-vacancy F-center. Due to the perspective, not all Au atoms are visible. A snapshot of a CO molecule approaching an adsorbed oxygen molecule is shown. (After [431])

In addition, a snapshot along the CO oxidation path according to an abstraction or so-called *Eley–Rideal mechanism* is shown. A CO molecule approaches an adsorbed O_2 molecule and reacts spontaneously to form a weakly bound ($\sim 0.2\,\text{eV}$) CO_2 molecule that can directly desorb plus an adsorbed oxygen atom. Recall that although the CO oxidation is strongly exothermic, it is hindered by a large activation barrier in the gas phase. Another reaction pathway of the Langmuir-Hinshelwood type where the two reactants are initially coadsorbed on the top-facet of the Au_8 cluster has also been found with a similarly small barrier. Through these reaction channels the low-temperature CO oxidation down to 90 K can proceed. As far as the higher-temperature oxidation is concerned, further channels have been identified at the periphery of the gold cluster. Their barriers are much smaller at the Au_8 cluster adsorbed above the F-center than on the perfect surface giving an explanation for the enhanced activity of the clusters on the defect-rich substrate.

9.3 Biologically Relevant Systems

Molecules relevant in biology, biochemistry and biological surface science [432] are usually much too complex to be fully treated by ab initio electronic structure methods. Typically, classical force field methods dominate in the simulation of biomolecular systems. However, these methods do not give any information about the electronic structure. These information can be obtained by mixed quantum-classical embedding schemes in which the active center treated by electronic structure methods is embedded in a classical po-

Fig. 9.3. Most favourable adsorption configuration for a LL-cysteine dimer adsorbed on a four-atom-vacancy on the Au(110) surface in the missing-row reconstruction determined by DFT-GGA calculations [435]. The Au atoms of the deeper layers are shaded in darker grey

tential of the remaining atoms at the periphery. In particular, the QM/MM (Quantum Mechanics/Molecular Mechanics) hybrid method [433] which represents an extension of the Car-Parrinello scheme (see p. 48) has been used successfully for, e.g., the simulation of enzymatic systems [55].

However, simple adsorbed biomolecular systems can already be fully treated by quantum electronic structure methods [434]. The adsorption of cysteine (HS-CH_2-$CH(NH_2)$-$COOH$), an amino acid, on Au(110) has been investigated by both STM experiments and DFT calculations [435]. Cysteine exists in two different so-called enantiomeric forms, L-cysteine and D-cysteine, i.e. two forms that are each others' mirror image with different chirality. The STM experiments have found a high stereoselectivity in the dimerization of adsorbed cysteine molecules on the Au(110) surface which reconstructs in the missing-row structure. Only either LL pairs or DD pairs have been identified.

These findings have been rationalized by DFT-GGA calculations [435]. The calculated most favourable adsorption configuration for a LL-cysteine dimer on Au(110) is shown in Fig. 9.3. The presence of sulfur causes the formation of vacancies on the gold rows due to the tendency of sulfur to bind to low-coordinated atoms. Since sulfur prefers the bridge site, the LL dimer is slightly rotated. In this configuration, three bonds are formed which mainly stabilize this structure: sulphur–gold, amino–gold and carboxylic–carboxylic. In any possible LD dimer adsorption structure, at least one of these bonds is lost, making the LD dimer energetically unfavourable. This explains the high selectivity observed in the STM experiments and fits into the picture that chiral recognition might be in general driven by the formation of three-point contacts [436].

9.4 Industrial Applications

Quantum chemistry methods based on Hartree-Fock theory have been an integral part of research and development in the chemical and pharmaceutical industry for some decades. Companies which manufacture products for which surface structures are relevant have been much more reluctant to employ first-principles electronic structure methods. This is caused by the fact that surface structures that are employed for industrial purposes are usually far away from being perfect. Cluster or slab calculations containing in the order of 100 atoms are therefore often not directly relevant for the research and development process of new catalysts or semiconductor devices. Still, although it is impossible to create new products theoretically from the scratch (and probably will remain impossible for a long time), electronic structure calculations can still add valuable information to the research and development process, in particular for properties where measurements are much more time consuming or not possible [437].

This has already been realized by some manufacturers in the semiconductor and chemical industry. One of the first examples of a successful collaboration between fundamental academic research, both experimental and theoretical, and industrial development has led to the design of a new catalyst for the steam-reforming process [438]. Here I will focus on the contribution of electronic structure calculations [438, 439] to the design of the catalyst; the corresponding experiments have already been reviewed in detail [440].

In the steam-reforming process, hydrocarbon molecules (mainly CH_4) and water are converted into H_2 and CO. This is a important process of great technological relance since it is the first step for several large scale chemical processes such as ammonia synthesis, methanol production or reactions that need H_2 [440]. The catalysts usually used for this reaction are based on Ni. However, during the catalyzed reaction also an unwanted by-product, namely graphite, is formed. A graphite overlayer on the Ni surface leads to a poisoning of the reaction, i.e., it lowers the activity of the catalyst. Such poisoning processes are very costly since they reduce the time the catalyst can be used so that they require a more frequent maintenance of the reactor unit in the chemical plant.

One way of changing the reactivity of metal surfaces is to modify their chemical composition by alloying them with other metals. Some metals that are immiscible in the bulk may still be able to form alloys at the surface. Au and Ni is such a system. The rate-limiting process in the steam-reforming process on Ni is the dissociation of CH_4 into CH_3 and H. DFT calculations by Kratzer et al. showed that this process is hindered by a relatively high barrier of 1.1 eV on Ni(111) [439]. If a Ni atom on the (111) surface has one or two Au atoms as neighbors, this barrier is even increased by 165 meV and 330 meV, respectively. Due to the fact that Au is a noble metal, the CH_4 dissociation barrier over the Au atom is even much higher [438]. An analysis of the calculated electronic structure revealed that the presence of neighboring Au atoms leads to a downshift of the d states at the Ni atom which reduces the reactivity at the Ni atoms [173]. Hence alloying a Ni surface with Au atoms leads to a reduced activity of the catalyst. However, DFT calculations also demonstrated [438] that the presence of the Au atoms lowers the chemisorption energy considerably for C atoms on Ni. If carbon is less strongly bound to the surface, the formation of CO becomes more likely which prevents the building up of a graphite layer.

Altogether, the DFT calculations showed that the lowering of the C chemisorption energy by alloying Ni with Au is much more effective than the increase of the CH_4 dissociation barrier. Hence one ends up with a catalyst that is slightly less reactive but much more robust and stable due to its higher resistance to graphite formation. These fundamental theoretical results together with experimental studies have led to the design of a new catalyst that is now patented [440].

Besides this application for the development of better catalysts, DFT calculations have for example also contributed to the research and development process with respect to the equipment simulation in the electronic industry [441] and in the manufacturing of discharge fluorescent lamps [442]. It can be expected that further applications will follow.

References

1. M. Born and J. R. Oppenheimer, Über die Theorie der Molekeln, Ann. Phys. **84**, 457 (1927).
2. H. Hellmann, *Einführung in die Quantenchemie*, Deuke, Leipzig, 1937.
3. R. P. Feynman, Forces in molecules, Phys. Rev. **56**, 340 (1939).
4. J. Callaway, *Quantum Theory of the Solid State*, Academic Press, Boston, second edition, 1991.
5. M. Born and K. Huang, *Dynamical Theory of Crystal Lattices*, Clarendon Press, Oxford, 1954.
6. M. Tinkham, *Group Theory and Quantum Mechanics*, McGraw-Hill, New York, 1964.
7. J. Sakurai, *Modern Quantum Mechanics*, Benjamin/Cummings, Menlo Park, 1985.
8. V. Heine, *Group Theory in Quantum Mechanics*, Pergamon Press, London, 1960.
9. W. Ludwig and C. Falter, *Symmetries in Physics*, Springer, Berlin, second edition, 1996.
10. N. W. Ashcroft and N. D. Mermin, *Solid State Physics*, Saunders College, Philadelphia, 1976.
11. C. Kittel, *Introduction to Solid State Physics*, John Wiley & Sons, New York, sixth edition, 1986.
12. M. Desjonquères and D. Spanjaard, *Concepts in Surface Physics*, Springer, Berlin, second edition, 1996.
13. G. Binnig, H. Rohrer, C. Gerber, and E. Weibel, Surface studies by scanning tunneling microscopy, Surf. Sci. **131**, L379 (1983).
14. D. R. Hartree, Proc. Cambridge. Philos. Soc **24**, 328 (1928).
15. V. A. Fock, Z. Phys. **15**, 126 (1930).
16. W. H. Press, S. A. Teukolsky, W. T. Vetterling, and B. P. Flannery, *Numerical Recipes in Fortran 90. The Art of Parallel Scientific Computing*, Cambridge University Press, second edition, 1996.
17. G. Baym, *Lectures on Quantum Mechanics*, Benjamin/Cummings, Menlo Park, 1973.
18. D. M. Ceperley and B. J. Alder, Ground state of the electron gas by a stochastic method, Phys. Rev. Lett. **45**, 566 (1980).
19. A. Szabo and N. S. Ostlund, *Modern quantum chemistry: introduction to advanced electronic structure theory*, McGraw-Hill, New York, 1989.
20. K. Raghavachari and J. Anderson, Electron correlation effects in molecules, J. Phys. Chem. **100**, 12960 (1996).
21. A. A. Hasanein and M. W. Evans, *Computational Methods in Quantum Chemistry*, World Scientific, Singapore, 1996.
22. J. A. Pople, Nobel Lecture: Quantum chemical models, Rev. Mod. Phys. **71**, 1267 (1999).

23. C. Møller and M. Plesset, Note on an approximation treatment for many-electron systems, Phys. Rev. **46**, 618 (1934).
24. J. L. Whitten and H. Yang, Theory of chemisorption and reactions on metal surfaces, Surf. Sci. Rep. **24**, 55 (1996).
25. M. Head-Gordon, Quantum chemistry and molecular processes, J. Phys. Chem. **100**, 13213 (1996).
26. P. Hohenberg and W. Kohn, Inhomogeneous electron gas, Phys. Rev. **136**, B864 (1964).
27. W. Kohn and L. Sham, Self-consistent equations including exchange and correlation effects, Phys. Rev. **140**, A1133 (1965).
28. R. M. Dreizler and E. K. U. Gross, *Density Functional Theory: An Approach to the Quantum Many-Body Problem*, Springer, 1990.
29. M. C. Payne, M. P. Teter, D. C. Allan, T. A. Arias, and J. D. Joannopoulos, Iterative minimization techniques for ab initio total-energy calculations: molecular dynamics and conjugate gradients, Rev. Mod. Phys. **64**, 1045 (1992).
30. W. Kohn, Nobel Lecture: Electronic structure of matter—wave functions and density functionals, Rev. Mod. Phys. **71**, 1253 (1999).
31. E. H. Lieb, Thomas-Fermi and related theories of atoms and molecules, Rev. Mod. Phys. **53**, 603 (1981).
32. A. D. Becke, Density-functional exchange-energy approximation with correct asymptotic behavior, Phys. Rev. A **38**, 3098 (1988).
33. C. Lee, W. Yang, and R. Parr, Development of the Colle-Salvetti correlation-energy formula into a functional of the electron density, Phys. Rev. B **37**, 785 (1988).
34. J. P. Perdew, J. A. Chevary, S. H. Vosko, K. A. Jackson, M. R. Pederson, D. J. Singh, and C. Fiolhais, Atoms, molecules, solids, and surfaces: Applications of the generalized gradient approximation for exchange and correlation, Phys. Rev. B **46**, 6671 (1992).
35. J. P. Perdew, K. Burke, and M. Ernzerhof, Generalized gradient approximation made simple, Phys. Rev. Lett. **77**, 3865 (1996).
36. B. Hammer, L. B. Hansen, and J. K. Nørskov, Improved adsorption energetics within density-functional theory using revised Perdew-Burke-Ernzerhof functionals, Phys. Rev. B **59**, 7413 (1999).
37. T. Grabo, T. Kreibich, S. Kurth, and E. K. U. Gross, Orbital functionals in density functional theory: the optimized effective potential method, in *Strong Coulomb correlations in electronic structure: Beyond the Local Density Approximation*, edited by V. I. Anisimov, Gordon and Breach, Tokyo, 1998.
38. J. C. Phillips and L. Kleinman, New method for calculating wave functions in crystals and molecules, Phys. Rev. **116**, 287 (1959).
39. M. L. Cohen and J. R. Chelikowsky, *Electronic Structure and Optical Properties of Semiconductors*, Springer, Berlin, 1988.
40. D. R. Hamann, M. Schlüter, and C. Chiang, Norm-conserving pseudopotentials, Phys. Rev. Lett. **43**, 1494 (1979).
41. G. B. Bachelet, D. R. Hamann, M. Schlüter, and C. Chiang, Pseudopotentials that work: From H to Pu, Phys. Rev. B **26**, 4199 (1982).
42. N. Troullier and J. L. Martins, Efficient pseudopotentials for plane-wave calculations, Phys. Rev. B **43**, 1993 (1991).
43. N. Troullier and J. L. Martins, Efficient pseudopotentials for plane-wave calculations II. operators for fast iterative diagonalization, Phys. Rev. B **43**, 8861 (1991).
44. D. Vanderbilt, Soft self-consistent pseudopotentials in a generalized eigenvalue formalism, Phys. Rev. B **41**, 7892 (1990).

45. H. J. Monkhorst and J. D. Pack, Special points for Brillouin-zone integrations, Phys. Rev. B **13**, 5188 (1976).
46. G. Kresse and J. Furthmüller, Efficiency of ab-initio total energy calculations for metals and semiconductors using a plane-wave basis set, Comput. Mater. Sci. **6**, 15 (1996).
47. J. C. Slater, Wave functions in a periodic potential, Phys. Rev. **51**, 846 (1937).
48. O. K. Andersen, Linear methods in band theory, Phys. Rev. B **12**, 3060 (1975).
49. B. Kohler, S. Wilke, M. Scheffler, R. Kouba, and C. Ambrosch-Draxl, Force calculation and atomic-structure optimization for the full-potential linearized augmented plane-wave code WIEN, Comput. Phys. Commun. **94**, 31–48 (1996).
50. P. E. Blöchl, Projector augmented-wave method, Phys. Rev. B **50**, 17953 (1994).
51. G. Kresse and D. Joubert, From ultrasoft pseudopotentials to the projector augmented-wave method, Phys. Rev. B **59**, 1758 (1999).
52. R. Car and M. Parrinello, Unified approach for molecular dynamics and density-functional theory, Phys. Rev. Lett. **55**, 2471 (1985).
53. P. Bendt and A. Zunger, Simultaneous relaxation of nuclear geometries and electric charge densities in electronic structure theories, Phys. Rev. Lett. **50**, 1684 (1983).
54. D. Marx and J. Hutter, Ab initio Molecular Dynamics: Theory and Implementation, in *Modern Methods and Algorithms of Quantum Chemistry*, edited by J. Grotendorst, volume 3 of *NIC series*, pages 329–477, John von Neumann-Institute for Computing, Jülich, 2000.
55. P. Carloni and U. Röthlisberger, Simulations of Enzymatic Systems: Perspectives from Car-Parrinello Molecular Dynamics Simulations, in *Theoretical Biochemistry – Processes and Properties of Biological Systems*, edited by L. Eriksson, page 215, Elsevier, Amsterdam, 2001.
56. W. M. C. Foulkes, L. Mitas, R. J. Needs, and G. Rajagopal, Quantum Monte Carlo simulations of solids, Rev. Mod. Phys. **73**, 33 (2001).
57. S. Goedecker and C. J. Umrigar, Natural orbital functional for the many-electron problem, Phys. Rev. Lett. **81**, 866 (1998).
58. J. C. Slater and G. F. Koster, Simplified LCAO method for the periodic potential problem, Phys. Rev. **94**, 1498 (1954).
59. C. M. Goringe, D. R. Bowler, and E. Hernández, Tight-binding modelling of materials, Rep. Prog. Phys. **60**, 1447 (1997).
60. S. Goedecker, Linear scaling electronic structure methods, Rev. Mod. Phys. **71**, 1085 (1999).
61. P. Löwdin, J. Chem. Phys. **18**, 365 (1950).
62. A. P. Sutton, M. W. Finnis, D. G. Pettifor, and Y. Ohta, The tight-binding bond model, J. Phys. C: Solid State Phys. **21**, 35 (1988).
63. M. J. Mehl and D. A. Papaconstantopoulos, Applications of a tight-binding total-energy method for transition and noble metals: Elastic constants, vacancies, and surfaces of monatomic metals, Phys. Rev. B **54**, 4519 (1996).
64. N. D. Lang and W. Kohn, Theory of metal surfaces: charge density and surface energy, Phys. Rev. B **1**, 4555 (1970).
65. N. D. Lang and W. Kohn, Theory of metal surfaces: work function, Phys. Rev. B **3**, 1215 (1971).
66. R. Hoffmann, A chemical and theoretical way to look at bonding on surfaces, Rev. Mod. Phys. **60**, 601 (1988).
67. M. Lischka and A. Groß, Hydrogen adsorption on an open metal surface: H_2/Pd(210), Phys. Rev. B **65**, 075420 (2002).

68. R. Smoluchowski, Anisotropy of the electronic work function of metals, Phys. Rev. **60**, 661 (1941).
69. A. Euceda, D. M. Bylander, and L. Kleinman, Self-consistent electronic structure of 6- and 18-layer Cu(111) films, Phys. Rev. B **28**, 528 (1983).
70. A. Euceda, D. M. Bylander, L. Kleinman, and K. Mednick, Self-consistent electronic structure of 7- and 19-layer Cu(001) films, Phys. Rev. B **27**, 659 (1983).
71. P. M. Echenique and J. B. Pendry, Theory of image states at metal-surfaces, Prog. Surf. Sci. **32**, 111 (1990).
72. U. Höfer, I. L. Shumay, C. Reuß, U. Thomann, W. Wallauer, and T. Fauster, Time-resolved coherent photoelectron spectroscopy of quantized electronic states on metal surfaces, Science **277**, 1480 (1997).
73. W. R. Tyson and W. A. Miller, Surface free-energies of solid metals – estimation from liquid surface-tension measurements, Surf. Sci. **62**, 267 (1977).
74. M. Methfessel, D. Hennig, and M. Scheffler, Trends of the surface relaxations, surface energies, and work functions of the 4d transition metals, Phys. Rev. B **46**, 4816 (1992).
75. L. Vitos, A. V. Ruban, H. L. Skriver, and J. Kollár, The surface energy of metals, Surf. Sci. **411**, 186 (1998).
76. S. Sakong and A. Groß, unpublished.
77. S. Lindgren, L. Walldén, J. Rundgren, and P. Westrin, Low-energy electron diffraction from Cu(111): Subthreshold effect and energy-dependent inner potential; surface relaxation and metric distances between spectra, Phys. Rev. B **29**, 576 (1984).
78. S. Walter, V. Blum, L. Hammer, K. Heinz, and M. Giesen, The role of an energgy -dependent inner potential in quantitative low-energy electron diffraction, Surf. Sci. **458**, 155 (2000).
79. S. Liem, G. Kresse, and J. Clarke, First principles calculations of oxygen adsorption and reconstruction of Cu(110) surface, Surf. Sci. **415**, 194 (1998).
80. D. L. Adams, H. B. Nielsen, and J. N. Andersen, Oscillatory relaxation of the Cu(110) surface, Surf. Sci. **128**, 294 (1983).
81. D. Spiv̌ák, Structure and energetics of Cu(100) vicinal surfaces, Surf. Sci. **489**, 151 (2001).
82. S. Parkin, P. Watson, R. McFarlane, and K. Mitchell, A revised LEED determination of the relaxations present at the (311) surface of copper, Solid State Commun. **78**, 841 (1991).
83. M. Albrecht, R. Blome, H. L. Meyerheim, W. Moritz, and I. K. Robinson, Surf. Sci., in press.
84. G. Kresse and J. Furthmüller, Efficient iterative schemes for ab initio total-energy calculations using a plane-wave basis set, Phys. Rev. B **54**, 11169 (1996).
85. M. W. Finnis and V. Heine, Theory of lattice contraction at aluminium surfaces, J. Phys. Chem. B **105**, L37 (1973).
86. P. Krüger and J. Pollmann, Dimer reconstruction of diamond, Si, and Ge (001) surfaces, Phys. Rev. Lett. **74**, 1155 (1995).
87. M. Rohlfing, P. Krüger, and J. Pollmann, Efficient scheme for GW quasi-particle band-structure calculations with applications to bulk Si and to the Si(001)–(2×1) surface, Phys. Rev. B **52**, 1905 (1995).
88. R. I. G. Uhrberg, G. V. Hansson, J. M. Nicholls, and S. A. Flodström, Experimental studies of the dangling- and dimer-bond-related surface electron bands on Si(100) (2×1), Phys. Rev. B **24**, 4684 (1981).

References 253

89. L. S. O. Johansson, R. I. G. Uhrberg, P. Mårtensson, and G. V. Hansson, Surface-state band structure of the Si(100) 2×1 surface studied with polarization-dependent angle-resolved photoemission on single-domain surfaces, Phys. Rev. B **42**, 1305 (1990).
90. L. Hedin, New method for calculating 1-particle Greens function with application to electron-gas problem, Phys. Rev. **139**, A796 (1965).
91. L. Hedin and S. Lundqvist, in *Advances in research and application*, edited by F. Seitz, D. Turnbull, and H. Ehrenreich, volume 23 of *Solid State Physics*, page 1, Academic Press, New York, 1969.
92. M. Rohlfing, Theory of excitons in low-dimensional systems, phys. stat. sol. (a) **188**, 1243 (2001).
93. M. Rohlfing and S. G. Louie, Electron-hole excitations and optical spectra from first principles, Phys. Rev. B **62**, 4927 (2000).
94. P. Chiaradia, A. Cricenti, S. Selci, and G. Chiarotti, Differential reflectivity of Si(111)2×1 surface with polarized-light: A test for surface-structure, Phys. Rev. Lett. **52**, 1145 (1984).
95. J. Dabrowski and M. Scheffler, Self-consistent study of the electronic and structural-properties of the clean Si(001)(2×1) surface, Appl. Surf. Sci. **56**, 15 (1992).
96. J. Northrup, Electronic structure of Si(100)c(4×2) calculated within the GW approximation, Phys. Rev. B **47**, 10032 (1993).
97. R. M. Tromp, R. J. Hamers, and J. E. Demuth, Si(001) dimer structure observed with scanning tunneling microscopy, Phys. Rev. Lett. **55**, 1303 (1985).
98. R. A. Wolkow, Direct observation of an increase in buckled dimers on Si(001) at low-temperature, Phys. Rev. Lett. **68**, 2636 (1992).
99. K. Takayanagi, Y. Tanishiro, M. Takahashi, and S. Takahashi, Structure-analysis of S(111)-7×7 reconstructed surface by transmission electron-diffraction, Surf. Sci. **164**, 367 (1985).
100. A. Zangwill, *Physics at Surfaces*, Cambridge University Press, Cambridge, 1988.
101. H. Lüth, *Surfaces and Interfaces of Solids*, volume 15 of *Springer Series in Surface Sciences*, Springer, Berlin, 1993.
102. K. D. Brommer, M. Needels, B. Larson, and J. D. Joannopoulos, Ab initio theory of the Si(111)-(7×7) surface reconstruction: A challenge for massively parallel computation, Phys. Rev. Lett. **68**, 1355 (1992).
103. T. Ohno, Energetics of As dimers on GaAs(001) As-rich surfaces, Phys. Rev. Lett. **70**, 631 (1993).
104. J. Northrup and S. Froyen, Energetics of GaAs(100)-(2×4) and (4×2) reconstructions, Phys. Rev. Lett. **71**, 2276 (1993).
105. N. Moll, A. Kley, E. Pehlke, and M. Scheffler, GaAs equilibrium crystal shape from first principles, Phys. Rev. B **54**, 8844 (1996).
106. N. Chetty and R. M. Martin, Determination of integrals at surfaces using the bulk crystal symmetry, Phys. Rev. B **44**, 5568 (1991).
107. J. Northrup and S. Froyen, Structure of GaAs(001) surfaces: The role of electrostatic interactions, Phys. Rev. B **50**, 2015 (1994).
108. R. E. Watson, J. W. Davenport, M. L. Perlman, and T. K. Sham, Madelung effects at crystal surfaces: Implications for photoemission, Phys. Rev. B **24**, 1791 (1981).
109. X. G. Wang, A. Chaka, and M. Scheffler, Effect of the environment on α Al_2O_3 (0001) surface structures, Phys. Rev. Lett. **84**, 3650 (2000).
110. C. Verdozzi, D. R. Jennison, P. A. Schultz, and M. P. Sears, Sapphire(0001) surface, clean and with d-metal overlayers, Phys. Rev. Lett. **82**, 799 (1999).

111. R. Di Felice and J. E. Northrup, Theory of the clean and hydrogenated Al_2O_3 (0001)-(1×1) surfaces, Phys. Rev. B **60**, R16287 (1999).
112. I. Batyrev, A. Alavi, and M. W. Finnis, *Ab initio* calculations on the Al_2O_3(0001) surface, Faraday Discuss. **114**, 33–43 (1999).
113. P. Guénard, G. Renaud, A. Barbier, and M. Gautier-Soyer, Determination of the α-Al_2O_3(0001) surface relaxation and termination by measurements of crystal truncation rods, Surf. Rev. Lett. **5**, 321 (1998).
114. J. Toofan and P. R. Watson, The termination of the α-Al_2O_3(0001) surface: a LEED crystallography determination, Surf. Sci. **401**, 162 (1998).
115. G. Renaud, Oxide surfaces and metal/oxide interfaces studied by grazing incidence X-ray scattering, Surf. Sci. Rep. **32**, 1 (1998).
116. E. Soares, M. A. Van Hove, C. F. Walters, and K. F. McCarty, Structure of the α-Al_2O_3(0001) surface from low-energy electron diffraction: Al termination and evidence for anomalously large thermal vibrations, Phys. Rev. B **65**, 195405 (2002).
117. W. Mackrodt, R. Davey, S. Black, and R. Docherty, The morphology of α-Al_2O_3 and α-Fe_2O_3 – the importance of surface relaxation, J. Cryst. Growth **80**, 441 (1987).
118. I. Vilfan, T. Deutsch, F. Lancon, and G. Renaud, Structure determination of the $(3\sqrt{3} \times 3\sqrt{3})$ reconstructed α-Al_2O_3(0001), Surf. Sci. **505**, L215 (2002).
119. R. Wiesendanger and H. Güntherodt, editors, *Scanning Tunneling Microscopy III*, Springer Series in Surface Science, Springer, Berlin, 1993.
120. J. Tersoff and D. R. Hamann, Theory and application for the scanning tunneling microscope, Phys. Rev. Lett. **50**, 1998 (1983).
121. J. Tersoff and D. R. Hamann, Theory of the scanning tunneling microscope, Phys. Rev. B **31**, 805 (1985).
122. M. Tsukada, K. Kobayashi, N. Isshiki, and H. Kageshima, 1st-principles theory of scanning tunneling microscopy, Surf. Sci. Rep. **13**, 265 (1991).
123. H. Ness and A. Fisher, Nonperturbative evaluation of STM tunneling probabilities from ab initio calculations, Phys. Rev. B **56**, 12469 (1997).
124. D. Tománek, S. G. Louie, H. J. Mamin, D. W. Abraham, R. E. Thomson, E. Ganz, and J. Clark, Theory and observation of highly asymmetric atomic structure in scanning-tunneling-microscopy images of graphite, Phys. Rev. B **35**, 7790 (1987).
125. Lord Rayleigh, Proc. London Math. Soc. **17**, 4 (1885).
126. J. Fritsch and U. Schröder, Density functional calculations of semiconductor surface phonons, Phys. Rep. **309**, 209 (1999).
127. S. Baroni, S. de Gironcoli, A. Dal Corso, and P. Giannozzi, Phonons and related crystal properties from density-functional perturbation theory, Rev. Mod. Phys. **73**, 515 (2001).
128. J. Fritsch and P. Pavone, Ab initio calculation of the structure, electronic states, and the phonon dispersion of the Si(100) surface, Surf. Sci. **344**, 159 (1995).
129. J. R. Dutcher, S. Lee, B. Hillebrands, G. J. McLaughlin, B. G. Nickel, and G. I. Stegeman, Surface-grating-induced zone folding and hybridization of surface acoustic modes, Phys. Rev. Lett. **68**, 2464 (1992).
130. D. C. Allan and E. J. Mele, Surface vibrational excitations on Si(001) 2×1, Phys. Rev. Lett. **53**, 826 (1984).
131. J. B. Hudson, *Surface Science: An Introduction*, Butterworth-Heinemann, Stoneham, 1992.
132. J. Lennard-Jones, Trans. Faraday Soc. **28**, 333 (1932).

133. K. Autumn, Y. A. Liang, S. T. Hsieh, W. Zesch, W. P. Chan, T. W. Kenny, R. Fearing, and R. J. Full, Adhesive force of a single gecko foot-hair, Nature **405**, 681 (2000).
134. S. Holloway and J. Nørskov, *Bonding at Surfaces*, Liverpool University Press, 1991.
135. E. Zaremba and W. Kohn, van der Waals interaction between an atom and a solid surface, Phys. Rev. B **13**, 2270 (1976).
136. E. Zaremba and W. Kohn, Theory of helium adsorption on simple and noble-metal surfaces, Phys. Rev. B **15**, 1769 (1977).
137. M. Petersen, S. Wilke, P. Ruggerone, B. Kohler, and M. Scheffler, Scattering of rare-gas atoms at a metal surface: Evidence of anticorrugation of the helium-atom potential-energy surface and the surface electron density, Phys. Rev. Lett. **76**, 995 (1996).
138. M. I. Trioni, F. Montalenti, and G. P. Brivio, Ab-initio adiabatic noble gas-metal interaction: the role of the induced polarization charge, Surf. Sci. **401**, L383 (1998).
139. M. Scheffler and C. Stampfl, Theory of adsorption on metal substrates, in *Electronic Structure*, edited by K. Horn and M. Scheffler, volume 2 of *Handbook of Surface Science*, pages 286–356, Elsevier, Amsterdam, 2000.
140. W. Kohn, Y. Meir, and D. Makarov, van der Waals energies in density functional theory, Phys. Rev. Lett. **80**, 4153 (1998).
141. E. Hult, H. Rydberg, B. Lundqvist, and D. Langreth, Unified treatment of asymptotic van der Waals forces, Phys. Rev. B **59**, 4708 (1999).
142. H. B. G. Casimir and D. Polder, The Influence of Retardation on the London-van der Waals Forces, Phys. Rev. **73**, 360 (1948).
143. M. Bordag, U. Mohideen, and V. M. Mostepanenko, New developments in the Casimir effect, Phys. Rep. **353**, 1 (2001).
144. F. Shimizu, Specular reflection of very slow metastable neon atoms from a solid surface, Phys. Rev. Lett. **86**, 987 (2001).
145. T. B. Grimley, Indirect interaction between atoms or molecules adsorbed on metals, Proc. Phys. Soc. London Sect. A **90**, 751 (1967).
146. D. Newns, Self-consistent model of hydrogen chemisorption, Phys. Rev. **178**, 1123 (1969).
147. P. W. Anderson, Localized magnetic states in metals, Phys. Rev. **124**, 41 (1961).
148. R. Brako and D. M. Newns, Theory of electronic processes in atom scattering from surfaces, Rep. Prog. Phys. **52**, 655 (1989).
149. G. P. Brivio and M. I. Trioni, The adiabatic molecule-metal surface interaction: theoretical approaches, Rev. Mod. Phys. **71**, 231 (1999).
150. W. Brenig and K. Schönhammer, On the theory of chemisorption, Z. Phys. **267**, 201 (1974).
151. M. Scheffler and A. M. Bradshaw, The electronic structure of adsorbed layers, in *Adsorption at Solid Surfaces*, edited by D. A. King and D. P. Woodruff, volume 2 of *The Chemical Physics of Solid Surfaces*, pages 165–257, Elsevier, Amsterdam, 1983.
152. N. D. Lang and A. Williams, Theory of atomic chemisorption on simple metals, Phys. Rev. B **18**, 616 (1978).
153. J. Callaway, t matrix and phase shifts in solid-state scattering theory, Phys. Rev. **154**, 515 (1967).
154. H. Hjelmberg, Hydrogen chemisorption by the spin-density functional formalism. II. Role of the sp-conduction electrons of metal surfaces, Physica Scripta **18**, 481 (1978).

155. J. Nørskov and N. Lang, Effective-medium theory of chemical binding: Application to chemisorption, Phys. Rev. B **21**, 2131 (1980).
156. J. Nørskov, Chemisorption on metal surfaces, Rep. Prog. Phys. **53**, 1253–1295 (1990).
157. M. Puska, R. Nieminen, and M. Manninen, Atoms embedded in an electron gas: Immersion energies, Phys. Rev. B **24**, 3037 (1981).
158. J. Nørskov, Covalent effects in the effective-medium theory of chemical binding: hydrogen heats of solution in the $3d$ metals, Phys. Rev. B **26**, 2875 (1982).
159. P. Nordlander, S. Holloway, and J. Nørskov, Hydrogen adsorption on metal surfaces, Surf. Sci. **136**, 59 (1984).
160. A. Christensen, A. V. Ruban, P. Stoltze, K. W. Jacobsen, H. L. Skriver, J. K. Nørskov, and F. Besenbacher, Phase diagrams for surface alloys, Phys. Rev. B **56**, 5822 (1997).
161. M. S. Daw and M. I. Baskes, Embedded-atom method: derivation and applications to impurities, surfaces, and other defects in metals, Phys. Rev. B **29**, 6443 (1984).
162. S. M. Foiles, M. I. Baskes, and M. S. Daw, Embedded-atom-method functions for the fcc metals Cu, Ag, Au, Ni, Pd, Pt and their alloys, Phys. Rev. B **33**, 7983 (1986).
163. M. S. Daw, S. M. Foiles, and M. I. Baskes, The embedded-atom method – a review of theory and applications, Mater. Sci. Rep. **9**, 252 (1993).
164. M. I. Baskes, Application of the embedded-atom method to covalent materials: A semiempirical potential for silicon, Phys. Rev. Lett. **59**, 2666 (1987).
165. P. van Beurden and G. J. Kramer, Parametrization of modified embedded-atom-method potentials for Rh, Pd, Ir, and Pt based on density functional theory calculations, with applications to surface properties, Phys. Rev. B **63**, 165106 (2001).
166. S. E. Wonchoba and D. G. Truhlar, Embedded diatomics-in-molecules potential energy function for methyl radical and methane on nickel surfaces, J. Phys. Chem. B **102**, 6842 (1998).
167. K. Fukui, Role of frontier orbitals in chemical reactions, Science **218**, 747 (1982).
168. P. J. Feibelman and D. R. Hamann, Electronic-structure of a poisoned transition-metal surface, Phys. Rev. Lett. **52**, 61 (1984).
169. P. J. Feibelman and D. R. Hamann, Modification of transition-metal electronic-structure by P, S, Cl, and Li adatoms, Surf. Sci. **149**, 48 (1985).
170. J. Harris and S. Andersson, H_2 dissociation at metal surfaces, Phys. Rev. Lett. **55**, 1583 (1985).
171. M. H. Cohen, M. V. Ganduglia-Pirovano, and J. Kudrnovský, Orbital symmetry, reactivity, and transition metal surface chemistry, Phys. Rev. Lett. **72**, 3222 (1994).
172. B. Hammer and J. K. Nørskov, Electronic factors determining the reactivity of metal surfaces, Surf. Sci. **343**, 211 (1995).
173. B. Hammer and J. K. Nørskov, Why gold is the noblest of all the metals, Nature **376**, 238 (1995).
174. V. Pallassana, M. Neurock, L. B. Hansen, B. Hammer, and J. K. Nørskov, Theoretical trends of hydrogen chemisorption on Pd(111), Re(0001) and $Pd_{ML}/Re(0001)$, $Re_{ML}/Pd(111)$ pseudomorphic overlayers, Phys. Rev. B **60**, 6146 (1999).
175. S. Wilke, D. Hennig, and R. Löber, Ab initio calculations of hydrogen adsorption on (100) surfaces of palladium and rhodium, Phys. Rev. B **50**, 2548 (1994).

176. A. Roudgar and A. Groß, submitted.
177. V. Ledentu, W. Dong, P. Sautet, A. Eichler, and J. Hafner, H-induced reconstructions on Pd(110), Phys. Rev. B **57**, 12482 (1998).
178. W. Dong, V. Ledentu, P. Sautet, A. Eichler, and J. Hafner, Hydrogen adsorption on palladium: a comparative theoretical study of different surfaces, Surf. Sci. **411**, 123 (1998).
179. E. Kampshoff, N. Waelchli, A. Menck, and K. Kern, Hydrogen-induced missing-row reconstructions of Pd(110) studied by scanning tunneling microscopy, Surf. Sci. **360**, 55 (1996).
180. J. Yoshinobu, H. Tanaka, and M. Kawai, Elucidation of hydrogen-induced (1×2) reconstructed structures on Pd(110) from 100 to 300 K by scanning tunneling microscopy, Phys. Rev. B **51**, 4529 (1995).
181. D. Tománek, S. Wilke, and M. Scheffler, Hydrogen induced polymorphism of Pd(110), Phys. Rev. Lett. **79**, 1329 (1997).
182. K. Christmann, Interaction of hydrogen with solid surfaces, Surf. Sci. Rep. **9**, 1 (1988).
183. S. Wilke and M. Scheffler, Potential-energy surface for H_2 dissociation over Pd(100), Phys. Rev. B **53**, 4926 (1996).
184. A. Dedieu, Theoretical studies in palladium and platinum molecular chemistry, Chem. Rev. **100**, 543 (2000).
185. A. Eichler, G. Kresse, and J. Hafner, Ab-initio calculations of the 6D potential energy surfaces for the dissociative adsorption of H_2 on the (100) surfaces of Rh, Pd and Ag, Surf. Sci. **397**, 116 (1998).
186. W. Dong and J. Hafner, H_2 dissociative adsorption on Pd(111), Phys. Rev. B **56**, 15396 (1997).
187. V. Ledentu, W. Dong, and P. Sautet, Ab initio study of the dissociative adsorption of H_2 on the Pd(110) surface, Surf. Sci. **412**, 518 (1998).
188. G. Blyholder, Molecular orbital view of chemisorbed carbon monoxide, J. Phys. Chem. **68**, 2772 (1964).
189. H. Aizawa and S. Tsuneyuki, First-principles study of CO bonding to Pt(111): validity of the Blyholder model, Surf. Sci. **399**, L364 (1998).
190. P. J. Feibelman, B. Hammer, J. K. Nørskov, F. Wagner, M. Scheffler, R. Stumpf, R. Watwe, and J. Dumesic, The CO/Pt(111) puzzle, J. Phys. Chem. B **105**, 4018 (2001).
191. M. Ø. Pedersen, M. L. Bocquet, P. Sautet, E. Laegsgaard, I. Stensgaard, and F. Besenbacher, CO on Pt(111): binding site assignment from the interplay between measured and calculated STM images, Chem. Phys. Lett. **299**, 403 (1999).
192. B. Hammer, O. H. Nielsen, and J. K. Nørskov, Structure sensitivity in adsorption: CO interaction with stepped and reconstructed Pt surfaces, Catal. Lett. **46**, 31 (1997).
193. K. D. Rendulic, G. Anger, and A. Winkler, Wide range nozzle beam adsorption data for the systems H_2/nickel and H_2/Pd(100), Surf. Sci. **208**, 404 (1989).
194. M. L. Burke and R. J. Madix, Hydrogen on Pd(100)-S: The effect of sulfur on precursor mediated adsorption and desorption, Surf. Sci. **237**, 1 (1990).
195. S. Wilke and M. Scheffler, Mechanism of poisoning the catalytic activity of Pd(100) by a sulfur adlayer, Phys. Rev. Lett. **76**, 3380 (1996).
196. C. M. Wei, A. Groß, and M. Scheffler, Ab initio calculation of the potential energy surface for the dissociation of H_2 on the sulfur-covered Pd(100) surface, Phys. Rev. B **57**, 15572 (1998).

197. B. Hammer, Coverage dependence of N_2 dissociation at an N, O, or H precovered Ru(0001) surface investigated with density functional theory, Phys. Rev. B **63**, 205423 (2001).
198. A. Logadottir, T. H. Rod, J. K. Nørskov, B. Hammer, S. Dahl, and C. J. H. Jacobsen, The Brønsted-Evans-Polanyi relation and the volcano plot for ammonia synthesis over transition metal catalysts, J. Catal. **197**, 229 (2001).
199. M. Mavrikakis, B. Hammer, and J. K. Nørskov, Effect of strain on the reactivity of metal surfaces, Phys. Rev. Lett. **81**, 2819 (1998).
200. M. Gsell, P. Jakob, and D. Menzel, Effect of substrate strain on adsorption, Science **280**, 717 (1998).
201. P. Jakob, M. Gsell, and D. Menzel, Interactions of adsorbates with locally strained substrate lattices, J. Chem. Phys. **114**, 10075 (2001).
202. P. K. Schmidt, K. Christmann, G. Kresse, J. Hafner, M. Lischka, and A. Groß, Coexistence of atomic and molecular chemisorption states: H_2/Pd(210), Phys. Rev. Lett. **87**, 096103 (2001).
203. A. Alavi, P. Hu, T. Deutsch, P. L. Silvestrelli, and J. Hutter, CO oxidation on Pt(111): An ab initio density functional theory study, Phys. Rev. Lett. **80**, 3650 (1998).
204. A. Eichler and J. Hafner, Reaction channels for the catalytic oxidation of CO on Pt(111), Surf. Sci. **433-435**, 58 (1999).
205. B. Hammer, Reactivity of a stepped surface – NO dissociation on Pd(211), Faraday Discuss. **110**, 323 (1998).
206. Z.-P. Liu and P. Hu, General trends in CO dissocation on transition metal surfaces, J. Chem. Phys. **114**, 8244 (2001).
207. B. Hammer, The NO+CO reaction catalyzed by flat, stepped, and edged Pd surfaces, J. Catal. **199**, 171 (2001).
208. K.-H. Allers, H. Pfnür, P. Feulner, and D. Menzel, Fast reaction-products from the oxidation of CO on Pt(111): Angular and velocity distributions of the CO_2 product molecules, J. Chem. Phys. **100**, 3985 (1994).
209. G. D. Billing, *Dynamics of Molecule Surface Interactions*, Wiley, New York, 2000.
210. L. Verlet, Computer "experiments" on classical fluids. I. Thermodynamical properties of Lennard-Jones molecules, Phys. Rev. **159**, 98 (1967).
211. D. Heermann, *Computer Simulation Methods in Theoretical Physics*, Springer, Berlin, second edition, 1990.
212. S. A. Adelman and J. D. Doll, Generalized Langevin equation approach for atom-solid-surface scattering: general formulation for classical scattering off harmonic solids, J. Chem. Phys. **64**, 2375 (1976).
213. S. Nosé, A unified formulation for the constant temperature molecular dynamics method, J. Chem. Phys. **81**, 511 (1984).
214. W. G. Hoover, *Molecular Dynamics*, volume 258 of *Lecture Notes in Physics*, Springer, Berlin, 1986.
215. R. Newton, *Scattering Theory of Waves and Particles*, Springer, New York, second edition, 1982.
216. W. Brenig, T. Brunner, A. Groß, and R. Russ, Numerically stable solution of coupled channel equations: The local reflection matrix, Z. Phys. B **93**, 91 (1993).
217. W. Brenig and R. Russ, Numerically stable solution of coupled channel equations: The local transmission matrix, Surf. Sci. **315**, 195 (1994).
218. W. Brenig, , A. Groß, and R. Russ, Numerically stable solution of coupled channel equations: The wave-function, Z. Phys. B **97**, 311 (1995).
219. J. A. Fleck, J. R. Morris, and M. D. Feit, Time-dependent propagation of high-energy laser-beams through atmosphere, Appl. Phys. **10**, 129 (1976).

220. M. D. Feit, J. A. Fleck, and A. Steiger, Solution of the Schrödinger-equation by a spectral method, J. Comput. Phys. **47**, 412 (1982).
221. H. Tal-Ezer and R. Kosloff, An accurate and efficient scheme for propagating the time-dependent Schrödinger-equation, J. Chem. Phys. **81**, 3967 (1984).
222. M. Hand and J. Harris, Recoil effects in surface dissociation, J. Chem. Phys. **92**, 7610 (1990).
223. A. Groß and W. Brenig, NO/Ag(111) revisited, Surf. Sci. **302**, 403 (1994).
224. M. Dohle and P. Saalfrank, Surface oscillator models for dissociative sticking of molecular hydrogen at non-rigid surfaces, Surf. Sci. **373**, 95 (1997).
225. H. Guo, P. Saalfrank, and T. Seideman, Theory of photo-induced surface reactions of admolecules, Prog. Surf. Sci. **62**, 239 (1999).
226. J. C. Tully, Chemical dynamics at metal surfaces, Annu. Rev. Phys. Chem. **51**, 153 (2000).
227. G. Lindblad, Generators of quantum dynamical semigroups, Commun. Math. Phys. **48**, 119 (1976).
228. A. G. Redfield, Adv. Magn. Reson. **1**, 1 (1966).
229. A. De Vita, I. Štich, M. J. Gillan, M. C. Payne, and L. J. Clarke, Dynamics of dissociative chemisorption: Cl_22/Si(111)-(2×1), Phys. Rev. Lett. **71**, 1276 (1993).
230. A. Groß, M. Bockstedte, and M. Scheffler, Ab initio molecular dynamics study of the desorption of D_2 from Si(100), Phys. Rev. Lett. **79**, 701 (1997).
231. A. J. R. da Silva, M. R. Radeke, and E. A. Carter, Ab initio molecular dynamics of H_2 desorption from Si(100)-2×1, Surf. Sci. **381**, L628 (1997).
232. M. R. Radeke and E. A. E.A. Carter, Ab initio dynamics of surface chemistry, Annu. Rev. Phys. Chem. **48**, 243 (1997).
233. G. R. Darling and S. Holloway, The dissociation of diatomic molecules at surfaces, Rep. Prog. Phys. **58**, 1595 (1995).
234. A. Groß, Reactions at surfaces studied by ab initio dynamics calculations, Surf. Sci. Rep. **32**, 291 (1998).
235. G. Wiesenekker, G.-J. Kroes, and E. J. Baerends, An analytical six-dimensional potential energy surface for dissociation of molecular hydrogen on Cu(100), J. Chem. Phys. **104**, 7344 (1996).
236. A. Groß, S. Wilke, and M. Scheffler, Six-dimensional quantum dynamics of adsorption and desorption of H_2 at Pd(100): steering and steric effects, Phys. Rev. Lett. **75**, 2718 (1995).
237. M. Kay, G. R. Darling, S. Holloway, J. A. White, and D. M. Bird, Steering effects in non-activated adsorption, Chem. Phys. Lett. **245**, 311 (1995).
238. A. Groß, B. Hammer, M. Scheffler, and W. Brenig, High-dimensional quantum dynamics of adsorption and desorption of H_2 at Cu(111), Phys. Rev. Lett. **73**, 3121 (1994).
239. A. D. Kinnersley, G. R. Darling, S. Holloway, and B. Hammer, A comparison of quantum and classical dynamics of H_2 dissociation on Cu(111), Surf. Sci. **364**, 219 (1996).
240. A. E. DePristo and A. Kara, Molecule surface scattering and reaction dynamics, Adv. Chem. Phys. **77**, 163 (1991).
241. P. Kratzer, The dynamics of the H+D/Si(001) reaction: a trajectory study based on ab initio potentials, Chem. Phys. Lett. **288**, 396 (1998).
242. H. F. Busnengo, A. Salin, and W. Dong, Representation of the 6D potential energy surface for a diatomic molecule near a solid surface, J. Chem. Phys. **112**, 7641 (2000).
243. G. Kresse, Dissociation and sticking of H_2 on the Ni(111), (100) and (110) substrate, Phys. Rev. B **62**, 8295 (2000).

244. T. B. Blank, A. W. Brown, S. D. Calhoun, and D. J. Doren, Neural-network models of potential-energy surfaces, J. Chem. Phys. **103**, 4129 (1995).
245. K. T. No, B. H. Chang, S. Y. Kim, M. S. Jhon, and H. A. Scheraga, Description of the potential energy surface of the water dimer with an artificial neural network, Chem. Phys. Lett. **271**, 152 (1997).
246. S. Lorenz, A. Groß, and M. Scheffler, APS Bulletin **43**, 235 (1998).
247. A. Groß, M. Scheffler, M. J. Mehl, and D. A. Papaconstantopoulos, Ab initio based tight-binding hamiltonian for the dissocation of molecules at surfaces, Phys. Rev. Lett. **82**, 1209 (1999).
248. A. Groß, C.-M. Wei, and M. Scheffler, Poisoning of hydrogen dissociation at Pd(100) by adsorbed sulfur studied by ab initio quantum dynamics and ab initio molecular dynamics, Surf. Sci. **416**, L1095 (1998).
249. I. Estermann and O. Stern, Beugung von Molekularstrahlen, Z. Phys. **61**, 95 (1930).
250. R. Frisch and O. Stern, Anomalien bei der spiegelnden Reflexion und Beugung von Molekularstrahlen an Kristallspaltflächen. I)., Z. Phys. **84**, 430 (1933).
251. E. Hulpke, editor, *Helium Atom Scattering from Surfaces*, volume 27 of *Springer Series in Surface Sciences*, Springer, Berlin, 1992.
252. M. F. Bertino and D. Farías, Probing gas-surface potential energy surfaces with diffraction of hydrogen molecules, J. Phys.: Condens. Matter **14**, 6037 (2002).
253. R. G. Rowe and G. Ehrlich, Rotationally inelastic diffraction of molecular beams: H_2, D_2, HD from (001) MgO, J. Chem. Phys. , 4648 (1975).
254. G. Brusdeylins and T. J. P., Simultaneous excitation of rotations and surface phonons in the scattering of D_2 from a NaF crystal, Surf. Sci. **126**, 647 (1983).
255. K. B. Whaley, C.-F. Yu, C. S. Hogg, J. C. Light, and S. Sibener, Investigation of the spatially anisotropic component of the laterally averaged molecular hydrogen/Ag(111) physisorption potential, J. Chem. Phys. **83**, 4235 (1985).
256. A. Groß and M. Scheffler, Scattering of hydrogen molecules from a reactive surface: strong off-specular and rotationally inelastic diffraction, Chem. Phys. Lett. **263**, 567 (1996).
257. W. Brenig, Microscopic theory of gas-surface interaction, Z. Phys. B **48**, 127 (1982).
258. W. Brenig, Dynamics and kinetics of gas-surface interaction: sticking, desorption and elastic scattering, Physica Scripta **35**, 329 (1987).
259. W. Brenig, *Nonequilibrium Thermodynamics*, Springer, Berlin, 1990.
260. M. F. Bertino, F. Hofmann, and J. P. Toennies, The effect of dissociative chemisorption on the diffraction of D_2 from Ni(110), J. Chem. Phys. **106**, 4327 (1997).
261. D. Cvetko, A. Morgante, A. Santaniello, and F. Tommasini, Deuterium scattering from Rh(11O) surface, J. Chem. Phys. **104**, 7778 (1996).
262. B. Gumhalter, Single- and multiphonon atom-surface scattering in the quantum regime, Phys. Rep. **351**, 1 (2001).
263. A. Siber, G. B., J. Braun, A. P. Graham, M. F. Bertino, J. P. Toennies, D. Fuhrmann, and C. Wöll, Combined He-atom scattering and theoretical study of the low-energy vibrations of physisorbed monolayers of Xe on Cu(111) and Cu(001), Phys. Rev. B **59**, 5898 (1999).
264. B. Baule, Theoretische Behandlung der Erscheinungen in verdünnten Gasen, Ann. Physik **44**, 145 (1914).
265. E. K. Grimmelmann, J. C. Tully, and M. J. Cardillo, Hard-cube model analysis of gas-surface energy accomodation, J. Chem. Phys. **72**, 1039 (1980).

266. E. W. Kuipers, M. G. Tenner, M. E. M. Spruit, and A. W. Kleyn, Differential trapping probabilities and desorption of physisorbed molecules: application to NO/Ag(111), Surf. Sci. **205**, 241 (1988).
267. C. T. Rettner and C. B. Mullins, Dynamics of the chemisorption of O_2 on Pt(111): dissociation via direct population of a molecularly chemisorbed precursor at high-incidence kinetic-energy, J. Chem. Phys. **94**, 1626 (1991).
268. C. R. Arumainayagam, G. R. Schoofs, M. C. McMaster, and R. J. Madix, Dynamics of molecular adsorption of ethane with Pt(111): a supersonic molecular-beam study, J. Phys. Chem. **95**, 1041 (1991).
269. J. A. Stinnett, R. J. Madix, and J. C. Tully, Stochastic simulations of the trapping of ethane on Pt(111) from a realistic potential: The roles of energy transfer processes and surface corrugation, J. Chem. Phys. **104**, 3134 (1996).
270. H. Schlichting, D. Menzel, T. Brunner, W. Brenig, and J. C. Tully, Quantum effects in the sticking of Ne on a flat metal surface, Phys. Rev. Lett. **60**, 2515 (1988).
271. H. Schlichting, D. Menzel, T. Brunner, and W. Brenig, Sticking of rare gas atoms on the clean Ru(001), J. Chem. Phys. **97**, 4453 (1992).
272. R. W. Fuller, S. M. Harris, and E. L. Slaggie, S-matrix solution for forced harmonic oscillator, Am. J. Phys. **31**, 431 (1963).
273. G. B. Arfken and H. J. Weber, *Mathematical Methods for Physicists*, Academic Press, San Diego, 4th edition, 1995.
274. W. Brenig, Theory of inelastic atom-surface scattering: average energy loss and energy distribution, Z. Phys. B **36**, 81 (1979).
275. J. Böheim and W. Brenig, Theory of inelastic atom-surface scattering: examples of energy-distributions, Z. Phys. B **41**, 243 (1981).
276. R. Sedlmeir and W. Brenig, Inelastic atom-surface scattering: a comparison of classical and quantum treatments, Z. Phys. B **36**, 245 (1980).
277. W. Brenig, Low-energy limit of reflection and sticking coefficients in atom surface scattering: 1. short-range forces, Z. Phys. B **36**, 227 (1980).
278. I. A. Yu, J. M. Doyle, J. C. Sandberg, C. L. Cesar, D. Kleppner, and T. J. Greytak, Evidence for universal quantum reflection of hydrogen from liquid ^{4}He, Phys. Rev. Lett. **71**, 1589 (1993).
279. C. Eltschka, H. Friedrich, and M. J. Moritz, Quantum reflection, Phys. Rev. Lett. **86**, 2693 (2001).
280. H. Friedrich, G. Jacoby, and C. G. Meiter, Quantum reflection by Casimir–van der Waals potential tails, Phys. Rev. B **65**, 032902 (2002).
281. K. D. Rendulic and A. Winkler, Adsorption and desorption dynamics as seen through molecular beam techniques, Surf. Sci. **299/300**, 261 (1994).
282. G.-J. Kroes, Six-dimensional quantum dynamics of dissociative chemisorption of H_2 on metal surfaces, Prog. Surf. Sci. **60**, 1 (1999).
283. W. Diño, H. Kasai, and A. Okiji, Orientational effects in dissociative adsorption/associative desorption dynamics of $H_2(D_2)$ on Cu and Pd, Prog. Surf. Sci. **63**, 63 (2000).
284. G.-J. Kroes, E. J. Baerends, and R. C. Mowrey, Six-dimensional quantum dynamics of dissociative chemisorption of ($v = 0, j = 0$) H_2 on Cu(100), Phys. Rev. Lett. **78**, 3583 (1997).
285. D. A. McCormack, G.-J. Kroes, R. A. Olsen, J. A. Groeneveld, J. N. P. van Stralen, E. J. Baerends, and R. C. Mowrey, Molecular knife throwing: aiming for dissociation at specific surface sites through state-selection, Chem. Phys. Lett. **328**, 317 (2000).
286. C. T. Rettner, D. J. Auerbach, and H. A. Michelsen, Role of vibrational and translational energy in the activated dissociative adsorption of D_2 on Cu(111), Phys. Rev. Lett. **68**, 1164 (1992).

287. D. A. McCormack and G.-J. Kroes, A classical study of rotational effects in dissociation of H_2 on Cu(100), Phys. Chem. Chem. Phys. **1**, 1359 (1999).
288. Y. Miura, H. Kasai, and W. Diño, Dynamical quantum filtering in the scattering of H_2 on Cu(001), J. Phys.: Condens. Matter **14**, L479 (2002).
289. B. Hammer, M. Scheffler, K. Jacobsen, and J. Nørskov, Multidimensional potential energy surface for H_2 dissociation over Cu(111), Phys. Rev. Lett. **73**, 1400 (1994).
290. J. A. White, D. M. Bird, M. C. Payne, and I. Stich, Surface corrugation in the dissociative adsorption of H_2 on Cu(100), Phys. Rev. Lett. **73**, 1404 (1994).
291. H. A. Michelsen and D. J. Auerbach, A critical-examination of data on the dissociative adsorption and associative desorption of hydrogen at copper surfaces, J. Chem. Phys. **94**, 7502 (1991).
292. M. Karikorpi, S. Holloway, N. Henriksen, and J. K. Nørskov, Dynamics of molecule surface interactions, Surf. Sci. **179**, L41 (1987).
293. J. C. Polanyi and W. H. Wong, Location of energy barriers. I. Effect on the dynamics of reactions A+BC, J. Chem. Phys. **51**, 1439 (1969).
294. G. R. Darling and S. Holloway, Dissociation thresholds and the vibrational excitation process in the scattering of H_2, Surf. Sci. **307-309**, 153 (1994).
295. S. Holloway, Dynamics of gas-surface reactions, Surf. Sci. **299/300**, 656 (1994).
296. A. Groß, Surface temperature effects in dissociative adsorption: D_2/Cu(111), Surf. Sci. **314**, L843 (1994).
297. A. Eichler, J. Hafner, A. Groß, and M. Scheffler, Trends in the chemical reactivity of surfaces studied by ab initio quantum-dynamics calculations, Phys. Rev. B **59**, 13297 (1999).
298. A. Groß, Hydrogen dissociation on metal surfaces – a model system for reactions on surfaces, Appl. Phys. A **67**, 627 (1998).
299. D. A. King, Kinetics of adsorption, desorption, and migration at single-crystal metal surfaces, CRC Crit. Rev. Solid State Mater. Sci. **7**, 167 (1978).
300. A. Groß and M. Scheffler, Role of zero-point effects in catalytic reactions involving hydrogen, J. Vac. Sci. Technol. A **15**, 1624 (1997).
301. C. Crespos, H. F. Busnengo, W. Dong, and A. Salin, Analysis of H_2 dissociation dynamics on the Pd(111) surface, J. Chem. Phys. **114**, 10954 (2001).
302. A. Groß and M. Scheffler, Influence of molecular vibrations on dissociative adsorption, Chem. Phys. Lett. **256**, 417 (1996).
303. H. A. Michelsen, C. T. Rettner, D. J. Auerbach, and R. N. Zare, Effect of rotation on the translational and vibrational-energy dependence of the dissociative adsorption of D_2 on Cu(111), J. Chem. Phys. **98**, 8294 (1993).
304. D. Wetzig, M. Rutkowski, H. Zacharias, and A. Groß, Vibrational and rotational population distribution of D_2 associatively desorbing from Pd(100), Phys. Rev. B **63**, 205412 (2001).
305. M. Beutl, M. Riedler, and K. D. Rendulic, Strong rotational effects in the adsorption dynamics of H_2/Pd(111): evidence for dynamical steering, Chem. Phys. Lett. **247**, 249 (1995).
306. M. Gostein and G. O. Sitz, Rotational state-resolved sticking coefficients for H_2 on Pd(111): testing dynamical steering in dissociative adsorption, J. Chem. Phys. **106**, 7378 (1997).
307. G. R. Darling and S. Holloway, Rotational motion and the dissociation of H_2 on Cu(111), J. Chem. Phys. **101**, 3268 (1994).
308. A. Dianat and A. Groß, Rotational quantum dynamics in a non-activated adsorption system, Phys. Chem. Chem. Phys. **4**, 4126 (2002).

309. D. Wetzig, M. Rutkowski, R. Etterich, W. David, and H. Zacharias, Rotational alignment in associative desorption of H_2 from Pd(100), Surf. Sci. **402**, 232 (1998).
310. D. Wetzig, R. Dopheide, M. Rutkowski, R. David, and H. Zacharias, Rotational alignment in associative desorption of D_2 (v=0 and 1) from Pd(100), Phys. Rev. Lett. **76**, 463 (1996).
311. G. R. Darling and S. Holloway, The role of parallel momentum in the dissociative adsorption of H_2 at highly corrugated surfaces, Surf. Sci. **304**, L461 (1994).
312. A. Groß, The role of lateral surface corrugation for the quantum dynamics of dissociative adsorption and associative desorption, J. Chem. Phys. **102**, 5045 (1995).
313. G. Anger, A. Winkler, and K. D. Rendulic, Adsorption and desorption kinetics in the systems H_2/Cu(111), H_2/Cu(110) and H_2/Cu(100), Surf. Sci. **220**, 1 (1989).
314. K. W. Kolasinski, Dynamics of hydrogen interactions with Si(100) and Si(111) surfaces, Int. J. Mod. Phys. B **9**, 2753 (1995).
315. K. Sinniah, M. G. Sherman, L. B. Lewis, W. H. Weinberg, J. T. Yates, and K. C. Janda, New mechanism for hydrogen desorption from covalent surfaces: the monohydride phase on Si(100), Phys. Rev. Lett. **62**, 567 (1989).
316. U. Höfer, L. Li, and T. F. Heinz, Desorption of hydrogen from Si(100)2×1 at low coverages: the influence of π-bonded dimers on the kinetics, Phys. Rev. B **45**, 9485 (1992).
317. D. J. Doren, Kinetics and dynamics of hydrogen adsorption and desorption on silicon surfaces, Adv. Chem. Phys. **95**, 1 (1996).
318. K. W. Kolasinski, W. Nessler, K.-H. Bornscheuer, and E. Hasselbrink, Beam investigations of D_2 adsorption on Si(100): on the importance of lattice excitations in the reaction dynamics, J. Chem. Phys. **101**, 7082 (1994).
319. P. Bratu, K. L. Kompa, and U. Höfer, Optical second-harmonic investigations of H_2 and D_2 adsorption on Si(100)2×1: the surface temperature dependence of the sticking coefficient, Chem. Phys. Lett. **251**, 1 (1996).
320. W. Brenig, A. Groß, and R. Russ, Detailed balance and phonon assisted sticking in adsorption and desorption of H_2/Si, Z. Phys. B **96**, 231 (1994).
321. K. W. Kolasinski, W. Nessler, A. de Meijere, and E. Hasselbrink, Hydrogen adsorption on and desorption from Si: considerations on the applicability of detailed balance, Phys. Rev. Lett. **72**, 1356 (1994).
322. P. Bratu and U. Höfer, Phonon-assisted sticking of molecular hydrogen on Si(111)-(7×7), Phys. Rev. Lett. **74**, 1625 (1995).
323. P. Bratu, W. Brenig, A. Groß, M. Hartmann, U. Höfer, P. Kratzer, and R. Russ, Reaction dynamics of molecular hydrogen on silicon surfaces, Phys. Rev. B **54**, 5978 (1996).
324. C. J. Wu, I. V. Ionova, and E. A. Carter, Ab initio H_2 desorption pathways for H/Si(100): the role of $SiH_{2(a)}$, Surf. Sci. **295**, 64 (1993).
325. Z. Jing, G. Lucovsky, and J. L. Whitten, Mechanism of H_2 desorption from monohydride Si(100)2×1H, Surf. Sci. **296**, L33 (1993).
326. P. Nachtigall, K. D. Jordan, and C. Sosa, Theoretical study of the mechanism of recombinative hydrogen desorption from the monohydride phase of Si(100): the role of defect migration, J. Chem. Phys. **101**, 8073 (1994).
327. Z. Jing and J. L. Whitten, Multiconfiguration self-consistent-field treatment of H_2 desorption from Si(100)2×1H, J. Chem. Phys. **102**, 3867 (1995).
328. P. Kratzer, B. Hammer, and J. K. Norskøv, The coupling between adsorption dynamics and the surface structure: H_2 on Si(100), Chem. Phys. Lett. **229**, 645 (1994).

329. E. Pehlke and M. Scheffler, Theory of adsorption and desorption of H_2/Si(001), Phys. Rev. Lett. **74**, 952 (1995).
330. P. Nachtigall, K. D. Jordan, A. Smith, and H. Jónsson, Investigation of the reliability of density functional methods: Reaction and activation energies for Si-Si bond cleavage and H_2 elimination from silanes, J. Chem. Phys. **104**, 148 (1996).
331. E. Penev, P. Kratzer, and M. Scheffler, Effect of the cluster size in modeling the H_2 desorption and dissociative adsorption on Si(001), J. Chem. Phys. **110**, 3986 (1999).
332. A. Vittadini and A. Selloni, Density functional study of H_2 desorption from monohydride and dihydride Si(100) surfaces, Chem. Phys. Lett. **235**, 334 (1995).
333. E. Pehlke, Highly reactive dissociative adsorption of hydrogen molecules on partially H-covered Si(001) surfaces: A density-functional studys, Phys. Rev. B **62**, 12932 (2000).
334. F. M. Zimmermann and X. Pan, Interaction of H_2 with Si(001)–(2×1): Solution of the barrier puzzle, Phys. Rev. Lett. **85**, 618 (2000).
335. P. Kratzer, E. Pehlke, M. Scheffler, M. B. Raschke, and U. Höfer, Highly site-specific H_2 adsorption on vicinal Si(001) surfaces, Phys. Rev. Lett. **81**, 5596 (1998).
336. E. Pehlke and P. Kratzer, Density-functional study of hydrogen chemisorption on vicinal Si(001) surfaces, Phys. Rev. B **59**, 2790 (1999).
337. A. Biedermann, E. Knoesel, Z. Hu, and T. F. Heinz, Dissociative adsorption of H_2 on Si(100) induced by atomic H, Phys. Rev. Lett. **83**, 1810 (1999).
338. A. Groß and M. Scheffler, Ab initio quantum and molecular dynamics of the dissociative adsorption of hydrogen on Pd(100), Phys. Rev. B **57**, 2493 (1998).
339. P. Hänngi, P. Talkner, and M. Borkovec, Reaction-rate theory: fifty years after Kramers, Rev. Mod. Phys. **62**, 251 (1990).
340. G. Mills, H. Jónsson, and G. K. Schenter, Reversible work transition state theory: application to dissociative adsorption of hydrogen, Surf. Sci. **324**, 305 (1995).
341. H. Jónsson, Theoretical studies of atomic-scale processes relevant for crystal growth, Annu. Rev. Phys. Chem. **51**, 623 (2000).
342. G. L. Kellogg and P. J. Feibelman, Surface self-diffusion on Pt(001) by an atomic exchange mechanism, Phys. Rev. Lett. **64**, 3143 (1990).
343. C. Chen and T. T. Tsong, Displacement distribution of atomic jump direction in diffusion of Ir atoms on the Ir(001) surface, Phys. Rev. Lett. **64**, 3147 (1990).
344. P. J. Feibelman, Diffusion path for an Al adatom on Al(001), Phys. Rev. Lett. **65**, 729 (1990).
345. P. Ruggerone, C. Ratsch, and M. Scheffler, Density-functional theory of epitaxial growth of metals, in *Growth and Properties of Ultrathin Epitaxial Layers*, edited by D. A. King and D. P. Woodruff, volume 8 of *The Chemical Physics of Solid Surfaces*, pages 490–544, Elsevier, Amsterdam, 1997.
346. B. D. Yu and M. Scheffler, Physical origin of exchange diffusion on fcc(100) metal surfaces, Phys. Rev. B **56**, R15569 (1997).
347. T.-Y. Fu and T. T. Tsong, Atomic processes in self-diffusion of Ni surfaces, Surf. Sci. **454**, 571 (2000).
348. C. M. Chang, C. M. Wei, and J. Hafner, Self-diffusion of adatoms on Ni(100) surfaces, J. Phys.: Condens. Matter **13**, L321 (2001).
349. H. J. Kreuzer and S. H. Payne, Theoretical approaches to the kinetics of adsorption, desorption and reactions at surfaces, in *Computational Methods in Colloid and Interface Science*, edited by M. Borowko, Marcel Dekker, New York, 1999.

350. H. J. Kreuzer, Theory of sticking: The effect of lateral interactions, J. Chem. Phys. **104**, 9593 (1996).
351. H. J. Kreuzer, Kinetic lattice gas model with precursors, Surf. Sci. **238**, 305 (1990).
352. C. Stampfl, H. J. Kreuzer, S. H. Payne, H. Pfnür, and M. Scheffler, First-principles theory of surface thermodynamics and kinetics, Phys. Rev. Lett. **83**, 2993 (1999).
353. A. Böttcher, H. Niehus, S. Schwegmann, H. Over, and G. Ertl, CO oxidation reaction over oxygen-rich Ru(0001) surfaces, J. Phys. Chem. B **101**, 11185 (1997).
354. C. Stampfl, H. J. Kreuzer, S. H. Payne, and M. Scheffler, Challenges in predicitive calculations of processes on surfaces: surface thermodynamics and catalytic reactions, Appl. Phys. A **69**, 471 (1999).
355. G. Ehrlich and F. G. Hudda, Atomic view of surface self-diffusion: tungsten on tungsten, J. Chem. Phys. **44**, 1039 (1966).
356. R. L. Schwoebel and E. J. Shipsey, Step motion on crystal surfaces, J. Appl. Phys. , 3682 (1966).
357. H. Brune, Microscopic view of epitaxial metal growth: nucleation and aggregation, Surf. Sci. Rep. **31**, 121 (1998).
358. J. A. Venables, Nucleation calculations in a pair-binding model, Phys. Rev. B **36**, 4153 (1987).
359. R. Stumpf and M. Scheffler, Ab initio calculations of energies and self-diffusion on flat and stepped surfaces of Al and their implications on crystal growth, Phys. Rev. B **53**, 4958 (1996).
360. P. Ruggerone, A. Kley, and M. Scheffler, Microscopic aspects of homoepitaxial growth, Prog. Surf. Sci. **54**, 331 (1997).
361. T. A. Witten and L. M. Sander, Diffusion-limited aggregation, a kinetic critical phenomenon, Phys. Rev. Lett. **47**, 1400 (1981).
362. T. A. Witten and L. M. Sander, Diffusion-limited aggregation, Phys. Rev. B **27**, 5686 (1983).
363. S. Ovesson, A. Bogicevic, and B. Lundqvist, Origin of compact triangular islands in metal-on-metal growth, Phys. Rev. Lett. **83**, 2608 (1999).
364. T. Michely, M. Hohage, M. Bott, and G. Comsa, Inversion of growth speed anisotropy in two dimensions, Phys. Rev. Lett. **70**, 3943 (1993).
365. S. Esch, M. Hohage, T. Michely, and G. Comsa, Origin of oxygen induced layer-by-layer growth in homoepitaxy on Pt(111), Phys. Rev. Lett. **72**, 518 (1994).
366. M. Kalff, G. Comsa, and T. Michely, How sensitive is epitaxial growth to adsorbates, Phys. Rev. Lett. **81**, 1255 (1998).
367. P. J. Feibelman, Energetics of steps on Pt(111), Phys. Rev. B **52**, 16845 (1995).
368. P. J. Feibelman, Interlayer self-diffusion on stepped Pt(111), Phys. Rev. Lett. **81**, 168 (1998).
369. P. J. Feibelman, Self-diffusion along step bottoms on Pt(111), Phys. Rev. B **60**, 4972 (1999).
370. P. Kratzer and M. Scheffler, Reaction-limited island nucleation in molecular beam epitaxy of compound semiconductors, Phys. Rev. Lett. **88**, 036102 (2002), for an animation of the island nucleation process, see http://netserver.aip.org/cgi-bin/epaps?ID=E-PRLTAO-87-031152.
371. G. R. Bell, M. Itoh, T. S. Jones, and B. A. Joyce, Nanoscale effects of arsenic kinetics on GaAs(001)-(2×4) homoepitaxy, Surf. Sci. **423**, L280 (1999).

372. P. Stoltze and J. K. Nørskov, Bridging the "pressure gap" between ultrahigh-vacuum surface physics and high-pressure catalysis, Phys. Rev. Lett. **55**, 2502 (1985).
373. C. Stampfl, M. V. Ganduglia-Pirovano, K. Reuter, and M. Scheffler, Catalysis and corrosion: the theoretical surface-science context, Surf. Sci. **500**, 368 (2002).
374. A. Nettesheim, H. H. von Oertzen, A. an Rotermund, and G. Ertl, Reaction-diffusion patterns in the catalytic CO oxidation on Pt(110): front propagation and spiral waves, J. Chem. Phys. **98**, 9977 (1993).
375. R. Imbihl, Oscillatory reactions on single-crystal surfaces, Prog. Surf. Sci. **44**, 185 (1993).
376. R. Imbihl and G. Ertl, Oscillatory kinetics in heterogeneous catalysis, Chem. Rev. **95**, 697 (1995).
377. H. H. Rotermund, Imaging of dynamic processes on surfaces by light, Surf. Sci. Rep. **29**, 265 (1997).
378. G. Ertl, Oscillatory kinetics and spatio-temporal self-orgnization in reactions at solid surfaces, Science **254**, 1750 (1991).
379. D. Barkley, A model for fast computer-simulation of waves in excitable media, Physica D **49**, 61 (1991).
380. M. Bär, M. Gottschalk, N. Eiswirth, , and G. Ertl, Spiral waves in a surface reaction: model calculations, J. Chem. Phys. **100**, 1202 (1994).
381. M. Bär, M. Eiswirth, H. H. Rotermund, and G. Ertl, Solitary-wave phenomena in an excitable surface reaction, Phys. Rev. Lett. **69**, 945 (1992).
382. A.-L. Barabási and H. E. Stanley, *Fractal Concepts in Surface Growth*, Cambridge University Press, 1995.
383. T. Greber, Charge-transfer induced particle emission in gas surface reactions, Surf. Sci. Rep. **28**, 3 (1997).
384. W. Diño, H. Kasai, and A. Okiji, Dynamical phenomena including many body effects at metal surfaces, Surf. Sci. **500**, 105 (2002).
385. E. Runge and E. Gross, Density-functional theory for time-dependent systems, Phys. Rev. Lett. **52**, 997 (1984).
386. E. K. U. Gross, J. F. Dobson, and M. Petersilka, Density functional theory of time-dependent phenomena, in *Density functional theory*, edited by R. F. Nalewajski, volume 181 of *Topics in Current Chemistry*, page 81, Springer, Berlin, 1996.
387. A. Zangwill and P. Soven, Density-functional approach to local-field effects in finite systems: Photoabsorption in the rare gases, Phys. Rev. A **21**, 1561 (1980).
388. B. Champagne, E. A. Perpète, S. J. A. van Gisbergen, E.-J. Baerends, J. G. Snijders, C. Soubra-Ghaoui, K. A. Robins, and B. Kirtman, Assessment of conventional density functional schemes for computing the polarizabilities and hyperpolarizabilities of conjugated oligomers: An ab initio investigation of polyacetylene chains, J. Chem. Phys. **109**, 10489 (1998).
389. M. Petersilka, U. J. Gossmann, and E. K. U. Gross, Excitation energies from time-dependent density-functional theory, Phys. Rev. Lett. **76**, 1212 (1996).
390. B. Gergen, H. Nienhaus, W. H. Weinberg, and E. W. McFarland, Chemically induced electronic excitations at metal surfaces, Science **294**, 2521 (2001).
391. H. Nienhaus, Electronic excitations by chemical reactions on metal surfaces, Surf. Sci. Rep. **45**, 1 (2002).
392. B. Gergen, S. J. Weyers, H. Nienhaus, W. H. Weinberg, and E. W. McFarland, Observation of excited electrons from nonadiabatic molecular reactions of NO and O_2 on polycrystalline Ag, Surf. Sci. **488**, 123 (2001).

393. J. R. Trail, M. C. Graham, and D. M. Bird, Electronic damping of molecular motion at metal surfaces, Comp. Phys. Comm. **137**, 163 (2001).
394. J. R. Trail, M. C. Graham, D. M. Bird, M. Persson, and S. Holloway, Energy loss of atoms at metal surfaces due to electron-hole pair excitation: first-principles theory of chemicurrents, Phys. Rev. Lett. **88**, 166802 (2002).
395. H. Nienhaus, H. S. Bergh, B. Gergen, A. Majumdar, W. H. Weinberg, and E. W. McFarland, Electronic excitations by chemical reactions on metal surfaces, Phys. Rev. Lett. **82**, 466 (1999).
396. M. Head-Gordon and J. C. Tully, Molecular dynamics with electronic friction, J. Chem. Phys. **103**, 10137 (1995).
397. J. T. Kindt, J. C. Tully, M. Head-Gordon, and M. A. Gomez, Electron-hole pair contribution to scattering, sticking and surface diffusion: CO on Cu(100), J. Chem. Phys. **109**, 3629 (1998).
398. J. C. Tully, Mixed quantum classical dynamics, Faraday Discuss. **110**, 407 (1998).
399. J. C. Tully, Molecular dynamics with electronic transitions, Faraday Discuss. **93**, 1061 (1990).
400. M. S. Topaler, T. C. Allison, D. W. Schwenke, and D. G. Truhlar, What is the best semiclassical method for photochemical dynamics of systems with conical intersections?, J. Chem. Phys. **109**, 3321 (1998).
401. C. Bach and A. Groß, Semiclassical treatment of charge transfer in molecule-surface scattering, J. Chem. Phys. **114**, 6396 (2001).
402. F. M. Zimmermann and W. Ho, State-resolved studies of photochemical dynamics at surfaces, Surf. Sci. Rep. **22**, 127 (1995).
403. D. Menzel and R. Gomer, Desorption from metal surfaces by low-energy electrons, J. Chem. Phys. **41**, 3311 (1964).
404. P. A. Redhead, Interaction of slow electrons with chemisorbed oxygen, Can. J. Phys. **42**, 886 (1964).
405. R. Gomer, Mechanisms of electron-stimulated desorption, in *Desorption induced by electronic transitions DIET I*, edited by N. H. Tolk, M. M. Traum, J. C. Tully, and T. E. Madey, volume 24 of *Springer Series in Chemical Physics*, page 40, Springer, Berlin, 1983.
406. P. R. Antoniewicz, Model for electron-stimulated and photon-stimulated desorption, Phys. Rev. B **21**, 3811 (1980).
407. J. W. Gadzuk, L. J. Richter, S. Buntin, D. S. King, and R. R. Cavanagh, Laser-excited hot-electron induced desorption: A theoretical model applied to NO/Pt(111), Surf. Sci. **235**, 317 (1990).
408. J. W. Gadzuk, Resonance-assisted, hot-electron-induced desorption, Surf. Sci. **342**, 345 (1995).
409. H. Guo and T. Seideman, Quantum-mechanical studies of photodesorption of ammonia from a metal-surface: Isotope effects, final-state distributions, and desorption mechanisms, J. Chem. Phys. **103**, 9062 (1995).
410. P. Saalfrank, S. Holloway, and G. Darling, Theory of laser-induced desorption of ammonia from Cu(111): State-resolved dynamics, isotope effects, and selective surface photochemistry, J. Chem. Phys. **103**, 6720 (1995).
411. W. Nessler, K.-H. Bornscheuer, T. Hertel, and E. Hasselbrink, Internal quantum state distributions of NH_3 photodesorbed from Cu(111) at 6.4 eV, Chem. Phys. **205**, 205 (1996).
412. P. Saalfrank, Photodesorption of neutrals from metal surfaces: a wave packet study, Chem. Phys. **193**, 119 (1995).
413. T. Klüner, H.-J. Freund, V. Staemmler, and R. Kosloff, Theoretical investigation of laser induced desorption of small molecules from oxide surfaces: A first principles study, Phys. Rev. Lett. **80**, 5208 (1998).

414. H. Kuhlenbeck, G. Odörfer, R. Jaeger, G. Illing, M. Menges, T. Mull, H.-J. Freund, M. Pöhlchen, V. Staemmler, S. Witzel, C. Scharfschwerdt, K. Wennemann, T. Liedte, and M. Neumann, Molecular adsorption on oxide surfaces: Electronic-structure and orientation of NO on NiO(100)/Ni(100) and on NiO(100) as determined from electron spectroscopies and ab initio cluster calculations, Phys. Rev. B **43**, 1969 (1991).
415. T. Mull, B. Baumeister, M. Menges, H.-J. Freund, D. Weide, C. Fischer, and P. Andresen, Bimodal velocity distributions after ultraviolet-laser-induced desorption of NO from oxide surfaces: Experiments and results of modelcalculations, J. Chem. Phys. **96**, 7108 (1992).
416. P. Saalfrank, Stochastic wave packet vs direct density matrix solution of Liouville-von Neumann equations for photodesorption problems, Chem. Phys. **211**, 265 (1996).
417. P. Finger and P. Saalfrank, Vibrationally excited products after the photodesorption of NO from Pt(111): A two-mode open-system density matrix approach, Chem. Phys. Lett. **268**, 291 (1997).
418. T.-C. Shen, C. Wang, G. C. Abeln, J. R. Tucker, J. W. Lyding, P. Avouris, and R. E. Walkup, Atomic-scale desorption through electronic and vibrational-excitation mechanisms, Science **268**, 1590 (1995).
419. Y. Miyamoto and O. Sugino, First-principles electron-ion dynamics of excited systems: H-terminated Si(111) surfaces, Phys. Rev. B **62**, 2039 (2000).
420. O. Sugino and Y. Miyamoto, Density-functional approach to electron dynamics: Stable simulation under a self-consistent field, Phys. Rev. B **59**, 2579 (1999).
421. A. Liebsch, *Electronic Excitations at Metal Surfaces*, Plenum Press, New York, 1997.
422. C. B. Duke and E. W. Plummer, editors, *Frontiers in Surface and Interface Science*, North-Holland, 2002.
423. D. M. Kolb, An atomistic view of electrochemistry, Surf. Sci. **500**, 722 (2002).
424. M. T. M. Koper and R. A. van Santen, Interaction of halogens with Hg, Ag and Pt surfaces: a density functional study, Surf. Sci. **422**, 118 (1999).
425. R. Raybaud, M. Digne, R. Iftimie, W. Wellens, P. Euzen, and H. Toulhoat, Morphology and surface properties of boehmite (γ-AlOOH): A density functional theory study, J. Catal. **201**, 236 (2001).
426. S. Izvekov, A. Mazzolo, K. Van Opdorp, and G. A. Voth, Ab initio molecular dynamics simulation of the Cu(110)-water interface, J. Chem. Phys. **114**, 3284 (2001).
427. S. K. Desai, V. Pallassana, and M. Neurock, A periodic density functional theory analysis of the effect of water molecules on deprotonation of acetic acid over Pd(111), J. Phys. Chem. B **114**, 10954 (2001).
428. H.-J. Freund, Clusters and islands on oxides: from catalysis via electronics and magnetism to optics, Surf. Sci. **500**, 271 (2002).
429. M. Haruta, Size- and support-dependency in the catalysis of gold, Catalysis Today **36**, 153 (1997).
430. M. Valden, X. Lai, and D. W. Goodman, Onset of catalytic activity of gold clusters on titania with the appearance of nonmetallic properties, Science **281**, 1647 (1998).
431. A. Sanchez, S. Abbet, U. Heiz, W.-D. Schneider, H. Häkkinen, R. N. Barnett, and U. Landman, When gold is not noble: nanoscale gold catalysts, J. Phys. Chem. A **103**, 9573 (1999).
432. B. Kasemo, Biological surface science, Surf. Sci. **500**, 656 (2002).

433. M. Eichinger, P. Tavan, J. Hutter, and M. Parrinello, A hybrid method for solutes in complex solvents: Density functional theory combined with empirical force fields, J. Chem. Phys. **110**, 10452 (1999).
434. T. H. Rod and J. K. Nørskov, The surface science of enzymes, Surf. Sci. **500**, 678 (2002).
435. A. Kühnle, T. R. Linderoth, B. Hammer, and F. Besenbacher, Chiral recognition in dimerization of adsorbed cysteine observed by scanning tunneling microscopy, Nature **415**, 891 (2002).
436. T. D. Booth, D. Wahnon, and I. Wainer, Is chiral recognition a three-point process?, Chirality **9**, 96 (1997).
437. A. Groß, The virtual chemistry lab for reactions at surfaces: Is it possible? Will it be helpful?, Surf. Sci. **500**, 347 (2002).
438. F. Besenbacher, I. Chorkendorff, B. S. Clausen, B. Hammer, A. M. Molenbroek, J. K. Nørskov, and I. Stensgaard, Design of a surface alloy catalyst for steam reforming, Science **279**, 1913 (1998).
439. P. Kratzer, B. Hammer, and J. K. Norskøv, A theoretical study of CH_4 dissociation on pure and gold-alloyed Ni(111) surfaces, J. Chem. Phys. **105**, 5595 (1996).
440. J. H. Larsen and I. Chorkendorff, From fundamental studies of reactivity on single crystals to the design of catalysts, Surf. Sci. Rep. **35**, 165 (1999).
441. M. Hierlemann, C. Werner, and A. Spitzer, Equipment simulation of SiGe heteroepitaxy – model validation by ab initio calculations of surface diffusion processes, J. Vac. Sci. Technol. B **15**, 935 (1997).
442. J. Almanstötter, T. Fries, and B. Eberhard, Electronic structure of fluorescent lamp cathode surfaces: BaO/W(001), J. Appl. Phys. **86**, 325 (1997).

Index

absorption 128
acetic acid 241
acoustic phonons 88
active sites 139
adatom resonance 107
adatoms 17
adiabatic connection formula 103
adiabatic representation 233
adlayer
– commensurate 16, 161
– incommensurate 16, 161
adsorption
– atomic 163
– dissociative 171
$Al_2O_3(0001)$ 82
Al(100)
– exchange diffusion 197
Al(111) 211
– diffusion barriers 212
ammonia synthesis 217
An_n/MgO(100) 243
Anderson–Grimley–Newns model 104
Antoniewicz model 235
APW method 46
asymmetric dimer model 73
attempt frequency 193
Au(110) 245

backdonation 134
backfolding 11
band gap 73
band structure 52, 75
barrier
– late 174
Baule formula 163, 189
Bethe-Salpeter equation 76
biologically relevant systems 244
Bloch functions 13
Bloch sum 50
Bloch theorem 13
Blyholder model 134
bonding competition 142

Born–Oppenheimer approximation 7, 18, 89
Born–Oppenheimer energy surface 7, 48, 88, 238
Bravais lattice
– three-dimensional 9
– two-dimensional 13, 14
Brillouin zone
– first 10
Brønsted–Evans–Polanyi relation 138

C_2H_6/ Pt(111) 165
car exhaust catalyst 134, 136, 142
Car–Parrinello method 49
Casimir effect 104
Casimir-van der Waals potential 104, 171
CH_4/Ni(111) 246
chemical potential 77
chemical turbulence 220
chemicurrent 229–230
chemisorption
– ionic 112
– strong 106
– weak 106
chiral recognition 245
chirality 245
classical dynamics 147–149
cluster 243
CO oxidation 142–144, 218, 243
CO/Cu(111)
– sticking probability 230
CO/Pt(111) 134
CO/Ru(0001) 140
coadsorbates 136, 138
cohesive energy 66, 69
configuraton interaction 34
correlation
– energy 31
corrugation 165
– energetic 185
– geometric 185

coupled cluster theory 34
coupled-channel equations 150
critical cluster size 208
crystal orbital overlap population (COOP) 135
crystal-momentum 12
Cu surfaces 68

d-band model 125, 138
dangling bonds 31
density functional perturbation theory (DFPT) 89
density functional theory (DFT) 36–41, 74
– time-dependent (TDFT) 225–228, 237
density matrix 239
– reduced 151
density of states 61–63
– local 62
– projected 62
deposition 207
– flux 209
desorption induced by electronic transitions (DIET) 234–237
desorption induced by multiple electronic transitions (DIMET) 237
detailed balance 160, 179, 199
diabatic representation 233
diffraction 157
– rotationally inelastic 159
diffusion 195
– exchange 197
– hopping 197
diffusion coefficient
– chemical 197
– tracer 196
diffusion limited aggregation 213
diffusion quantum Monte Carlo 49
dissipation 148, 151, 229, 231
dynamical image plane 99
dynamical matrix 88
dynamical trapping 177
dynamics 147, 162, 171

effective medium theory 116
– atomic chemisorption energies 120
Ehrlich–Schwoebel barrier 207
elastic force constants 88
electrocatalysis 241
electrochemistry 241
electron
– correlation 31
– counting principle 80
– exchange 26
electron affinity 107
electron-hole pairs 228
electron-stimulated desorption (ESD) 234
electronic friction 230
electronically non-adiabatic processes 225
Eley–Rideal mechanism 244
embedded atom method 121
embedded-diatomics-in-molecules 123
embedding energy 116
energy transfer 157, 164, 168, 171
– rotation to translation 182
– to electron-hole pairs 231
– to parallel motion 167
– to phonons 157
– to rotations 167
– vibration to translation 179
equilibrium shape
– of a crystal 67
– of an island 211
exchange
– hole 29
exchange-correlation
– hole 39
exchange-correlation energy 6, 31
exciton 76
exponentiated distorted-wave Born approximation 161

F-center 243
Fermi energy 27
Fermi vector 27
fewest-switches algorithm 233, 239
Fick's law 197
fluctuation-dissipation theorem 230
forced oscillator model 169
Franck–Condon transitions 234
Frank–van der Merwe growth 205
Friedel oscillations 58
frontier orbitals 123
frozen-phonon technique 89
fuel cells 241

GaAs 77
– (100) 77
generalized gradient approximation (GGA) 40
generalized Langevin equation 149
graphite(0001) 86
group theory 9
growth 205–217

- direction 214
- modes 205
- of fcc(111) surfaces 210
GW approximation 74

$H_2/Cu(100)$ 172
$H_2/Pd(100)$ 160, 174
$H_2/S/Pd(100)$ 136
$H_2/Si(100)$ 186
$H/Pd(110)$ 130
$H/Si(111)$ 237
Hückel method 50
hard-cube model 163–166, 189, 190
Hartree energy 25
Hartree equations 23
Hartree theory 21–25
Hartree–Fock equations 26, 53
Hartree–Fock theory 25–31
He interaction with jellium surfaces 101
helium atom scattering 159, 161
Hellmann–Feynman theorem 7, 233
highest occupied molecular orbital (HOMO) 123, 134
Hohenberg–Kohn theorem 36
hole model 173
homogeneous electron gas 26

image charge 98
image potential 65, 107, 108
image potential states 65
industrial applications 246
ionization energy 107
island density 209
island nucleation 209
islands 213
isotope effect 189, 235, 240

Jahn–Teller effect 73
jellium model 26, 57
jellium surface 101, 110

k-space 10, 52
kinetic lattice gas model 198
kinetic Monte Carlo 210
kinetics 191
- of adsorption and desorption 199
- of diffusion 195
- of growth 205
kinks 17
Kohn–Sham equations 38

Langmuir kinetics 202
Langmuir–Hinshelwood mechanism 218
lattice gas 198
- Hamiltonian 198
lattice with a basis 9
Lindblad operator 152
linear response 90
linearized augmented plane waves (LAPW) method 47
- full potential (FP-LAPW) 48
Liouville–von Neumann equation 152
Lippmann–Schwinger equation 111
local density approximation (LDA) 39, 49, 74
- overbinding 40, 120
lowest unoccupied molecular orbital (LUMO) 123, 134

Markov approximation 152, 199
master equation 196, 199
mean-field method 232
Menzel–Gomer–Redhead (MGR) model 234, 240
microfacets 211
microscopic reversibility 199, 233
Miller indices 11
mixed quantum-classical dynamics 232
modified embedded atom method 123
molecular dynamics 147–149
- with electronic friction 230
- with electronic transitions 233
muffin-tin potential 46
multi-reference methods 34
Møller–Plesset theory 32–33

$N_2/Ru(0001)$ 138
nanostructured surfaces 243
nanotechnology 243
natural orbital functional theory 49
nearly-free electron model 60
Newns–Anderson model 104–110
- chemisorption function 107
$NH_3/Cu(111)$ 235
NO/Ag 229
$NO/NiO(100)$ 235
$NO/Pt(111)$ 235, 237
Nosé thermostat 149
nucleation theory 209
nudged elastic band method 195

$O/Ru(0001)$ 140, 203
open system dynamics 151
optical potentials 153
orbital functionals 41, 49

$Pd(110)$ 130
$Pd(111)$ 241

Pd(210)   62
phase shift
– in resonance scattering   113–114
photoemission electron microscopy (PEEM)   218
photon-stimulated desorption (PSD)   234
physisorption   97–104
plane waves   44
– augmented   46
– linerarized augmented   47
poisoning   136
polarizability   99
potential energy surface   95
precursor   175
prepairing mechanism   185
pressure gap   218
projected augmented waves (PAW)   48
projected bulk band structure   61, 73, 75
pseudopotential   41–44
– norm-conserving   43
– ultra-soft   44
Pt(100)-hex   139
Pt(11,7,5)   139
Pt(111)   134, 139, 215
Pulay forces   48

quantum chemistry methods   31–35, 225
quantum dynamics   149–153
quantum Monte Carlo (QMC)   31, 49
quantum reflection   171

rare gas sticking   168
rare gase   98, 101
rate equations   208
rates   191
Rayleigh waves   88, 162
Rayleigh–Ritz variational principle   22
reciprocal space   10
reconstruction   15
– adsorbate-induced   130
– missing row   16, 131, 219, 245
– pairing row   132
Redfield approach   152
reflection   157
relaxation   70
rotation
– cartwheel   167
– helicopter   167
rotational alignment   184
rotational hindering   181
roughness   223

Ru(0001)   140, 168
Runge–Gross theorem   225

sagittal plane   88
scanning tunneling microscope (STM)   85, 237
scattering   157
– elastic   157
– inelastic   157
– multiphonon   161
– rotationally inelastic   160
– singlephonon   161
Schrödinger equation   5
selective adsorption resonances   158
self interaction   23
self-consistency   24
self-consistent field (SCF) approximation   24
Shockley surface state   61, 64
Si
– (100)   72
– (111)-(7×7)   76
single-reference methods   32
size extensivity   34
slab model   45
Slater determinant   25
Smoluchowski smoothing   64, 71
sodium chloride structure   80
solid-liquid interface   241
spatiotemporal self-organization   218
specular reflection   158
steam reforming   246
steering   176
step flow   206
sticking probability   163
– normal energy scaling   165
– total energy scaling   165
STM-induced desorption   237
Stranski–Krastanov growth   205
structure gap   136
supercell approach   45
surface
– ionic   80–85
– metal   63–72
– non-polar   80
– polar   80
– semiconductor   72–80
surface band structure   72
surface dimer   72
surface energy   66, 205
surface exciton   76
surface phonons   87, 161
surface relaxation   70
surface resonance   61

surface stoichiometry   77
surface-hopping method   232

Tamm surface state   55, 61, 64
temperature programmed desorption (TPD)   143, 204
Tersoff–Hamann picture   86
Thomas–Fermi equation   36
Thomas–Fermi theory   36
tight-binding   50–52
time-reversal symmetry   179, 199
trajectory approximation   169
trajectory calculations   89
transition state theory   191–195, 210
– multi-dimensional rate   194
– one-dimensional rate   193

unit cell   10, 15
unit vectors   9

vacancy   17
van der Waals constant   100
van der Waals interaction   98–103
variational Monte Carlo   49
Verlet algorithm   147, 189
vibrational efficacy   173
vibrational temperature   180
vibrationally enhanced dissociation   173
vicinal surface   17
Volmer–Weber growth   205

wetting   205
Wigner–Seitz cell   10
Wigner–Seitz radius   28
work function   58, 218
Wulff construction   67, 211

Xe/Cu(100)   161
Xe/Cu(111)   161

Printing (Computer to Film): Saladruck Berlin
Binding: Stürtz AG, Würzburg